# Crossing Boundaries 2021



IOP Conference Series: Earth and Environmental Science
Volume 855

Online
24 – 25 March 2021

ISBN: 978-1-7138-4819-6
ISSN: 1755-1307

**Printed from e-media with permission by:**

Curran Associates, Inc.
57 Morehouse Lane
Red Hook, NY 12571

**Some format issues inherent in the e-media version may also appear in this print version.**



Institute of Physics
Dirac House, Temple Back
Bristol BS1 6BE UK

Phone: 44 1 17 929 7481
Fax: 44 1 17 920 0979

techtracking@iop.org

**Additional copies of this publication are available from:**

Curran Associates, Inc.
57 Morehouse Lane
Red Hook, NY 12571 USA
Phone: 845-758-0400
Fax: 845-758-2633
Email: curran@proceedings.com
Web: www.proceedings.com

# TABLE OF CONTENTS

The conference 'CROSSING BOUNDARIES, Towards a circular, sustainable and vital built environment' was held online on 24 – 25 March 2021. Initially the conference was scheduled to take place in the Rolduc Abbey in the Netherlands but due to COVID-19 pandemic and travel restriction the local organizing committee decided to held the conference online.

One of the major themes in architecture and construction is the move towards a zero energy and circular built environment. In the case of zero energy, most countries have legislation in place to realize this ambition. In the case of circularity, there is a different situation. Many countries do have ambitions in this field, but legislation is not yet embedded in policies. Collaterally, most buildings are still designed, constructed, operated and demolished in a traditional linear process. Currently less than three percent of all residual material is reused after the demolition of a building, which means that vast amounts of resources are wasted in the process. The small amount of materials that is recycled, is often used in inefficient ways, with collateral environmental impacts, that go across national boundaries. Worldwide, there is numerous ongoing research and exemplary projects in the field of circularity in the built environment, so, crossing boundaries between different fields of disciplines and nations is necessary to meet the ambitions of full circularity. The conference 'CROSSING BOUNDARIES, Towards a circular, sustainable and vital built environment' aims to create a platform on which knowledge on the subject can be exchanged on a scientific level, to study the environmental and economic implications and to share experiences on the latest developments in the field of circular building.

The Parkstad region in the south of the Netherlands is coping with a steady decline in population and therefore has a large housing surplus, making it an interesting testing ground for the practice of circular building. The Parkstad region has the ambition to develop itself into the (inter)national centre of circularity in the built environment. This ambition meets, and is supported by the agenda of IBA Parkstad, the International Building Exhibition in the region that is presenting its results in 2020-2021 in which sustainable and circular building is also an important theme.

Fitting in these developments, the goal of this conference is make the next step in the framework for enabling circularity in the built environment. The aim is to have a scientific conference on circularity in the built environment. The conference will result in a publication of a selection of best papers, and aims at giving guidance towards a circular built environment.

The conference is a collaboration between Zuyd University of Applied Sciences, IBA Parkstad and the City Region of Parkstad organization. List of committees:

**Organisation committee:**

- Michiel Ritzen, chair
- Nurhan Abuijdi, chair scientific committee
- Renske Bouten, member
- Paul Consten, member
- Wim Bakens, member

**Scientific committee:**

Ad Straub, TU Delft, the Netherlands
Agnes Schuurmans, SGS Intron, the Netherlands
Ahmed Z. Khan, ULB Brussels, Belgium
Ali Vahidi, TU Delft, the Netherlands
Anatolijs Borodinecs, Riga Technical University, Bulgary

Anne Paduart, VITO, Belgium
Antonín Lupíšek, Czech Technical University, Czech Republic
Caroline Newton, TU Delft, the Netherlands
Cecilia Furlan, TU Delft, the Netherlands
Charlotte Cambier, VUB Brussels, Belgium
Chris Geurts, TNO, the Netherlands
Christoph Maria Ravesloot, Rotterdam UAS, the Netherlands
Dario Cottafava, University of Turin, Italy
Dimitra Papadaki, National and Kapodistrian University of Athens, Greece
Elettra Agliardi, University of Bologna, Italy
Elke Knapen, UHasselt, Belgium
Eric Willems, Huygen Installatie Advisors, the Netherlands
Dirk Franco, Uhasselt/PXL, Belgium
Gino van Strijdonck, Zuyd UAS, the Netherlands
Griet Verbeeck, UHasselt, Belgium
Haico van Nunen, Rotterdam UAS, the Netherlands
Holger Wallbaum, Chalmers University of Technology, Sweden
Jacques Kimman, RVO, the Netherlands
John van Oorschot, Zuyd UAS, the Netherlands
Joop van Duren, Zuyd UAS, the Netherlands
Karen Allacker, KU Leuven, Belgium
Liane Thuvander, Chalmers University of Technology, Sweden
Lisanne Havinga, TU Eindhoven, the Netherlands
Luís Bragança, University of Minho, Portugal
Manuela Almeida, University of Minho, Portugal
Marc van den Berg, University of Twente
Marijke Steeman, Ghent University, Belgium
Marijn van de Weijer, Zuyd UAS, the Netherlands
Mark Kelly, GMIT, Ireland
Nikos Kalogeras, Zuyd UA & Wageningen University & Research, the Netherlands
Pascal Buskens, TNO, the Netherlands
Patrick Daly, TU Dublin, Ireland
Perica Savanovic, Avans UAS, the Netherlands
Peter-Paul van den Berg, Kamp C, Belgium
Petr Hajek, Czech Technical University, Czech Republic
Philippe Lemarchand, TU Dublin, Ireland
Rene Kemp, Maastricht University, the Netherlands
Richard Lorch, Buildings & Cities, UK
Rolf André Bohne, Norwegian University of Science and Technology, Norway
Ronald Rovers, TU Eindhoven, the Netherlands
Ruben Vrijhoef, TU Delft, the Netherlands
Shady Attia, Université de Liège, Belgium
Silu Bhochhibhoya, University of Twente
Sultan Çetin, TU Delft, the Netherlands
Targo Kalamees, Tallinn University of Technology, Estonia
Thomas Lützkendorf, KIT, Germany
Vanessa Gomes, Universidade Estadual de Campinas, Brazil
Veronique Vasseur, Maastricht University, the Netherlands

IOP Conf. Series: Earth and Environmental Science **855** (2021) 011001 doi:10.1088/1755-1315/855/1/011001

Wilfried van Sark, University of Utrecht, the Netherlands
Yvonne van der Meer, Maastricht University, the Netherlands

During the two day program we had 3 plenary speakers, 18 scientific parallel session, hosted the online closing conference of Super Circular Estate and hosted one special session of innovative district heating and cooling and seasonal storage. We are proud to look back at this interactive international online conference with a booming live chat, very interesting speakers and a great online closing conference of the Super Circular Estate project!

IOP Conf. Series: Earth and Environmental Science **855** (2021) 011002 doi:10.1088/1755-1315/855/1/011002

# Peer review declaration

All papers published in this volume of IOP Conference Series: Earth and Environmental Science have been peer reviewed through processes administered by the Editors. Reviews were conducted by expert referees to the professional and scientific standards expected of a proceedings journal published by IOP Publishing.

- **Type of peer review:** Double-blind
- **Conference submission management system:** Insight - https://insight.klinkhamergroup.com/crossingboundariesconference/account/closed
- **Number of submissions received:** 42
- **Number of submissions sent for review:** 42
- **Number of submissions accepted:** 34
- **Acceptance Rate (Number of Submissions Accepted / Number of Submissions Received X 100):** 81%
- **Average number of reviews per paper:** 2
- **Total number of reviewers involved:** 55
- **Any additional info on review process:** n/a
- **Contact person for queries:**
  Name : Nadia Debie
  Affiliation: Klinkhamer Group
  Email : nadia@klinkhamergroup.com

IOP Conf. Series: Earth and Environmental Science **855** (2021) 012001 doi:10.1088/1755-1315/855/1/012001

# Building energy simulations for different building types equipped with a high performance thermochromic smart window

D Mann[1,2], C Yeung[1,2], R Habets[1,2], Z Vroon[1,2,3] and P Buskens[1,2,4]

[1] The Netherlands Organisation for Applied Scientific Research (TNO), High Tech Campus 25, 5656AE Eindhoven, The Netherlands.
[2] Brightlands Materials Center, Urmonderbaan 22, 6167RD Geleen, The Netherlands.
[3] Zuyd University of Applied Sciences, Nieuw Eyckholt 300, 6400AN Heerlen, The Netherlands.
[4] Hasselt University (UHasselt), Institute for Materials Research (IMO), DESINe group, Martelarenlaan 42, 3500 Hasselt, Belgium

**Abstract.** With constantly progressing climate change and global warming, we face the challenge to reduce our energy consumption and $CO_2$ emission. To increase the energy-efficiency in buildings, we developed a thermochromic coating for smart windows which is optimized for intermediate climates. Here we present a building energy simulation study for the use of our smart window in the four main residential building types in the Netherlands. In the study we show that for all building types energy savings between 15-30% can be achieved. Hereby the impact of the windows on energy consumption is dependent on the window surface area as well as the total floor space. Furthermore we show that by the use of our new smart window, where the thermochromic coating is combined with a standard low-e coating, annual cost savings for energy between 220-445 € for a single household can be achieved. The thermochromic coating usually accounts for half of these cost savings, that is an addition in cost savings between 6-7.5 €/m$^2$ glass. Due to the low material and processing costs for the thermochromic coating, a return on invest within 7 years should be feasible with these annual cost savings.

With constantly progressing climate change and global warming, we face the challenge to reduce our energy consumption and $CO_2$ emission. In Europe, more than one third of our total energy consumption and $CO_2$ emission result from the building sector with more than 50% of this energy being used for Heating, Ventilation and Air-Conditioning (HVAC) systems [1]. A large part of this energy consumption results from heat loss through windows in cold conditions, or from cooling due to solar heat gain through windows. Therefore, windows account for more than 30% of the total energy consumption in buildings [2]. To improve the energy efficiency of windows and reduce the energy consumption of buildings, there are already several different energy-efficient glazing systems on the market. Nevertheless, the European building stock is still mostly equipped with dated, inefficient glazing. A recent study, commissioned by Glass for Europe, analyzing the energy and $CO_2$ emission savings potential, has shown that by equipping all buildings in Europe with energy-efficient glazing, the total energy consumption of the building sector can be reduced by over 30% for most countries in Europe. That means, if this transition were completed in the whole European Union, by 2030 annual energy saving of 75.5 Mt oil equivalents (with 1 kt of oil equivalent equals 11.6 GWh) and annual emission reduction of 94.5 Mt $CO_2$ would be achieved [3,4]. This shows the huge energy savings and emission reduction potential of not only the introduction of already available energy-efficient glazing systems, but also of the development of new materials and technologies already outperforming the commercial ones.

Current energy efficient glazing systems are either static infrared (IR) blockers, to reduce the transmission of solar heat into a building, low emissivity glazing, to increase insulation and reduce radiator heat loss through the window or a combination of both. These systems are suited best for either hot climates, where solar heat has to be blocked the whole year, or cold climates, where insulation and reduction of radiator heat loss is desired constantly. Current research and development focuses on adaptive systems, such as thermochromics [5], which can switch their solar heat gain properties depending on the building's needs. These properties are required in intermediate climates with hot summers and cold winters.

Thermochromics can switch their IR modulating properties from transmissive to blocking depending on the glazing temperature [5]. In a recent study we have shown that smart windows, where our thermochromic coating is combined with a standard low-e coating, are perfectly optimized for intermediate regions, where heating and cooling are equally important over the year [6]. In this study we have analyzed a free standing residential building in 10 different climate zones. For intermediate climates energy savings of up to 22% could be achieved with our smart window, outperforming all compared commercial energy-efficient glazing systems. Due to the nature of the study and to be able to compare the impact of the window in different climate regions, the rest of the building envelop was kept constant and not adapted to each location. For a more detailed analysis of the performance of the thermochromic smart widow in the Netherlands, we here present a study on the 4 main building types. In 2015, 23% of homeowners in the Netherlands lived in free-standing buildings, 19.6% were living in duplex houses, 42.5% lived in terrace houses and 15% in apartments [7]. Therefore we chose those 4 building types for a building energy simulation study.

In our labs we have developed a thermochromic coating with excellent optical properties for application in smart windows. The functional material in the coating is $VO_2$,which changes its crystal structure and therefore its optical properties at 68°C [5]. By use of metal ion doping, the switching temperature can be reduced to temperatures suitable for application, e.g. at 20°C [6,8]. The coating developed in our labs shows high visible transmission ($T_{vis}$) of 74% combined with a high solar modulation ($\Delta T_{sol}$) of 10% [9] (Figure 1). This combination of high $T_{vis}$ and $\Delta T_{sol}$ is unrivaled by thermochromic $VO_2$ based coating reported in literature to date [10].

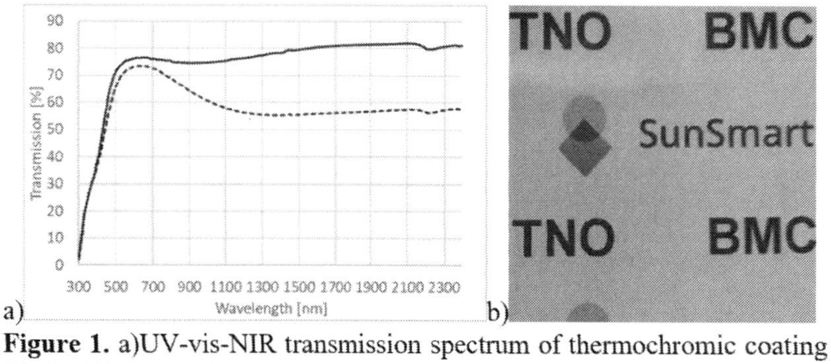


**Figure 1.** a)UV-vis-NIR transmission spectrum of thermochromic coating in cold (solid line) and hot (dashed line) stae. b) picture of coated glass plate.

For the building energy simulation study we integrated the thermochromic coated glass in an insulating glazing unit (IGU) with a low-e coated glass (Saint-Gobain ECLAZ) and a 13 mm argon filled gap, reaching a U value of 1.1 $W \cdot m^{-2} \cdot K^{-1}$. The selected buildings were designed according to current new building construction projects in the Netherlands found on the internet (Figure 2). For duplex and terrace houses as well as for apartments, one unit was choses instead of the total building complex. For terrace houses one of the central units was chosen. Wall material was selected according to current

Dutch regulations reaching an *Rc* value (thermal resistivity of the composite) of 4.7 $m^2K \cdot W^{-1}$. We used the software Energy Plus version 9.2.0 [11] for the simulations. The selected simulation period was one year with data points gathered monthly. All other parameters were used as reported in our previous building energy simulation study [6].

**Figure 2.** Selected buildings for building energy simulation study on thermochromic smart window. a) free-standing, b) apartment, c) terrace house and d) duplex house.

As a reference point, we simulated the annual energy consumption of the 4 building units equipped with an IGU made of two clear glass panes (Pilkington Optiwhite) and a 13 mm argon filled gap. The energy demand of these reference buildings were then compared to the simulated annual energy consumption for the same buildings equipped with our thermochromic smart window. The selected free-standing building, duplex and terrace house had a similar ratio of $m^2$ window surface to $m^2$ living space. The most significant difference between those 3 buildings was the total floor area and the number of walls in contact with the outside environment. Here the free-standing building with the largest floor area and all 4 outside walls in contact with the environment showed the largest impact of the smart window on annual energy consumption. Here annual energy savings of 21.2% could be reached (Figure 3). With reduced floor area and either three or two outside walls being in contact with the environment, for duplex and terrace building respectively, the impact of the smart window on total energy consumption declined. Here annual energy savings of 16.1 and 15.4% were achieved for the duplex and terrace house, respectively (Figure 3). The apartment was a more special case, since it is a one floor unit with large window surfaces on the two outside walls in contact with the environment. Here the big window surface area in relation to the total floor area, leads to a big impact of the thermochromic smart window on annual energy consumption. Here annual energy savings of 29.4% could be reached in comparison to standard IGUs (Figure 3).

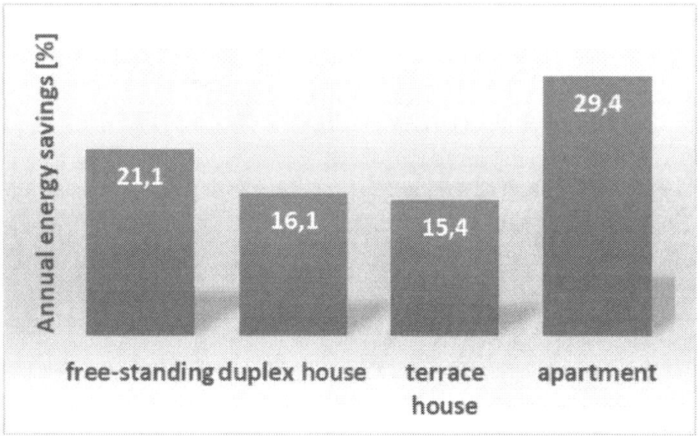


**Figure 3.** Annual energy savings of different building types in the Netherlands equipped with thermochromic smart windows in comparison to standard clear IGUs.

The simulated energy consumption and savings could be divided into heating, cooling and lighting contributions. Suggesting that natural gas was used for heating and electricity for cooling and lighting, the average cost price for gas and electricity in the Netherlands could be used to calculated annual cost savings for each building type. Here the annual cost savings are very much dependent on the total energy costs for each building type. Therefor the building type with the highest annual energy savings in %, doesn`t necessarily also have the highest total annual cost savings. Here the free-standing building showed with 445 € the largest annual cost savings of all building types (Figure 4), due to the highest total energy demand. When subtracting the cost savings, which can be attributed to the low-e coating in the smart window, the added cost savings per m$^2$ attributed to the thermochromic coating can be calculated. For the free-standing building, the duplex and the terrace house, the contribution of the thermochromic coating was approximately 50%, which leads to annual cost savings between 6-7.5 €/m$^2$ window. Due to the cheap material and processing cost a return on invest within 7 years will be possible with these attributed cost savings. Due to the big impact of the thermochromic smart window on cost savings for the apartment, here annual cost savings attributed to the thermochromic coating in the smart window of 14.5 €/m$^2$ window can be achieved, making the new technology very attractive for home owners.

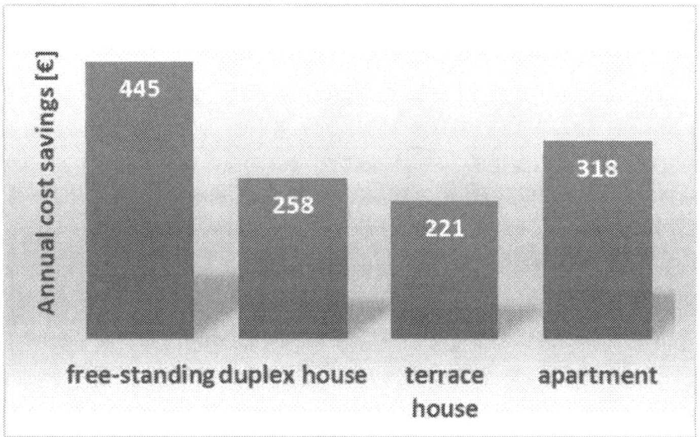


**Figure 4.** Annual cost savings of different building types in the Netherlands equipped with thermochromic smart windows in comparison to standard clear IGUs.

In this report, we showed the results of a building energy simulation study on the four main building types in the Netherlands. We compared the annual energy savings for use of a thermochromic smart window in a free-standing building, a duplex house, a terrace house and an apartment to standard clear

IGUs. The thermochromic coating developed in our labs shows a combination of high $T_{vis}$ and $\Delta T_{sol}$, which is unrivaled in scientific literature on thermochromic $VO_2$ based coatings. In our simulation study we could show that the use of a smart window, where our thermochromic coating is combined with a standard low-e coated glass, can lead to annual energy savings between 15-21% in free-standing buildings, duplex and terrace houses. For apartments with large window areas even higher energy savings of up to 30% can be reached. In total the thermochromic coating adds between 6-7.5 €/m² in cost savings to standard low-e glass, which is currently used in new built homes. Here due to low additional production costs, a return on invest within 7 years is feasible. For apartments with large window surface areas even higher cost savings are possible, up to 14.5 €/m² window.

**Acknowledgements**

The research work in this paper was carried out within Brightlands Materials Center, a joint research initiative between TNO and the province of Limburg. The work is funded by the Dutch government as well as the province of Limburg. Additionally, the work was partly performed within the Interreg Vlaanderen-Nederland project EnEf and within the Sia RAAK Pro project Window of the Future

**References**

[1] Mardiana A and Riffat SB 2015 *J. Earth Sci. Climate Change* **S3** 001.
[2] Feist W and Schnieders J 2009 *Eur. Phys. J. Spec. Top.* **167** 141–153.
[3] TNO. *Potential Impact of High-Performance Glazing on Energy and $CO_2$ Savings in Europe*; TNO: Delft, The Netherlands, 2019.
[4] https://glassforeurope.com/glazing-saving-potential-2030-2050/ (accessed on 24 November 2020).
[5] Cui Y, Ke Y, Liu C, Chen Z, Wang N, Zhang L, Zhou Y, Wang S, Gao Y and Long Y 2018 *Joule* **2** 1707–1746.
[6] Mann D, Yeung C, Habets R, Vroon Z and Buskens P 2020 *Energies* **13** 2842.
[7] Statistics Netherlands (CBS) 2016 https://www.cbs.nl/nl-nl/nieuws/2016/14/vier-op-de-tien-huishoudens-wonen-in-een-rijtjeshuis (accessed on 24 November 2020)
[8] Zhang J, He H, Xie Y and Pan B 2013 *J. Chem. Phys.* **138** 114705.
[9] $T_{vis}$ and $\Delta T_{sol}$ are defined as $T_{vis,sol} = \frac{\int \Phi_{vis,sol}(\lambda)T(\lambda)d\lambda}{\int \Phi_{vis,sol}(\lambda)d\lambda}$ and $\Delta T_{sol} = T_{sol,cold} - T_{sol,hot}$, where $T(\lambda)$ is the transmittance at wavelength $\lambda$, $\Phi_{vis}(\lambda)$ is the photoscopic spectral sensitivity of human vision, and $\Phi_{sol}(\lambda)$ as the AM 1.5 solar irradiance spectrum.
[10] Yeung C, Habets R, Leufkens L, Colberts F, Vroon Z, Mann D and Buskens P 2020 *submitted manuscript.*
[11] Energy Plus Building Energy Simulation Software, US Department of Energy. Available online: https://energyplus.net/ (accessed on 24 November 2020).

IOP Conf. Series: Earth and Environmental Science **855** (2021) 012002 doi:10.1088/1755-1315/855/1/012002

# Circular building with raw earth: a qualitative assessment of two cases in Belgium

**Pelicaen E[1], Janssens B[1], Knapen E[1]**

[1]Faculty of Architecture and Arts, Hasselt University, Belgium

erik.pelicaen@uhasselt.be

**Abstract**. The built environment puts high pressure on our planet, and a great deal is related to resource extraction, material production and waste generation. In the context of circular construction, buildings must be designed and built in order to keep our natural resources in closed material loops for as long as possible. Raw earth has regained attention in the building industry as an abundant, low-impact and highly recyclable building material. However, little is known and experienced about the implementation of raw earth in circular building design. Therefore, this research offers a better understanding of the circularity of earth architecture by assessing two contemporary Belgian cases. Based on literature, semi-structured interviews and the analysis of technical documents, the circularity of the two cases is qualitatively assessed at different scales and levels. It appears that circularity is highest on the material scale and lowest on the building scale for both cases. It is also found that earth as a building material does not easily fit in existing circular assessment frameworks. This investigation represents a contribution towards the development of design support for circular building with raw earth.

## 1. Introduction

Globally, the built environment is responsible for the largest shares of final energy use, carbon emissions, resource depletion and waste generation [1–4]. To counteract these issues, demands to include circular economy principles in the construction sector are increasing [5,6]. Another tendency is the growing interest in vernacular building materials with various environmental benefits such as raw earth [7,8]. Raw, crude or unbaked earth is considered to be one of the oldest and most widely used building materials [7,9,10]. Although its use was almost abandoned in more industrialized countries, it has resurfaced as a low-impact, widely available and highly recyclable building material [7,11–13].

Raw earth used for construction is a mix of subsoil fractions of different particle sizes including silts, sands and possibly gravels within a clayey binding matrix providing cohesion [13,14]. The most common earthen construction techniques are wattle and daub (pressing earth onto a woven lattice of wooden strips in a timber load-bearing structure), cob (layering of an earth-straw mix to form masonry walls), rammed earth (compacting of consecutive earth layers in a formwork), adobe (masonry of moulded or extruded earth bricks) and compressed earth blocks (masonry of manually or mechanically pressurized earth blocks) [7,12,15,16].

Considering the global challenges the built environment is facing today, the interest for applications of raw earth in circular construction can grow. However, little is known about the circularity of earth architecture. Therefore, it is essential to have a clear understanding of the potential of this material in the context of a circular built environment. The circular economy and circular building are broad terms to grasp as they are a sustainability paradigm requiring a systemic worldview [17]. Therefore, the scope of this research is limited to the environmental and technological dimensions situated at the micro-scale (resources, materials and components) of circular building research [6]. This research aims to start filling this gap by assessing the circularity of two contemporary earth buildings in Belgium. Within a more extensive research project [18], it contributes towards the development of design support for circular building with raw earth.

## 2. Methodology

A qualitative assessment of the circularity of earthen constructions was performed on two cases (**Figure 2**) of which the selection criteria were as follows: located in Belgium, contemporary architecture (built after 2000), containing load-bearing or self-supporting earthen building elements, and two different earthen construction techniques. The Bio Class, designed by BC architects in 2015, is a centre for nature education which was built with compressed earth blocks (CEBs) inside a warehouse located on the Fort V-site in Edegem. The Bee Hall, designed by DAM architects in 2017, has a similar function in addition to housing beehives and was built with rammed earth as part of the educational garden The Helix in Geraardsbergen. Both projects were realized with the same (sub)contractor: Het Leemniscaat.

**Figure 2.** *Left*: Load-bearing CEB-walls with exterior hempcrete insulation in the Bio Class (©Thomas Noceto). *Right*: Naked load-bearing rammed earth walls in the Bee Hall (©Ken Dupont).

### *2.1. Research strategy*

First, based on literature, a framework was developed to assess the circularity of the cases at different scales. Second, to assess the circularity of the two cases, three research methods were used: a literature review on earthen construction techniques in general, followed by case-specific semi-structured interviews and the analysis of technical documents. Literature was reviewed to assess the circular potential of earthen constructions. This was verified and complemented by data from the cases: semi-structured interviews with stakeholders and the analysis of technical documents. Data triangulation resulted in the verification of the circular potential of two earthen buildings.

### 2.2. *Data collection and analysis*

To develop a circular assessment matrix for this research (section **3.1**), literature was reviewed about circular building and existing circular assessment frameworks. In parallel, properties of earth as a building material and earthen constructions were inventoried through literature as well. Based on literature, a first general assessment of the circularity of earthen constructions was performed.

These findings were verified and complemented by a case study analysis of two earthen buildings. The respondents were selected based on their involvement in the design and construction of the cases: two architects and a contractor. The three semi-structured interviews took place at the office of the respondent, except for one which was conducted online due to COVID-19 measures. The recorded interviews were held in Dutch, took approximately 90 minutes, and were transcribed in their original language. The context of the interview was introduced with general open questions about their knowledge, opinion and experiences with earth construction and circular building. This was followed by case-specific questions, in which the cases were assessed on different scales. To stimulate the conversation and to avoid respondent's biases, a card-sorting method was used. For different scales and levels, a selection of circularity indicators (**Tables 3, 4** and **5**) was written on cards that were sorted by the respondents according to what extent it was applied in the case. By use of this method, the interviewees were encouraged to be critical about the circularity of their project. In the online interview, the card-sorting method was done with the *Miro* online visual collaboration platform. The transcribed interviews were coded and analyzed according to the circularity indicators and underlying principles.

Technical documents from the cases were obtained from the respondents and were analyzed using the circular assessment matrix. The most used documents were architectural and executive plans, construction details, visual media of the building process, and construction specifications.

## 3. Results

In the following paragraphs, a circular assessment matrix that will be used for this research is presented (section *3.1*) and the circularity of the two cases is assessed by means of this matrix on three different scales, i.e. the material scale, the element scale and the building scale (sections *3.2*, *3.3* and *3.4*).

### 3.1. *Developing a circular assessment matrix*

To assess the circularity of raw earth as a building material and earthen construction techniques from an environmental and technological perspective, the assessment must be done at different scales. Since the desired assessment scales for this research are seldom found together in one framework, a new circular assessment matrix was developed by combining existing frameworks found in literature.

First, the assessment is performed at three scales: the building material scale, corresponding to the resources or building products; the building element scale, corresponding to the parts of a building that fulfil a specific set of functions (e.g. an outer wall); and the building scale, corresponding to a composition of building elements offering space to people, activities and goods [19]. Second, each scale contains different levels at which the case can be assessed: the components level, the interfaces level, and the composition level. The scales and levels correspond to the rows and columns of the matrix in **Table 1,** respectively. The levels were adopted from the element and building scales of a qualitative assessment framework by OVAM [19] containing key principles of Design for Change, mainly inspired by Vandenbroucke [20]. To ensure coherence, the authors chose to maintain these levels at the material scale. At the material scale, the components refer to the extracted resources used for the building product, the interfaces refer to the bonds between the materials of the building product, and the composition refers to the assembly of different materials to manufacture the building product. At the element scale, the components refer to the design of the building products, the interfaces refer to the interaction of different components, and the composition considers the assembly of components in a building element. At the building scale, the components refer to the design of the building elements, the interfaces consider the interaction between different building elements, and the composition refers to the assembly of building elements [19,20].

Each field in the matrix of **Table 1** contains *circular principles* that will be assessed in the cases. At the element and building scale, these were adopted from [19]. At the material scale, these were selected from various sources. Sixteen 'circular design qualities' were summarized by Galle et al. [21], six of which apply to the material scale: reused, recycled, renewed, compostable, safe and healthy, and pure. The EMF [22] considers different 'material circularity indicators': recyclability, reuse, utility, scarcity, toxicity, and impact. A study to include 'circularity indicators' in BREEAM [23] indicates that material selection must be made keeping renewability, toxicity, scarcity and impact in mind. According to the WTCB [24], the integration of circular thinking in the selection of materials and products consists of choosing pure, non-toxic, renewable, reusable, recyclable, biodegradable or compostable, local materials with a low environmental impact. **Table 1** shows the resulting choice of circular principles.

At each assessment scale and level, the circular principles contain a set of indicators that were defined based on literature (**Tables 3, 4** and **5**). In the following sections, at all scale levels and for each indicator, the circular potential is assessed based on literature, and verified in the cases based on the interview and project data using the values described in **Table 2**.

**Table 1**. Circular assessment matrix with principles in each field (adapted from [19]).

| **Scale ↓** | **Level →** | Components | Interfaces | Composition |
|---|---|---|---|---|
| Material | | Resources<br>*Renewability*<br>*Scarcity*<br>*Toxicity* | Bonds between materials<br>*Biodegradability* | Assembly of materials<br>*Purity*<br>*Impact*<br>*Recovery* |
| Element | | Building products<br>*Durability*<br>*Manageability*<br>*Compatibility* | Joints between components<br>*Reversibility*<br>*Simplicity*<br>*Speed* | Assembly of components<br>*Independence*<br>*Pace-layering*<br>*Prefabrication* |
| Building | | Building elements<br>*Demountability*<br>*Reusability*<br>*Expandability* | Joints between elements<br>*Accessibility* | Assembly of elements<br>*Versatility* |

**Table 2.** Assessment values and corresponding meanings.

| **Symbol** | **Circular potential** | **Verification** |
|---|---|---|
| + | High | Potential is applied in case |
| – | Low | Potential is not applied in case |
| ? | Not found in literature | Not verifiable |

*3.2. Circular assessment at the material scale*

**Table 3** shows the circular assessment of raw earth at the material scale. The formation of raw earth is an ongoing chemical or physical weathering process of rocks that takes place constantly [15]. However, one may argue that it is a non-renewable resource as these processes largely exceed the human time scale and therefore, cannot be regarded as such [7]. Moreover, renewable resources are commonly referred to as biologically sourced or regrowable materials [21,22]. Then again, it can be argued that recyclable materials are considered renewable in the technical cycle of circularity [23]. Except for permafrost regions, naked bedrocks, and sand deserts, the availability of subsoil is pervasive [13]. However, the composition of the earth mix varies between construction techniques, and all components are not always locally available [25]. Earth is not considered to be a toxic material. On the contrary, multiple studies show that it improves indoor air quality [11], as it buffers relative humidity thanks to its hygroscopic behaviour [16]. The ability of raw earth to plasticize when water is added makes the recycling possibilities practically endless. By crushing, drying, and sieving, one can separate the earthen material and retrieve the virgin resources. However, the purity of the material affects its technical suitability when recycled as the presence of stabilizers or additives in the mix may significantly decrease the plasticizing properties [12]. Earthen building materials can be considered biodegradable since no chemical bonds hinder natural decomposition. However, if stabilizers or additives are used, rapid biodegradation of earthen materials is only possible in the case where these are organic [26]. Composting might be possible if organic additives can be separated from the earth, but no information was found in literature. Pure earthen construction wastes can be returned to the site without environmental hazards [7,12], but landscape transformation and changes in the soil composition are possible. Life Cycle Assessment (LCA) of CEB and rammed earth shows that the two processes with the largest impact are the transportation of soil and the use of stabilisers such as lime or cement [27]. Regarding material recovery, Schroeder [12] and Dethier [28] mention that the transformation of excavation soil into earthen building materials could result in lower demand for landfill space.

In the Bio Class, the earth mix contains 50% clay from a nearby quarry (~10 km) which was transported by truck, and 50% of two fractions of sand from a local distributor (~20 km) which was transported by boat from an unknown source and further transported by truck to the construction site. Wet clay lumps were mechanically crushed to dry on-site for four weeks with two ventilators naturally. Once dried out, the clay was mixed with the sand in an electric mixer, and the earth mix was compressed with a hydraulic press at 15 MPa into 19,000 blocks with a compressive strength of 3,3 MPa. Per hour, 200 CEBs were manufactured in a team of five to six volunteers during a three-week workshop. No additives nor stabilizers were used in the CEBs. However, the earth mortar was stabilized with 10% of cement, as it was designed in accordance with French earth building regulations for load-bearing CEB-walls. Recyclability, biodegradability and purity are therefore affected since, at the end-of-life phase of the building, the CEBs cannot easily be separated from the stabilized earth mortar.

In the Bee Hall, excavation soil – which is usually considered as waste – from the site was used in the rammed earth mix. However, a reformulation with additional clay, sand and gravel fractions from two distant quarries (~100 km) were needed to make the mix suitable for ramming. The exact proportions of the fractions are unknown. The mixing was done ex-situ and was transported in big-bags to the construction site by truck. No additives or stabilizers were used.

**Table 3.** Circular potential at the material scale of earth as a building material based on literature and verified in the Bio Class (BC) and in the Bee Hall (BH).

| Level | Principle | Indicator | Lit. | BC | BH |
|---|---|---|---|---|---|
| | Renewability | Rapid replenishment of virgin resource | – | ? | ? |
| | | Regrowth at least as fast as functional service life | – | ? | ? |
| | Scarcity | Abundancy in the Earth's crust | + | ? | ? |
| | | Local availability of resources | + | + | – |
| | Toxicity | No emittance of volatile organic compounds (VOC) | + | + | + |
| | | No exposure to health risks of labourers in the life cycle | + | + | + |
| | Biodegra-dability | Ability to naturally decompose rapidly | + | – | + |
| | | Ability to be used as mulch or compost | ? | – | – |
| | Purity | No contamination in the life cycle of the material | + | ? | ? |
| | | Selection of untreated/unfinished materials | + | + | + |
| | | Limitation of different materials combined in a product | + | + | + |
| | | Efficient separation of materials manually/mechanically | + | – | + |
| | Impact | Low environmental impact throughout the life cycle | + | + | + |
| | | Low embodied energy throughout the life cycle | + | + | + |
| | Recovery | A proportion of residual waste stream in the product | + | – | + |
| | | A proportion of recycled material in the product | + | – | – |

### *3.3. Circular assessment at the element scale*

**Table 4** shows the circular assessment of the two cases at the element scale. Little research was found about the circularity of earthen constructions at this scale. By far, most of the literature is related to durability as it has become an increasingly popular topic over the last two decades [14] and draws the most attention from the general public [29]. Water is the most detrimental environmental agent to durability [14]. The key erosion mechanism is related to the kinetic energy from raindrops impacting the wall surface [29]. Moreover, more frequent maintenance is needed with earth constructions compared to concrete or fired bricks [13]. The addition of stabilizers such as lime or cement considerably improves durability [9], yet according to [13], this process merely upgrades earth into a low-quality concrete at a high environmental cost. Unstabilized earthen walls are durable – even in wet climates – provided that the architectural design principles of a "good hat" (roof) and "good boots" (foundation) are respected [13]. However, the ability to pass durability tests is one of the main reasons why stabilization is used in everyday practice. Then again, many of the durability tests designed for conventional building materials are not representative of earthen walls due to a large discrepancy between lab tests and natural conditions [9,13,14]. Improvements are being made to attain conformity, e.g. by selecting adequate tests for case-specific conditions [30]. However, quantitative observations within a large time scale in natural conditions turn out to be the most reliable way to assess the durability of earthen constructions [31]. Moreover, calculated erosion can be used to improve the durability of exposed rammed earth walls. The walls are built a few centimetres thicker than initially planned, and so-called erosion checks (horizontal layers of protruding fired clay or trass-lime mortar) decelerate the water flow down the wall causing the abrasion [32,33]. According to OVAM [34], manageability and pace-layering of vertical load-bearing structures are independent of the materiality. Regarding the manageability of vertical load-bearing structures, massive blocks are characterized by small dimensions and are, therefore, easy to handle [34]. However, the low speed of (dis)assembly relative to larger components is often considered an adverse effect [34]. Massive walls have large dimensions and therefore have low manageability, with the result that they are only reusable in a similar design. Adobes are considered to have compatible dimensions [34], and therefore, the compatibility potential for CEBs can be deemed similar. Prefabricated rammed earth is also regarded as compatible by OVAM [34]. Apart from a load-bearing function, massive structures mostly have a space-dividing function and therefore structural interventions are needed when one wishes to change the laout of the building. Therefore, pace-layering potential of adobes and rammed earth walls are considered low. Although this has not yet been tested, the reversibility of a CEB wall is likely to be dependent on the cohesive strength of the blocks and the mortar used. Unstabilized earth mortar is considered reversible according to Rauch [33]. Fired bricks layered with lime mortar can be reclaimed undamaged with simple tools [34]. However, this is a labor-intensive process that may be limited through prefabrication of larger modules. Blocks cannot be removed separately and are therefore not considered independent as stated by OVAM [34]. In the case of a monolithic rammed earth wall, independence can be considered valid since the wall is the component itself. Regarding prefabrication, a higher degree of quality and facilitation of the construction process can be obtained by off-site industrial production of rammed earth elements. Mechanical prefabrication machines, as developed by Lehm Ton Erde in Austria, can decrease the labor-intensity and accelerate the production process [33].

In the Bio Class, all CEB-walls are load-bearing and are either single- or double-layered using a conventional masonry technique (cross bond). The 5,5 kg weighing CEBs of 29×14×7 cm were dimensioned according to the hydraulic press standard and did not correspond to standards of conventional fired bricks. The blocks were carried from pallets with a brick clamp and layered with the cement-stabilized earth mortar by two labourers. Most walls contain an arch-shaped glass door opening with a diameter of 3,5 m and are separated from a concrete plinth by an impermeable layer to prevent the capillary rise. The CEB-walls support flat timber rafters with external hempcrete insulation covering the whole building. On the interior, the blocks are left apparent without finishing, and on the exterior, a wooden frame is attached to guide the loose-fill hempcrete formwork. The self-supporting hempcrete facade is fixed to the plinth by a steel profile connected to the foundation. Glazed bricks were used in the sanitary unit due to the CEB's vulnerability to water.

Regarding the components (CEBs), the respondent from BC architects appointed the highest score of circularity in the card-sorting exercise to the ease of maintenance and repair (durability) along with architectural possibilities (compatibility). Lowest scores were given to the requirement for maintenance and damage associated with transportation (durability). To extend the service life of components, the architect highlights the importance of a good design to protect the earth from humidity and to prevent cracking due to point loads. During construction, the wet hempcrete humidified the clayey CEBs in its drying process and caused the arches to sag 1 or 2 cm due to a decrease in stiffness. Damage due to transportation is more common with CEBs than with other building materials. Maintenance has not occurred since the project was finished recently, except for final treatment with a sponge float of the interior wall surface to make the transition of the block edges and the mortar less sensitive. Regarding the interfaces (earth mortar joints and other connections), visible and accessible joints (speed) scored highest along with disassembly without damage (reversibility). Lowest scores of circularity were given to the number of joints (speed) and the use of standard joints and tools (reversibility). Although standard joints and tools are used, the architect mentions a somewhat higher technical complexity than fired brick masonry or concrete walls due to the CEB's limited strength. Regarding the composition (assembly of CEBs with other materials in the wall), the independence of components scored highest and the added value of prefabrication degree lowest. The hempcrete skin and CEB structure are separated, but the services are embedded in the skin and perforate the structure. The architect acknowledges the possibility of separating the services as well but states this was a design choice. To the question how the Bio Class could be disassembled and reassembled on another location, the architect suggested comparing the cost and time efficiency of two methods: the disassembly of CEBs in one wall and the demolition and remanufacturing of CEBs in another. In the first case, the time (and therefore cost) invested in recovering all blocks could be too high as a percentage of loss is inevitable due to disassembly and transportation. In the second case, there would be no loss, and the demolition would be faster, but the cost to remanufacture could be higher.

In the Bee Hall, the two zig-zag rammed earth walls are load-bearing and 35 cm thick. One wall is interrupted by a steel pivoting door and is perforated by a hexagon-shaped opening of about 60 cm in diameter. The two walls are connected on the interior by a window and a timber-framed partitioning wall. As in the Bio Class, the walls stand on a concrete plinth separated by an impermeable layer. The walls support a flat timber green roof with an overhang around the perimeter of the building to protect the rammed earth from precipitation. During two two-day workshops with four volunteers and one labourer, the rammed earth mix was manually shovelled from big-bags into the standardized concrete formwork. The earth was rammed layer by layer with a manual and a pneumatic rammer.

Regarding the components (in this case the rammed earth walls), the respondent from DAM architects appointed the highest score to cards with durability indicators except for the damage associated with transportation. Other lowest scores were given to manageability and standard dimensions (although the latter was a design choice). The architect believes maintenance is not needed as long as vandalism does not occur, since the roof design protects the rammed earth from incident rainfall. Regarding the interfaces (connections of the wall with other materials), the architect gave the highest score to speed and the lowest to reversibility. Reinforcement bars run 40 cm through the walls every meter, at the bottom through the concrete plinth and at the top through the timber divider beam. The latter is connected to the wall by an earth mortar without stabilizer. At the level of the windows, a timber column was rammed within the walls to connect the window frames to the supporting structure. The architect argues disassembly of the walls cannot be done without damaging the rammed earth unless performed with extreme care and therefore neither cost nor time-efficient. Regarding the composition (assembly of the wall), the architect appoints the highest score to independence and the lowest to prefabrication. The architect also argues the added value of prefabrication and the associated transportation damage.

**Table 4.** Circular potential at the element scale of compressed earth blocks (CEB) and rammed earth (RE), based on literature and verified in the Bio Class (BC) and in the Bee Hall (BH).

| Level | Principle | Indicator | CEB | | RE | |
|---|---|---|---|---|---|---|
| | | | Lit. | BC | Lit. | BH |
| | Durability | Long expected service life | + | + | + | + |
| | | Little maintenance required | – | – | – | – |
| | | Easy to repair or remanufacture | + | + | + | + |
| | | Little damage associated with transportation | ? | – | + | – |
| | Compatibility | Standard shapes and sizes of components | + | + | + | – |
| | | Exchange of components with other buildings | + | + | + | – |
| | | Capability to form new configurations | + | + | + | – |
| | Manageability | Small-sized components | + | + | – | – |
| | | Lightweight components | + | + | – | – |
| | | Ergonomically stackable components | + | + | – | – |
| + | Reversibility | Disassembly of components without damage | + | ? | ? | – |
| | | Reusability of the joints | + | – | + | + |
| | Simplicity | Standard joints and tools | + | + | + | + |
| | | No need for specialized training | + | + | + | + |
| | | A small number of different components | + | + | + | + |
| | Speed | A small number of joints | – | – | + | + |
| | | A small number of different types of joints | + | + | + | + |
| | | Visible and reachable joints | + | + | + | + |
| | Pace-layering | Layering of components according to lifespan | – | – | – | – |
| | | Layers with shorter lifespan closer to surface | – | – | – | – |
| | Independence | Geometry allows individual removal | – | – | + | + |
| | | Disassembly from multiple directions | – | – | + | + |
| | Prefabrication | Off-site preassembly into larger modules | ? | – | + | – |

### 3.4. *Circular assessment at the building scale*

**Table 5** shows the circular assessment of the two cases at the building scale. It is important to note that in this case, the component level tackles the building elements in their entirety, i.e. the CEB walls with hempcrete insulation in the case of BC and the rammed earth walls in the case of BH. The interfaces level deals with the connections between these walls, and the composition addresses the assembly of these walls in the building volume. No literature was found directly related to the circular principles on this scale, except for the low accessibility of the joints in cross-bonding brickwork [34].

The Bio Class is a rectangular single-storey volume positioned in oblique inside a rectangular warehouse. The building (surface area ~350 m$^2$) is subdivided into six square volumes of about 25 m$^2$. Except for one wall delimiting the bathroom area, all internal walls have the same arch-shaped openings as most external walls and some spaces are separated by glass doors. The architect appointed the highest score to the demountability of the infill and the ease of reinforcing supporting elements. The lowest score was given to movable building elements. The supporting elements cannot be disassembled as a whole, but the infill elements such as the glazed-brick partitioning walls can possibly be disassembled from the supporting structure. Reuse of the supporting elements as a whole is excluded, and reusability of the infill elements in other buildings depends on their compatibility. Although the CEBs are slightly over-dimensioned due to higher safety factors as chosen by the structural engineer, this was not a design choice with the prospect of expandability. The cross-bonding brickwork hinders the accessibility of supporting elements as one element cannot be separated from the other. The layout can be considered multipurpose thanks to the identical shapes and sizes of the spaces, but it is not transformable since the elements are not movable or demountable.

The Bee Hall (surface area ~50 m$^2$) is a single-storey volume comprising a hexagonal space in between two hexagon halves, open at one end and closed by a glass pane at the other. Three spaces are divided by a timber partitioning wall and a window. The architect appointed the highest score to the demountability and reusability of the infill and gave the lowest scores to the over-dimensioning of technical elements and the transformable lay-out. The supporting elements cannot be disassembled as a whole, but the infill elements such as the timber partitioning wall can be disassembled from the supporting structure. As in the Bio Class, reuse of the supporting elements as a whole is excluded, and reusability of the infill elements in other buildings depends on their compatibility. Like the Bio Class, the safety factors were taken higher by the structural engineer, but the design was not intended for expansion. For the same reason, technical elements were not over-dimensioned. Reinforcement of supporting elements seems a difficult option as the synergy between the existing and the added wall must be ensured. Accessibility of the supporting elements is complicated because of the reinforcement bars at the bottom and at the top, as well as the wooden columns inside the rammed earth walls connected to the window frames. Given the size and shape of the building, the layout is not considered multipurpose. As in the Bio Class, the immobile and undemountable supporting elements do not allow the layout to be transformed.

**Table 5.** Circular potential at the building scale of the Bio Class (BC) and the Bee Hall (BH), based on the interviews and the analysis of technical documents.

| Level | Principle | Indicator | BC | BH |
|---|---|---|---|---|
| | Demountability | Selective disassembly of the support | – | – |
| | | Selective disassembly of the infill | + | + |
| | Reusability | Reusable elements in the same building | – | – |
| | | Reusable elements in other buildings | – | – |
| | Expandability | Over-dimensioning of supporting elements | – | – |
| | | Over-dimensioning of technical elements | – | – |
| | | Ease of reinforcing supporting elements | + | – |
| | Accessibility | Visible and reachable joints | – | – |
| | | Reversible joints between elements | – | – |
| | Versatility | Multipurpose building layout | + | – |
| | | Transformable building layout | – | – |
| | | Movable building elements | – | – |

## 4. Discussion

In the following paragraphs, the results of the assessment are discussed on three scales by comparing theory and practice on the one hand and by comparing the two earthen construction techniques on the other. Additionally, improvement potential for the cases and particular potential compared to more conventional building materials are discussed. Furthermore, critical reflections are made for the methodology, the framework and the assessment; and recommendations are given for future research.

### *4.1. Interpretations and implications*

At the material scale, the results indicate that, based on findings in literature, the circular potential of earth as a building material is high for the majority of principles. The results also show that the circular potential that was identified based on literature, highly corresponds with the verification in the two cases. However, it must be said that some principles are independent of the studied cases, i.e. the replenishment of earth as a virgin resource by weathering of rocks and the abundance of these resources in the Earth's crust. Possible improvements in the Bio Class are the elimination of cement stabilizer in the earth mortar increasing the purity, biodegradability and reducing the impact; more local resourcing of the sand fraction reducing the impact and using a proportion of excavation soil or recycled earthen material. Improvement potential in the Bee Hall lies in the more local resourcing of the clay, sand and gravel fractions needed to reformulate the excavated soil to make it suitable for ramming. Particular potential lies in the low impact, low toxicity and low scarcity, and in the high purity and high biodegradability. Moreover, the use of excavation soil in the earth mix seems particularly promising in terms of circularity.

At the element scale, the results demonstrate that the circular potential is relatively high and mainly corresponds with the verification in the cases. By comparing the techniques, the overall manageability is high for CEB and low for RE, whereas the overall independence is low for CEB and high for RE. In addition, the speed of disassembly would be lower with CEB than with RE because of the many joints compared to none in a RE wall. Although pace-layering is ruled out for massive vertical load-bearing structures, the structure and skin are separated in the Bio Class, and the structure and skin are one and the same in the Bee Hall. Additionally, it must be noted that some principles are less related to the earthen material than others. For instance, manageability depends more on the component's geometry than on the materiality. Likewise, pace-layering depends more on the chosen structure design than on the materiality (columns and beams versus massive walls), although this is an indirect consequence of the material's properties. Possible improvements in the Bio Class lie again in the elimination of cement stabilizer in the earth mortar, potentially increasing the reversibility. In the Bee Hall, prefabrication of the rammed earth elements may increase the reversibility due to less damage with disassembly. Particular potential on this scale lies in the simplicity and the reversibility. The latter is due to the fact that unstabilized earth is able to plasticize when water is added. Moreover, the low adhesive strength of earth allows it to be easily separated from other materials. Furthermore, the remnants of earth on the other material can easily be washed away with water. It seems therefore fair to say that, in terms of reversibility, water may be seen as a friend rather than a foe, other than in terms of durability.

At the building scale, the results show a relatively low circular potential. Given the fact that the two studied cases were not intentionally designed to maximize circularity is a plausible reason for their low potential on this scale. For instance, versatility and expandability rely much on design choices and the prospect of future adaptation of the building. Thus, improvement and particular potential are out of place here. To the question how they would tackle a disassembly and reassembly of the discussed case, the interviewees had similar responses: they would choose between disassembly and reuse; demolition, recycling and reproducing; or demolition, disposal and reproducing by balancing cost, time and environmental impact. Transportation of the earthen material and the associated impact was one of the primary factors of consideration. Hence, some questions from a circular perspective may arise: When is it interesting for earthen building elements and components to be demountable and reusable? And when is it more interesting to demolish and dispose or to recycle and reproduce, given the fact that there are fewer negative externalities linked to these processes than with conventional building materials? In fact, disassembly and reuse of earthen building components or elements have possibly not yet been tested in practice. Therefore, the next step in research may be to physically test the reuse potential of earthen construction samples to give more insights from a technical perspective. From an environmental perspective, a comparative life cycle analysis of reuse versus recycling or disposal of earthen building components and elements may support decision-making processes.

### 4.2. *Limitations and recommendations*

Raw earth is an atypical building material in the context of circularity since it hovers in between the biological and the technical cycle of resource flows. Therefore, it does not fit well in existing circularity frameworks, and circular checklists are more challenging to use due to ambiguities of terms and definitions. Moreover, "reuse" is often confused with "recycle" in earth building literature and claims about reusability should therefore be treated with care. Furthermore, circular indicators specific to earth construction, such as calculated erosion and erosion checks to improve durability, may be needed to complement existing frameworks in order to perform a more rigorous assessment.

By reflecting on the assessment of the two cases, some remarks must be made. First, in this research project, a new circular assessment matrix was developed because a framework including the desired scales was not available in literature. However, the framework development was not the main focus of this study. It is also noteworthy that the qualitative assessment does not attribute weight factors to the principles, ignoring the possibility that reversibility may be more important than say manageability. Second, the selected cases are somewhat out of the ordinary. Since the Bio Class is situated inside a warehouse, it is protected from the weather, and therefore the durability of the earth may be significantly higher than if the building were exposed to precipitation. The Bee Hall is a small building for seasonal events and does not contain insulation, thereby presumably increasing simplicity and speed of disassembly. The cases are still very recent, and therefore maintenance, adaptation or disassembly have not yet occurred (and may never occur). Thus, statements made by the architects related to these phases are based on their intuition and not on practical experience. Besides, BC architects have extensive expertise with earth building and incorporated lessons learnt and insights from other projects in Bio Class. Another comment is the fact that no specific benchmarks were defined as more research is needed to state clear definitions of the indicators applied to the case of earth construction.

By reflecting on the methodology, a more in-depth analysis of the cases prior to the interviews would have allowed discussing some constructive features more in detail with the interviewees. Then again, case-specific probing questions are time-consuming in an interview, and this might be avoided by a questionnaire beforehand. A final remark to be made is the difficulty for the respondents to estimate all the indicators at each scale, level and principle correctly.

## 5. Conclusion

This research aimed to assess the circularity of two contemporary earth buildings in Belgium, in order to contribute to a better understanding of the circular potential of earth as a building material. Based on a qualitative assessment of the two cases, it can be concluded that the circular potential found in literature was generally confirmed in practice. The circularity of the two cases is highest on the material scale, relatively high on the building element scale and relatively low on the building scale, although the buildings were not intentionally designed as circular buildings. The main differences between the studied construction techniques (CEB and RE) are in terms of manageability, independence and speed of disassembly of components. Particular potential lies in the low impact, toxicity and scarcity, and the high purity, biodegradability, reversibility and simplicity. However, more cases must be studied to provide more evidence to support these statements. From a circular point of view, it remains unclear whether disassembly and reuse of earthen building components are more desirable than demolition and disposal or recycling and remanufacturing. Future research (lab tests, LCA) may support the decision-making processes for circular building with raw earth.

## References

[1] Solís-Guzmán J, Marrero M, Montes-Delgado M V and Ramírez-de-Arellano A 2009 A Spanish model for quantification and management of construction waste *Waste Manag.* **29** 2542–8

[2] IEA and UNEP 2019 *2019 Global Status Report for Buildings and Construction: Towards a zero-emissions, efficient and resilient buildings and constructi on sector* vol 224

[3] Krausmann F, Gingrich S, Eisenmenger N, Erb K H, Haberl H and Fischer-Kowalski M 2009 Growth in global materials use, GDP and population during the 20th century *Ecol. Econ.* **68** 2696–705

[4] Hansen K, Braungart M and Mulhall D 2013 Resource Repletion, Role of Buildings *Sustainable Built Environments* (Springer New York) pp 502–25

[5] Benachio G L F, Freitas M do C D and Tavares S F 2020 Circular economy in the construction industry: A systematic literature review *J. Clean. Prod.* **260** 121046

[6] Pomponi F and Moncaster A 2017 Circular economy for the built environment: A research framework *J. Clean. Prod.* **143** 710–8

[7] Pacheco-Torgal F and Jalali S 2012 Earth construction: Lessons from the past for future eco-efficient construction *Constr. Build. Mater.* **29** 512–9

[8] Morel J C, Mesbah A, Oggero M and Walker P 2001 Building houses with local materials: Means to drastically reduce the environmental impact of construction *Build. Environ.* **36** 1119–26

[9] Guettala A, Abibsi A and Houari H 2006 Durability study of stabilized earth concrete under both laboratory and climatic conditions exposure *Constr. Build. Mater.* **20** 119–27

[10] Avrami E C, Guillaud H and Hardy M 2008 *Terra Literature Review: An Overview of Research in Earthen Architecture Conservation* (Getty Conservation Institute)

[11] Egenti C and Khatib J M 2016 *Sustainability of compressed earth as a construction material* (Elsevier Ltd.)

[12] Schroeder H 2015 *Sustainable building with earth* (New York: Springer)

[13] Van Damme H and Houben H 2018 Earth concrete. Stabilization revisited *Cem. Concr. Res.* **114** 90–102

[14] Fabbri A, Morel J-C and Gallipoli D 2018 Assessing the performance of earth building materials: a review of recent developments *RILEM Tech. Lett.* **3** 46–58

[15] Röhlen U and Ziegert C 2011 *Earth Building Practice - Planning, Design, Building* (Beuth Verlag)

[16] Minke G 2013 *Building with Earth: Design and Technology of a Sustainable Architecture* (Basel, Switzerland: Birkhaeuser)

[17] Ellen Macarthur Foundation; McKinsey & Company 2015 Growth within: a circular economy vision for a competitive europe *Ellen MacArthur Found.* 100

[18] Pelicaen E 2020 Circular building with raw earth: developing earthen building solutions for reversible applications in contemporary sustainable architecture (conference poster) *DS2BE*

[19] Debacker W, Galle W, Vandenbroucke M, Wijnants L, Chung Lam W, Paduart A, Herthogs P, De Temmerman N, Trigaux D, De Troyer F and De Weerdt Y 2015 *Veranderingsgericht bouwen: ontwikkeling van een beleids- en transitiekader* (Mechelen)

[20] Vandenbroucke M 2016 *Design, dimensioning and evaluation of demountable building elements (PhD thesis)* (Vrije Universiteit Brussel)

[21] Galle W, Cambier C, Elsen S, Lanckriet W, Poppe J, Tavernier I and Vandervaeren C 2020 *Building a Circular Economy: Design Qualities to Guide and Inspire Building Designers and Clients* (Brussels: Vrije Universiteit Brussel)

[22] The Ellen MacArthur Foundation and Granta Design 2015 *Circularity Indicators: An Approach to Measuring Circularity - Methodology* (Routledge)

[23] Kubbinga B, Bamberger M, van Noort E, van den Reek D, Blok M, Roemers G, Hoek J and Faes K 2018 *A Framework for Circular Buildings*

[24] Romnée A and Vrijders J 2018 *Naar een circulaire economie in de bouw: Inleiding tot de principes van de circulaire economie in de bouwsector* (Brussels)

[25] Jiménez Delgado M C and Guerrero I C 2007 The selection of soils for unstabilised earth building: A normative review *Constr. Build. Mater.* **21** 237–51

[26] Laborel-Préneron A, Aubert J E, Magniont C, Tribout C and Bertron A 2016 Plant aggregates and fibers in earth construction materials: A review *Constr. Build. Mater.* **111** 719–34

[27] Fernandes J, Peixoto M, Mateus R and Gervásio H 2019 Life cycle analysis of environmental impacts of earthen materials in the Portuguese context: Rammed earth and compressed earth blocks *J. Clean. Prod.* **241**

[28] Dethier J 2020 *The Art of Earth Architecture: Past, Present, Future* (Thames & Hudson)

[29] Heathcote K A 1995 Durability of earthwall buildings *Constr. Build. Mater.* **9** 185–9

[30] Beckett C T S, Jaquin P A and Morel J C 2020 Weathering the storm: A framework to assess the resistance of earthen structures to water damage *Constr. Build. Mater.* **242** 118098

[31] Bui Q B, Morel J C, Venkatarama Reddy B V. and Ghayad W 2009 Durability of rammed earth walls exposed for 20 years to natural weathering *Build. Environ.* **44** 912–9

[32] Heringer A, Howe L B and Rauch M 2019 *Upscaling earth : material, process, catalyst* (gta Verlag)

[33] Rauch M 2015 *Refined earth construction & design with rammed earth* ed O Kapfinger and M Sauer (Detail)

[34] Vandenbroucke M and Brancart S 2020 *Catalogus veranderingsgericht bouwen: functionele lagen* (Mechelen)

# Tectonics of avoidance

**M A Thomassen[1], P Munch-Petersen[1]**

1 Royal Danish Academy – Architecture, Design, Conservation, Philip de Langes Allé 10, 1435 Copenhagen K

**Abstract**

This article unfolds as a dialogue between architectural and economics concerns on how a sustainable circular practice can be introduced and upscaled in construction. It will point towards absolute sustainable targets and will be analyze in juxtaposition to the (weak) economic drivers that can get us there (Brejnrod et al., 2017; Eberhardt et al., 2020). First, it is established that sustainable innovation is design strategies that simplify construction in order to avoid environmental impacts tied to building material overuse. These strategies are defined as *tectonics of avoidance* and consist of two, direct and indirect, approaches to architectural design. Common for both is that as innovation they must be understood more at applied complex knowledge more than a specific product.

In the second part of the article, we discuss how this kind of information (knowledge) can be applied in the economic circumstances that frame construction and architecture today. How it challenges the path-dependency of design methods today and how information and complex knowledge can be 'sold' in an 'open source' or 'closed' approach to the existing marked for construction.

## 1. Introduction

Is it possible to discuss economical and ecological sustainability at once? Or is the prevailing 'economic climate' incompatible with the 'ecological climate'? Today sustainability has become a selling point for almost all commodities and especially in architectural construction, so what kind of technology are we are selling and does that mean that economic growth and prosperity can align with achieving a balance with nature?

What we, for matters of simplicity, would like to propose is, that there are two kind of technologies: One kind of technology is about adding materials, intelligence etc. to existing solutions. We call this "adding technology". Another kind of technology is about rethinking human needs and building physics and based on this, avoiding using materials. We call this "avoidance technology".

In the first part of this conceptual paper, we will argue why we think avoidance technologies are essential for a sustainable future and what substantiates them. Today, we have to go beyond "reduce, reuse, recycle" – we have to "avoid" (*Affaldshierarkiet*, 2016). We will establish what avoidance technology can represent in the context of architecture today. What kind of innovation we are looking for and what sustainability, circular economy and the planetary boundaries, can mean in a practical sense in architecture (Webster, 2017)? When 'avoidance' is operationalized in an architectural setting we see an increased focus on the resources (material) of construction and this new focus on mitigation material environmental impacts and overuse of virgin materials drive major new approaches to architecture. We call this innovative architectural approach: *Tectonics of avoidance*.

Secondly, we will look at the economic circumstances that can lead the way or hinder *Tectonics of avoidance* in getting traction in architecture and construction today. Here we will discuss the challenges and potentials in finding a place in the marked for subtraction. In essence: how do we sell nothing?

## 2. Part 1 - The tectonics of avoidance

To substantiate the design method of 'tectonics of avoidance (TOA)' two different design approached of TOA needs to be investigated and the preconditioning motivation behind it needs to be acknowledged.

The goal of TOA is to identify architectural design solutions that exemplify *nothingness* – showcasing what can be avoided in common practice today. TOA can be viewed from two angels addressing two mayor challenges: First, the impacts tied to material use. Here the focus is on the overconsumption of (virgin) materials (mitigating raw material scarcity) and the array of environmental impacts building materials represent now and over time (OECD, 2019).

Secondly, we look at a different (passive) approach to indoor climate regulation that also represent a minimalist material approach to architectural design. Over the lifespan of a building the impacts of at building can be divided into two categories: Impacts concerning construction (Product / Manufacture- and Construction Process Stage, LCA stage A, EN 15978:2011) and impact concerning building life (Use, End-of-Life Stage and Benefits and Loads Beyond building life, EN 15978:2011). As above – a NOW and OVER TIME scenario (European Commission & Joint Research Centre, 2010).

The first reflects the impact loads tied to the manufacturing of building materials and the construction of the building. The second impact loads are tied to the operational impacts during the building lifespan. Here we talk about maintenance, repair, operational energy, demolition, disposal and more.

Historically though, these different impact loads have been addresses and understood as separate technical problems – if addressed formally at all. In the *Danish Building Code* there are demands for energy efficiency of the building but no requirement to the physical system that ensure the performance of the building (Falk Hansen & Hansen, 2018). Sideswiping the fact that energy and energy infrastructure have real material preconditions in and outside the building. As such the impacts of physical materials of the building that ensure the energy performance of the building and their environmental impacts are not addressed. TOA1 and TOA2 aim at addressing the totality of this dichotomy.

### *2.1. Tectonics of Avoidance 1: Avoiding excessive material use and embedded impacts*

Simply put the focus here is to minimize impacts tie to the material choices by choosing low impact materials instead of high impact materials. First challenge is to define what constitutes a low impact material'. And here a division can be made between embedded material impacts now and embedded material impacts over time (Munch-Petersen, 2020). The impacts tied to building construction is being addressed increasingly recently, and as *LCA (LifeCycleAssesment)* is becoming mandatory in many situations (certification etc.) a common awareness of the environmental impact of building materials is evolving. The understanding of the material impact loads is becoming well documented but its practical material implications are not obvious. Due to the urgency of the climate crises and for practical simplicity TOA can be divided into two meaningful approaches focusing on 'when' impact occurs, exemplified below:

TOA 1A: Construction material impacts - NOW

TOA 1B: Construction material impacts - OVER TIME

TOA 1A: NOW. Example: In *Det Åndbare Hus,* renewables materials make up the base of construction (Realdania, 2015). Virgin renewable materials are the driver and only very few virgin unrenewable materials were used. The construction is simple, less layered and less complex. The envelope is permeable construction making vapor barriers obsolete.

TOA 1B: OVER TIME. Example 2: In *Næste Skur*, the aim is to expand the functional lifespan of building materials (*Næste Skur*, 2020). Ensuring that material lifespan is not defined by the building component it is part of. This is showcased by using primarily reused materials in a design for further future re-use of materials (*Design for Disassembly*) (Crowther, 2000). Here the technical assembly is key. Maintaining the reversibility of the design is the priority.

Though TOA in both cases represent a simpler material composition the complexity of design rises. Knowledge on materials embedded impacts, building physics, raw material scarcity etc. needs to be an active part of the design process. This indicates that architecture in alignment to TOA1 are both more constraint (artistically) but also at different and more complex approach to design. The question then becomes if the 'free art' of architecture will be stifled or productively challenged or - if indeed – architecture has ever been a free art? Maybe the burden of climate change is the creative framework architecture as an artform needs to become relevant, revolutionary and transformative – to regain its potential to be an artform and an innovative force?

*2.2. Tectonics of Avoidance 2: Minimal design through passive indoor climate regulation.*

TOA2 is a holistic approach where a total minimal impact of the architectural design is the goal, where the point of departure is passive regulation of the indoor climate. As a by-product the total embedded material impacts (TOA1) also are reduced (see below).

TOA2 is a design strategy that seek to minimize impact loads through a passive approach to indoor climate – removing the technical infrastructure and mechanics of indoor environment regulation:

Example three: In Baumschlager Eberle's building *BE2226* the relation between the operational energy and construction materials impacts is exemplified (Eberle & Aicher, 2016). The aim was to establish a building that had no operational energy for heating, cooling and ventilation. And thus, no need to allocate space and construction for technical infrastructure in the building.

The project show that by designing in a way that eliminates operational energy/impacts (mechanical ventilation, cooling and heating) from the building the impacts of construction material was also minimized. As the technical building systems was no longer needed there was also no need for aggregate components and machinery, tech-flooring, radiators and the list goes on. Furthermore, as these common technical devises was no longer needed the need for shafts, suspended ceiling concealing the mechanical installations etc. was no longer needed and as a result the height of the walls and the resulting volume of the building decreased resulting in less need for building materials. This showcased how operational energy strategies and construction material use in synergy can lead to a simpler building with lower impact than that of conventional buildings (with indoor climate regulating systems).

The dilemma is that this too represent a more complex design process. Commonly, according to the design dilemma, the further into the process of design one gets, the more knowledge is gained. That means that the complicated interplay between operational energy, technical systems and the construction is difficult to address simultaneously like in BE2226. The problem is that the building energy performance is commonly calculated on the basis of the design and as such the calculation comes in the later stages of the design. Often so late that the overall premise of the design is difficult alter. Then the technical system 'function' as necessary regulators to achieve the indoor climate wanted – or symptom treatment of the design with a non-functional indoor climate.

In BE2226 the knowledge on passive indoor climate regulation was present at the early design stages and more importantly it was made a central design parameter. This made possible a situation where design decisions and in-depth technical knowledge went hand-in-hand but also had the effect that the array of possible results was limited in comparison to design that is not limited by the pursuit of passive indoor climate. Put simply the hypothesis could be: It takes complex thinking to make a simple (low impact) design and contrary, simple thinking lead to complex (high impact) design. Suggesting that the *tectonics of avoidance* is a *purposed driven innovative process* that requires a holistic approach to design that has a specific aim of subtraction – getting rid of 'unnecessary' symptom treatment of design, but also (or therefore) has a more limited array of design outputs.

### *2.3. TOA in architectural practice*

It is complex to design in accordance with TOA. Especially TOA2 is foreign to common practice simply due to high level of specialist indoor climate expertise needed in early design phases. In order to create the necessary synthesis between all architectural elements (building materials, energy balance, indoor climate and technical systems) that ensure that all unnecessary is identified and avoided the full complexity of architecture needs to be addressed simultaneously. For example, if cooling and mechanical ventilation is sought avoided then the energy performance of the building, orientation, window size and glass specification, the potential thermal mass properties of indoor materials and insulation properties, permeability of the envelope needs to be addressed simultaneously with the plan layout (user loads) and spacial and overall volume is design. This means that the cultural/functional, the measurable and the unmeasurable aspects of architecture needs to be addressed simultaneously. Furthermore, the material impacts, and reversible technical assembly (TOA1) needs to be addressed as well. It seems that *Tectonic of Avoidance*, though the goal is a simpler design, represent much more complex idea- and design process.

Conventional design follows a strict phase understanding of design and creative process well described and defined in 'Description of services for Building and Landscape' (YBL18). In common practice that means that the design goes from a sketch/abstract level which is a superficially layout of the building and moves along towards a realizable concrete drawing and model project.

That movement for abstract to literal is often defined by few people in early phases and more people in later stages (often also more experiences people from more specialized fields). This means that the level of knowledge increases as the design becomes complex – but without the possibility to inform the original idea. But what TOA suggest is that in order to make the proper sustainable choices the knowledge needs to be present in early design – and the collective ambition for a simpler low impact totality of design need to be present in the design team. As such TOA can be viewed as a deliberate goal oriented innovative process ultimately depended on the knowledge present and the knowledge it can produce.

TOA will represent a different pattern of work. In order to deal with the complexity of design in early phases the workload has to be massive. This goes for the knowledge level necessary as well. This could represent a complete mirroring of the tradition workload distribution – few people in early design and more in latter stages – to a to a model where early design is the bulk of the work and later phases can be done by few people.

## 3. Part 2 - The economics of avoidance

### *3.1. The path dependency of design practice in construction*

As argued in part I of the paper, tectonics of avoidance is not compliant with existing practices and structures found in construction. In order for tectonics of avoidance to be achieved, new design approaches, a more fluent division of labor between professions as well as different payment schemes has to be installed.

As a sector, construction is characterized by a project-oriented approach where multiple companies collaborate for a limited time-horizon (that is, the length of the project). This agility in establishing high performance teams is supported by regulation, voluntary agreements and informal rules shared on a sector level (Thomassen 2003) that allow persons and companies to interact without prior personal working relations. It is in other words beneficial to use one single way of organizing in the construction industry.

In economic terms, there are increasing returns from using the same approach to design, contracts etc. As pointed out by David (1985), this potentially creates *path dependency* in which a dominant design is prohibitively costly to change from any individual agent even if alternatives along the way turns out to have a better long-term potential. As observed by Utterback, after the formative period of an industry, the dominant design will be very difficult to challenge by new players.

So even though the existing design practice in construction might not reflect the knowledge we now have on planetary boundaries, the switching cost towards TOA are expected to be substantial and, from a single firm point of view, difficult to manage. Such a mis-match between private and social returns on sustainable investments would be consistent with empirical observations from other industries – see for instance Andries and Stephan (2019) on data on Flamish firms or Frondel et al (2008) on German firms. As we will discuss in the next sections, this adverse relation is likely to hold merit also for firms challenging the dominant design of construction by introducing TOA. And even if companies and persons are not necessarily driven by economic motives, in the long run the transformation towards TOA is difficult to uphold without some level of positive market response.

### *3.2. Innovation as an (imperfect) market for information*

This echoes deeper theoretical concerns on the incentives to invest in innovation (regardless of its purpose). As made clear in the seminal works of Arrow on informational economics, Invention (and the innovation following from this) can broadly be understood "as the invention of knowledge" (Arrow 1962, 609). To innovate is to generate knowledge. And thus, to benefit from innovation is to benefit from knowledge. And this is where its becomes tricky according to Arrow as the market for information is far from perfect. Firstly, information is indivisible; you either have it or do not have it and consequently demand and supply curves cannot adjust smoothly in order to ensure equilibrium. Secondly, it is inappropriable[1]. The cost of "transmitting a given body of information is frequently very low" (Arrow 1962 614). From a welfare economics point of view this a nice thing. But from a private perspective this is troublesome[1] as you can either sell it (knowing that soon everyone will know the same as you) or keep it for yourself (knowing that this dramatically reduces its total use).

The result of this is a "downward bias" to devote resources to invention (bid 617). This bias is particularly strong when the provided information is not embedded in a final output but serve as an input for further generation of knowledge. Basic research, Arrow points out, is "especially unlikely to

---

be rewarded". This encourages private individuals to engage in innovation activities that can be commoditized (ibid p 618).

This gives us a potential clue to understand why investments in TOA might be less attractive than investments in "adding technologies".

For TOA 2 on the absolute complexity of design, the implication of Arrows insight seems obvious. As a design strategy, which is about avoiding products, naturally there are no products to build hard-to-copy knowledge into. The key knowledge behind TOA 1 is the ability to handle complex multicriteria problem solving in the initial design phases. This knowledge can either be kept as tacit knowledge held by, say a few persons at offices like B&E in which case it is of limited use. Or be codified in work procedures, standard solutions for critical interfaces etc. in which case it is easy for competitors to imitate by hiring people from these offices.

For TOA 1 working by means of low impact materials and a higher degree of re-use, the implication is less straight forward. Knowledge build into low impact products are not per se more difficult to protect than knowledge embedded in high impact products. However, high impact products are often high-impact as a consequence of being more processed than simpler "dug-out-of-the ground" materials, which tend to make them more protectable as the process can be patented or build into a fine-tuned large scale production facility. Secondly, strategies for design for disassembly are often build on transparency. Consider for instance the nine tectonic strategies for a circular architectural practice proposed in Anne Beim et al 2019 (p. 28). These circular principles works by means of simplicity (few materials, non-processed materials), visibility (logic intuitive / easy to understand), replaceability (standardized dimensions, materials accessible) reversibility (no harmfull substances, non-processed materials). It is clear that these properties are beneficial for circularity and ensures the long-term value of the materials. But it is also clear that this recipe for circularity is close to a disaster for any producer hoping to protect his knowledge by embedding it into sealed products or advanced manufacturing processes.

From an investment point of view the implication seems discouraging for both TOA 1 and TOA 2 – it is easier to appropriate knowledge build into technologies that adds rather than reduces complexity. And even if technologies of avoidance is beneficial from a societal and ecological point of view, its business case is really bad.

However, if we loosen the assumption on perfectly maximizing agents just a bit, other motives for innovation than profit comes into play. For instance if the innovation contributes to a sustainable future or not. Plainly speaking, the possibility of circular innovations may attract persons and companies that will not otherwise be in the market. For them, the question is not to invest in (often) traditional adding technologies or circular avoidance technologies. For them, the question is if the benefit from investing in circular solutions is large enough to sustain further development and marketing – if the business case is strong enough not to kill all the good intentions so to speak.

As Winter (2006) points out, there are plenty of cases in which innovation takes place even if the inventor can attract only a marginal part of its overall outcome. What matters is if the inventor can attract enough income to cover costs and some (perhaps very limited) risk premium. The happy observation is that TOA innovations are not necessarily very costly to develop or produce. They require a good deal of thought, but with respect to materials, equipment and processing knowledge they are – by definition as seen in the *tectonics principles for circularity* cited above – cheap.

The more troublesome observation is that new solutions in construction requires a good deal of additional innovation (or at least adjustment) around the new solution as we discussed in the section on path dependency. Here, we have to do with what Teece in his 1986 article on "Profiting from Innovation" (Teece 2006) identified as complementary assets. What according to Teece determines the success of a core technological insight is not the insight as such, but the company's ability to align a wide range of arrangements around it. Thus, the innovator does not only have to develop the product. To some degree he has to develop the system or the infrastructure for the product. This makes innovation difficult as it widens and deepens the necessary activities.

The size of the company matters in this respect. The cost of obtaining information is the same whether it is used one or many time and hence it is more advantageous for a large company, with many potential ways of using it, to innovate. A smaller firm with less resources will be forced to tailormade its organization to a specific use of the new technology at an early stage. This is risky, in particular for innovations in their very early phases, where the most efficient way of applying it is extremely hard to predict (Arrow 1962, p. 618). In the wording of Teece, it is much more likely that the large company has access to the required complementary assets. Which all in all implies that "*There is a fundamental scale economy in capturing the returns from R&D/innovation and the "scale" that often matters is the scale of application of the innovating firm*" (Winter 2006, p. 13).

So even though tectonics of avoidance might be cheap when it comes to the technology as such, they are expensive when it comes to investments in supplementary assets. And this is not promising when we recall our previous conclusion that avoidance technologies have a low appropriability and thus have to be cheap in order to be sustained economically.

Circular solutions is undoubtedly obvious and mandatory to apply in construction taking its impact on the planetary boundaries into account. But is at the same time far from obvious taking the industry structure of construction into account. Circularity as a goal is straight forward. The way to get there, is as best, crooked.

## 4. Final discussion: Alternative Strategies for capturing the benefits of investment in tectonics of avoidance - Baumschlager Eberle as "youtuber" or "darth wader"?

Let us conclude by returning to one of the examples of avoidance technologies introduced in part I of the paper and say that we, very hypothetically, was given the job of helping B&E with a strategy for how they could expand their BE2226 solution; that is passively controlled, simple buildings with a clear tectonic design that only requires inhabitants in order to be well-tempered and with a healthy indoor climate. What we, in all humbleness and backed by the theoretical discussions presented in this paper, would suggest to B&E is that "You can do one of two. You can turn to the dark side and become darth vader. Or you can turn to a perhaps even more subtle force and become a youtuber".

*Darth Vader?* As discussed above, company size matters. If B&E were to expand their solution beyond a limited number of projects done for the clients of B&E an obvious answer was to grow their business into something much bigger and diversified. Bigger in order to being more financial robust for the long time horizons in bringing new products to the market as stressed by Schumpeter. Bigger to have economics of scale on information that is hard to get paid for unless you keep it within the closed doors of the company as stressed by Arrow. Bigger to have access to all the complementary assets needed for succeeding with a novel product or service as stressed by Teece and Winter.

But not only this. Our analysis will point out that it is not only or mainly about growing your business horizontally (as might appear to be sufficient when reading Arrow, Teece and Winter). It is also about going into other parts of the value chain; that is to grow vertically. If they should succeed encapsulating their knowledge in a way that will generate them a sustained income, they will have to build it into all the products and materials "surrounding" how they can provide a sound indoor climate which is close to say that B&E should provide the entire office building. B&E will have to turn into a turn key contractor for, in this instance, office buildings. A transformation towards absolutely control and capitalization of the final building that many within construction, we assume, would consider as being "turning to the dark side".

*Youtuber?* In 1973 Arrow explains the existence of firms as one of organizing on information costs. As there is costs of handling information, it may pay to invest in procedures and structures for handling information. These procedures and structures will partly reduce the amount of information exchanged, partly make the transfer of information easier to conduct. And they will be designed in order to optimize certain kind of information flows and is hence not part of an overall common structure; they are in short what constitute the firm and explains "…the internal structure of firms and more generally of other economic organizations (Arrow 1973, pp-20.21).

What is striking when reading Arrows description of information structure here 50 years later, is how much its sounds like the prescription not for firms, but for the internet. The internet is precisely a bunch of technologies and protocols that partly reduce the cost of transmitting information, partly reduce the amount of abundant information as provider and supplier of information are easily matched (Olesen 2012).

Now we, for better or worse, see the blossom of an economy of digital connectedness. We see something that we thought was almost impossible: That it pays to give away information for "free". Or more precisely to use the conceptual framework of Arrow and Winter: that the extent of digital empowered markets are so big, that information can be provided at a price very close to zero. In the context of B&E it would translate to that B&E gave away their information "for free". And perhaps even made investments in distributing it by tutorials, on-line training, establishing local communities of "avoidance freaks" and the like. All this would be provided at no (visible) cost. The "only" pay B&E would get, is the attention of this potentially global community with the revenues from data (and perhaps projects) following from this. B&E would essentially be a youtuber.

As these line of reasoning suggest, there might be multiple ways to sustain the emergence of strategies to handle increasingly more complex processes. The task ahead is not only to reduce and reuse materials but simply not to use them. This demands an interlinked knowledge level that is rare in construction today. The real lessons to be learned from the example of Part 1 is not about specific plans, projects or products they provide for different clients in different climatic zones. The real lesson is about the knowledge present to address complexity. How this knowledge is transferred and made widely applicable by transformation of education, academic traditions, economic drivers as well as legislative standards is, we believe, the fundamental sustainable issue to address.

IOP Conf. Series: Earth and Environmental Science **855** (2021) 012003 doi:10.1088/1755-1315/855/1/012003

## References

*Affaldshierarkiet*. (2016, July 7). Affaldshierarkiet - Videncenter for Cirkulær Økonomi i Byggeriet - VCØB. https://vcob.dk/byggeaffald/affaldshierarkiet/

Brejnrod, K. N., Petersen, S., & Birkved, M. (2017). The absolute environmental performance of buildings. *Building and Environment*, *119*, 87–98.

Crowther, P. (2000). *DEVELOPING GUIDELINES FOR DESIGNING FOR DECONSTRUCTION*. 11.

Eberhardt, L. C. M., Birkved, M., & Birgisdottir, H. (2020). Building design and construction strategies for a circular economy. *Architectural Engineering and Design Management*, 1–21. https://doi.org/10.1080/17452007.2020.1781588

Eberle, D., & Aicher, F. (Eds.). (2016). *be 2226 - die Temperatur der Architektur: Portrait eines energieautonomen Hauses = be 2226 - the temperature of architecture: portrait of an energy self-sufficient house*. Birkhäuser.

European Commission & Joint Research Centre. (2010). *ILCD handbook: General guide for life cycle assessment : detailed guidance*. Publications Office of the European Union. http://lct.jrc.ec.europa.eu/pdf-directory/ILCD-Handbook-General-guide-for-LCA-DETAIL-online-12March2010.pdf

Falk Hansen, C., & Hansen, M. L. (2018). *Bygningsreglement 2018(BR 18)*. Byggecentrum.

Munch-Petersen, P. (2020). *Circular Facade Design: The Tectonics of Circular Economy* [PhD-thesis]. KADK.

*Næste skur*. (2020). https://www.naeste.dk

OECD. (2019). *GLOBAL MATERIAL RESOURCES OUTLOOK TO 2060: Economic drivers and environmental consequences*. ORGANISATION FOR ECONOMIC CO-OPERATION AND DEVELOPMENT -ORGANIZATION FOR ECONOMIC.

Realdania. (2015). *Det åndbare hus*. https://realdania.dk/projekter/det-aandbare-hus

Webster, K. (2017). *The circular economy: A wealth of flows* (Second edition). Ellen MacArthur Foundation Publishing.

Andries, Petra; Stephan, Ute. 2019. "Environmental Innovation and Firm Performance: How Firm Size and Motives Matter", Sustainability 11, no. 13: 3585.

Arrow, Kenneth J. 1962. "Economic Welfare and the Allocation of Resources for Invention". University of Illinois at Urbana-Champaign's Academy for Entrepreneurial Leadership.

Arrow, Kenneth J. 1973. "Information and economic behavior", prepared for office of Naval Research, distributed bu National Technical Information service, U.S. Department of Commerce.

Beim, Anne; Ejstrup, Henriette; Frederiksen, Line K; Hildebrand, Linda; Madsen Ulrik S.; Munch-Petersen, Pelle and Skjöld, Simon R. 2019. "Circular Construction – Materials, Architecture, Tectonics", Cinark, KADK,

David, Paul A. 1985. "Clio and the Economics of QWERTY", The American Economic Review, Vol. 75, No. 2, Papers and Proceedings of the Ninety-Seventh Annual Meeting of the American Economic Association. (May, 1985), pp. 332-337.

Frondel, Manuel, Horbach, Jens and Rennings, Klaus. 2008. "What triggers environmental management and innovation? Empirical evidence for Germany". Ecological Economics, 2008, vol. 66, issue 1, 153-160

Olesen, Jens Mogensen. 2012. "Netværk". Aarhus Universitsforlag. Part of "Tænkepauser – viden til hverdagen".

Sidney Winter. 2006. "The Logic of Appropriability: From Schumpeter to Arrow to Teece," LEM Papers Series 2006/21, Laboratory of Economics and Management (LEM), Sant'Anna School of Advanced Studies, Pisa, Italy.

Teece, David J.. 2006. "Reflections on "Profiting from Innovation"," Research Policy, Elsevier, vol. 35(8), pages 1131-1146, October.

Thomassen, Mikkel A. 2004. "The economic organisation of building processes. On specialisation and coordination in interfirm relations", Technical University of Denmark, Ph.D. thesis

IOP Conf. Series: Earth and Environmental Science **855** (2021) 012004 doi:10.1088/1755-1315/855/1/012004

# Environmental impact evaluation of five circular concrete scenarios

**R E J van den Bogert[1] and M J Ritzen[1, 2]**

[1] Zuyd University of Applied Sciences, Smart Urban ReDesign research team, Academy Built Environment, Heerlen.

[2] VITO, Unit Smart Energy and Built Environment, Boeretang 200, 2400 Mol, Belgium

**Abstract.** In the Netherlands, the material concrete accounts for 40-60% of the total environmental impact of the Dutch construction industry. Cement accounts for 80-95% of this impact, while not being the main component expressed in mass.

Traditional methods of (re)using and recycling concrete in the construction industry have a high detrimental environmental impact, and are currently not fitting in the transition from a linear economy to a circular economy.

In the Dutch concrete treaty (DCT) (07-2018), the main goals of the 'Green Deal for a sustainable concrete industry' (2011-2015) became tangible. The goals concerning the (re)use of concrete in the DCT are:

1. Lowering $CO_2$ emissions by 30% relative to $CO_2$ emissions in 1990, with the ambition to lower $CO_2$ emissions by 49%, by 2030.
2. 100% high quality reuse of concrete from the demolition stock by 2030.

To achieve these goals, changing traditional use of concrete is necessary. Consequently, there is a demand for initiatives that boost the development of new, circular methods of (re)using concrete, resulting in a lower environmental impact. However, data on the actual environmental impact of these methods is still to be collected. In the European project Urban Innovation Actions (UIA), Super Circular Estate–SUPERLOCAL, this gap of knowledge is addressed, and the actual environmental impact of circular concrete methods is studied. In this project, 10-story high apartment blocks are deconstructed, and circular solutions for material re-use and their environmental impact are researched.

The aim of this study is the analysis and comparison of the actual environmental impact of different methods of how concrete, from the existing building stock, not designed for reuse, can be (re)used in a more circular manner.

This study is designed as a comparative analysis. Based on literature and project-expert consultation, five circular concrete scenarios and a baseline scenario have been formulated. These scenarios are considered promising and feasible methods of (re)using concrete from SUPERLOCAL, in a circular manner. These circular scenarios are compared to the baseline scenario, using 100% new materials and mainstream/traditional methods of pouring concrete on site. The five circular scenarios are:

1. Compartments; extracting whole building components from existing building stock to be reused in new buildings.
2. Slabs; extracting smaller building components that can be used for new buildings.

IOP Conf. Series: Earth and Environmental Science **855** (2021) 012004 doi:10.1088/1755-1315/855/1/012004

3. Prefabricated elements (with 100% coarse aggregates from granulated rubble).
4. Pouring concrete on site, with 100% coarse aggregates from granulated rubble, and
5. BRX; pre-manufactured concrete bricks with 100% coarse aggregates from granulated rubble.

Data for the calculations is extracted from literature study, from data gathered through real world executions of several circular concrete scenarios in the project SUPERLOCAL, and using widely accepted data bases on embodied energy and embodied carbon in materials.

The outcome of this study is a process description of each scenario and a calculation of the corresponding environmental impact. The environmental impact is expressed in the amount of embodied energy (MJ/m$^3$ and MJ/unit) and the amount of embodied carbon (kg$CO_2$/m$^3$ and kg$CO_2$/unit).

Although scenario 1 and 2 score the highest on savings on embodied energy (up to 70%) and embodied carbon (up to 70%) compared to baseline, this cannot be deemed as the most promising scenario currently. The risks during execution are too high and the shortcomings on quality of the existing building stock and current building regulations are too big of an obstacle to overcome to call it a feasible scenario.

The most promising scenarios with the highest feasibility of implementation in the built environment are scenario 3,4 and 5. The risks during execution of these scenarios are low and current building regulation can be met with relative ease. Furthermore, there is significant room for improvements. Optimizing manufacturing processes and material usage has directly significant influence on the environmental impact of these scenarios. The possibilities of end of life re-use scenarios are also significant factors in this.

## 1. Background

With the signing of the Paris Agreement [1], global warming is now widely acknowledged as a serious problem. Actions that reduce the risk and impact of climate change due to global warming must be taken by every country that has signed the agreement. One of the goals set in the Paris Agreement is reducing the emission of $CO_2$ drastically. Every sector must contribute to obtain these goals. For the Dutch construction industry these goals are set in the Dutch Concrete Treaty [2].

In the Netherlands, the material concrete accounts for 40-60% of the total environmental impact of the Dutch concrete industry. Cement accounts for 80-95% of this impact, while not being the main component expressed in mass.

Traditional methods of (re)using concrete in the building industry have a high detrimental environmental impact and are not fitting in the transition from a linear economy to a circular economy.

In the Dutch concrete treaty (DCT) [3], the main goals of the 'Green Deal for a sustainable concrete industry' (2011-2015) became tangible. The goals concerning the (re)use of concrete in the DCT are:

1. Lowering $CO_2$ emissions by 30% relative to $CO_2$ emissions in 1990, with the ambition to lower $CO_2$ emissions by 49%, by 2030, and
2. 100% high quality reuse of concrete demolition stock by 2030.

To achieve these goals, changing traditional use of concrete is necessary. Consequently, there is a demand for initiatives that boost the development of new, circular methods of (re)using concrete, resulting in a lower environmental impact. However, data on the actual environmental impact of these methods is still to be collected. In the European project Urban Innovation Actions (UIA), Super Circular Estate–SUPERLOCAL, an urban development project in the Netherlands, this gap of knowledge is addressed and the use of new circular materials and building methods and their social, economic and environmental effects is investigated. In SUPERLOCAL, 10-story high apartment blocks are deconstructed, and circular solutions for material re-use and their environmental impact are researched. Within this project, several demonstrator cases are executed that provide the opportunity to investigate circular concrete building methods and study their actual environmental impact. For example, the realisation of three circular model housing units, using only materials harvested from the 10-story high apartment block [4].

## 2. Aim
The aim of this study is the analysis and comparison of the actual environmental impact of different methods of how concrete, from the existing building stock, not designed for reuse, can be (re)used in a more circular manner.

## 3. Methodology
This study is designed as a comparative analysis. Based on literature and project-expert consultation, five circular concrete scenarios and a baseline scenario have been formulated. These circular scenarios are considered promising and feasible methods of (re)using concrete from SUPERLOCAL, in a circular manner. These circular scenarios are compared to the baseline scenario, using 100% new materials and mainstream/traditional methods of pouring concrete on site. The five circular scenarios are:

1. Compartments; extracting whole building components from existing building stock to be reused in new buildings (Figure 1).
2. Slabs; extracting smaller building components that can be used for new buildings (Figure 2).
3. Prefabricated elements (with 100% coarse aggregates from granulated rubble) (Figure 3).
4. Pouring concrete on site, with 100% coarse aggregates from granulated rubble (Figure 4), and
5. BRX; pre-manufactured concrete bricks with 100% coarse aggregates from granulated Rubble (Figure 5).

**Figure 1.** Complete compartment being hoisted out.

**Figure 2.** Damaged slabs intended for reuse.

IOP Conf. Series: Earth and Environmental Science **855** (2021) 012004 doi:10.1088/1755-1315/855/1/012004

**Figure 3.** Placing of prefabricated concrete elements.

**Figure 4.** On-site pouring of recycled concrete.

**Figure 5.** premanufactured BRX.

## 4. Data collection and outcome measures for calculating environmental impact of the scenarios.

Data for the calculations is extracted from literature study, from data gathered through real world executions of several circular concrete scenarios in the project in the project SUPERLOCAL and using widely accepted data bases on embodied energy and embodied carbon in materials [5].

There are several methods to account for the beneficial effects of recycling in an assessment. For each assessment, one must choose the fitting approach, based on the goal and scope of the study. In this study, the Recycled Content Approach (100:0 method) is applied. This method benefits in full of the materials being recycled, but does not account for possible end of life re-use scenarios [6].

In formulating the five circular concrete scenarios, end of life recycling was not included. The exclusion of end of life recycling, keeps the necessary associated assumptions to a minimum and increases the focus on the use of materials and building methods. By excluding the end of life recycling factor, using a recycled content approach is legit, for this approach appoints all the benefits of recycling to the materials being saved and thereby excludes assumptions for possible end of life re-use or recycling for materials and products with a long lifespan, like buildings. This leads to a clear demarcation.

A database containing information that can be used when working from a recycled content approach is the ICE database [5]. The range of the ICE database is cradle-to-gate. This includes: material extraction, material processing, transportation to the factory gate and manufacturing until the product is ready to leave the factory. End of life recycling is outside the range of cradle-to-gate. This makes the ICE database suitable to use in a recycled content approach study. The exception being materials where the demand is higher than the supply of virgin materials (e.g. steel). For these exceptions, adequate adjustments for the embodied energy and embodied carbon must be made.
The outcome of this study is a process description of each scenario and a calculation of the corresponding environmental impact. The environmental impact is expressed in the amount of embodied energy (MJ/$m^3$ and MJ/unit) and the amount of embodied carbon (kg$CO_2$/$m^3$ and kg$CO_2$/unit).
Because four out of five scenarios had a real world demonstrator, practical knowledge on these scenarios is available. Therefore, the calculated outcome of each scenario is also weighed against points of improvement obtained from this data.Lastly, a recommendation on the most promising scenario with the highest feasibility of implementation in the built environment is made.

## 5. Scenario description and methodology

*5.1 Baseline.*
In order to make the comparison between the different scenarios, a volume and a building method is set for the baseline unit as well as a unit for the outcome. The baseline volume is set at 38.07$m^3$ of concrete. This is the amount of concrete used in one of the housing units of the aforementioned demonstrator case. This unit is also used to calculate the outcome of scenario 1; compartments. The baseline scenario accounts for all the work and materials involved in putting up the baseline unit, using current mainstream methods of pouring concrete on site. This involves:

1. Materials used in and during the construction of the baseline unit.
2. Energy used during the construction of the baseline unit (e.g. transport and general use of energy on the building site), and
3. Labor performed during construction of the baseline unit and scenario specific activities.

The main materials being used are concrete and reinforcement steel (rebar). Because the amount of cement used in the concrete composition has a big influence on the environmental impact, it is of great consequence to pick a concrete composition that is representable for an average concrete composition that is used in the Dutch concrete industry. The same applies for the amount of steel used in the concrete construction. CE Delft has gathered data on this subject in their research on the environmental impact of the use of concrete in the Dutch building industry [7]. This data is used to determine the amounts of materials used in the baseline unit.

Data on the general use of energy during the build of the baseline unit is extracted from a chain analysis of the Dutch concrete industry [8]. When required, a surcharge for transport of materials, gate to building site, is set at 0,0031 MJ/kg and 0,0028kg$CO_2$/kg. This is based on transportation of the materials by truck; cargo 35 tons, average of 100km supply and discharge route and an average diesel consumption of 0,30L/km.
Labour required for constructing the baseline unit consists of: setting up the formwork, laying the reinforcement steel, pouring the concrete and demoulding. Data on this subject is extracted from a report on the feasibility of circular concrete scenario's [9].
Scenario specific activity for the baseline scenario is demolishing the existing building stock. Data on this subject is extracted from data gathered during real world execution of the fourth scenario in the project SUPERLOCAL; pouring concrete on site with 100% coarse aggregates from granulated rubble.

*5.2 Scenario 1: Compartments; extracting whole building components from existing building stock to be reused in new buildings*

For this scenario a real life demonstrator has been executed in the project SUPERLOCAL. This scenario involves sawing out a complete housing unit of an existing high apartment block and re-using it as the structural work for a new dwelling.

The data acquired through this demonstrator is used in the calculation of this scenario. Because this scenario has high fixed costs, it is relevant to research the influence of upscaling. Therefor a sub-scenario is also calculated, were the maximum capacity of the high apartment block of 50 apartments is being used. In the demonstrator case it was unclear what the occupancy rate of the machinery was. Therefor several sub-scenarios concerning the occupancy rate of the machinery have been calculated (100%-75%-50% and 25% occupancy rate).

The scenario specific work and materials included in the calculation for this scenario are as following:

Preparatory work; Improving the carrying capacity of the soil near the high apartment block for the big cranes involved in this scenario, putting up the cranes involved in hoisting out the apartments, stamping the high apartment block to ensure safety while performing the labor, removing the concrete balconies that are attached to the housing unit and designing and manufacturing a hoisting installation specifically for this project.

Work included during construction: Sawing out the housing units, hoisting out the housing units onto a low loader, moving to housing units with the low loader to their new location, hoisting the housing units onto their foundation.

*5.3 Scenario 2: Slabs, extracting smaller building components that can be used for new buildings.*

No demonstrator has been executed for this scenario. This theoretical scenario researches the effect of reducing the high fixed costs of scenario 1 by hoisting out smaller components (less than 3000kg per building component) instead of entire housing units. Using only one crane and making several high cost items redundant (e.g. specialized hoisting construction) . The high risks involved in hoisting out entire housing units, not designed for this proceeding, will be diminished drastically as well.

The fixed costs for this scenario are still relatively high. Therefore, the same sub-scenarios, including the occupancy rate of machinery scenarios, as for scenario 1 will be calculated.

The scenario specific work and materials included in the calculation for this scenario are similar to scenario 1. Amounts and unnecessary cost items are adjusted accordingly for this scenario.

*5.4 Scenario 3: Pouring circular concrete on site, with 100% coarse aggregates from granulated rubble.*

For this scenario a real life demonstrator in the project SUPERLOCAL has been executed as well. The foundation and stability walls for the model units in the demonstrator of scenario 1 have been constructed with this circular concrete. This scenario involves, harvesting, depositing and processing the coarse aggregates, mixing the concrete and pouring the concrete. All these proceedings are executed on site. The data acquired through this demonstrator is used in the calculation of this scenario.

The scenario specific material included in the calculation for this scenario is the circular concrete. The composition of this concrete is as following: 1600kg/m$^3$ coarse granulates, 385kg/m$^3$ CEM III,B and 97kg/m$^3$ reinforcement steel.

The scenario specific work included in the calculation for this scenario is as following: Harvesting, depositing and sorting debris for extraction of coarse aggregates, processing the debris to usable aggregates (0-4mm and 4-22mm), setting up the formwork, laying the reinforcement steel, setting up a mobile concrete plant, mixing and pouring the concrete with process control and lastly demoulding.

*5.5 Scenario 4 Prefabricated elements, with 100% coarse aggregates from granulated rubble.*

No demonstrator has been executed for this scenario in SUPERLOCAL, but information and exchange of experiences of a similar project using coarse aggregates from granulated rubble into prefabricated concrete structural work is available.

This scenario involves prefabricating concrete elements in a factory, using a concrete composition with 100% coarse aggregates from granulated rubble. Then transporting them to the building site and put them up with the use of a crane.
To determine the amounts of materials used in this scenario, the report of CE_Delft on the environmental impact of the use of concrete in the Dutch building industry is used again, but this time data on the composition of concrete products is used instead of concrete mortar [7]. Data on the coarse aggregates from granulated rubble is extracted from scenario 3 and adjusted accordingly to scenario 4. After testing serval concrete compositions, it is apparent that the sand needed for the concrete mortar can only be replaced by small amounts or not at all with the harvested granulates. Because of the coarse and inconsistent shape of recycled sand grains, the consistency of the mortar becomes unpredictable and will not flow as desired. In a factory process this is not workable, so new sand is used in the calculation of this scenario.
Furthermore a "precast concrete modification factor" from the ICE-Database has been added to this data to account for the work being done and energy used inside the factory gates.
A surcharge for transport of materials, gate to building site, is again based on transportation of the materials by truck and are set similar to baseline values. Lastly the consumption of diesel for the crane is set at an estimate of 75L/unit and labor on site is set at 0,95MJ/m$^3$ and 5,01kgCO$_2$/m$^3$ [9].

*5.6 Scenario 5 BRX; pre-manufactured concrete bricks with 100% coarse aggregates from granulated rubble.*
For this scenario a real life demonstrator in the project SUPERLOCAL has been executed as well. Several storage units are built using this method. This scenario involves producing stackable concrete blocks, in a dry manufacturing process. This method is developed by Pieter Scheer of Dusseldorp. Assembling of the BRX-system can be done in two ways. One method is to stack the bricks and glue them together using a cement mortar. Another method is to stack the bricks and connect them with steel tensile bars. Although the latter method is a better solution from a circular point of view (deconstruction without decreasing the value of the material is easily accomplished), this calculation is made with the first method because data on the method with tensile bars is not complete.
Specialized machinery is required to produce the concrete blocks. Unfortunately this machinery proved to be so delicate, that the goal of producing the blocks on site (like scenario 3) could not be achieved. Therefore, additional transportation of materials is accounted for in this analysis.
The concrete used in this process is made in the same way as in scenario 3 (pouring on site). The quality of the concrete for the "pouring on site scenario" is set at C20/25. For concrete blocks it is not necessary to have such a high quality. So it is realistic to use a lesser cement consumption for this scenario. The prescribed M10 (10N/mm$^2$) mortar for gluing the blocks affirms that this is a correct assumption. Cement consumption is set at 250kg/m$^3$ of concrete (CEM III / B42,5L LH-SR). M10 mortar is set at 42kg/m$^3$ of concrete. Labor is set at 1.02 hours of labor/m$^2$ and involves manufacturing the blocks, setting the profiles and gluing the blocks [9].

## 6. Results

Table 1 shows for each scenario the embodied energy [MJ/unit] and embodied carbon [kgCO$_2$/unit] and the percentage saved on embodied energy and embodied carbon compared to the baseline scenario. The same information is graphically represented in Figure 6 and 7. The complete calculation, including a more detailed process description can be found in Annex A: Calculation of the environmental impact of five circular concrete scenarios.

**Table 1.** embodied energy [EE] and embodied carbon [$ECO_2$] per scenario.

| | Embodied Energy [MJ/unit] | Embodied Carbon [$kgCO_2$/unit] | Percentage saved on EE compared to baseline. | Percentage saved on $ECO_2$ compared to baseline |
|---|---|---|---|---|
| **Baseline** | 143.088 | 21.194 | | |
| **Scenario 1** | 284.086 | 27.980 | -99% | -32% |
| **Scenario 1b** | 48.580 | 14.889 | 66% | 30% |
| **Scenario 2** | 192.118 | 43.071 | -34% | -103% |
| **Scenario 2b** | 41.865 | 6.436 | 70% | 70% |
| **Scenario 3** | 121.752 | 13.380 | 15% | 37% |
| **Scenario 4** | 165.726 | 20.867 | -16% | 2% |
| **Scenario 5** | 51.514 | 9.713 | 58% | 27% |

**Scenario 1b and 2b represent the outcome for upscaling to maximum potential of the scenario*

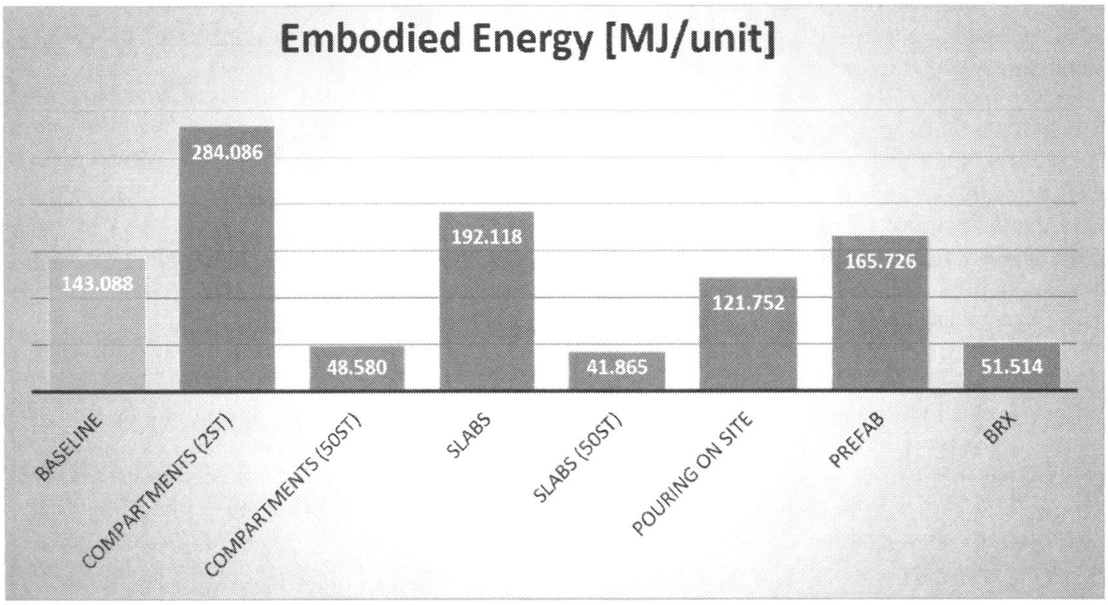


**Fig 6.** Embodied Energy [MJ/unit] per scenario.

IOP Conf. Series: Earth and Environmental Science **855** (2021) 012004 doi:10.1088/1755-1315/855/1/012004

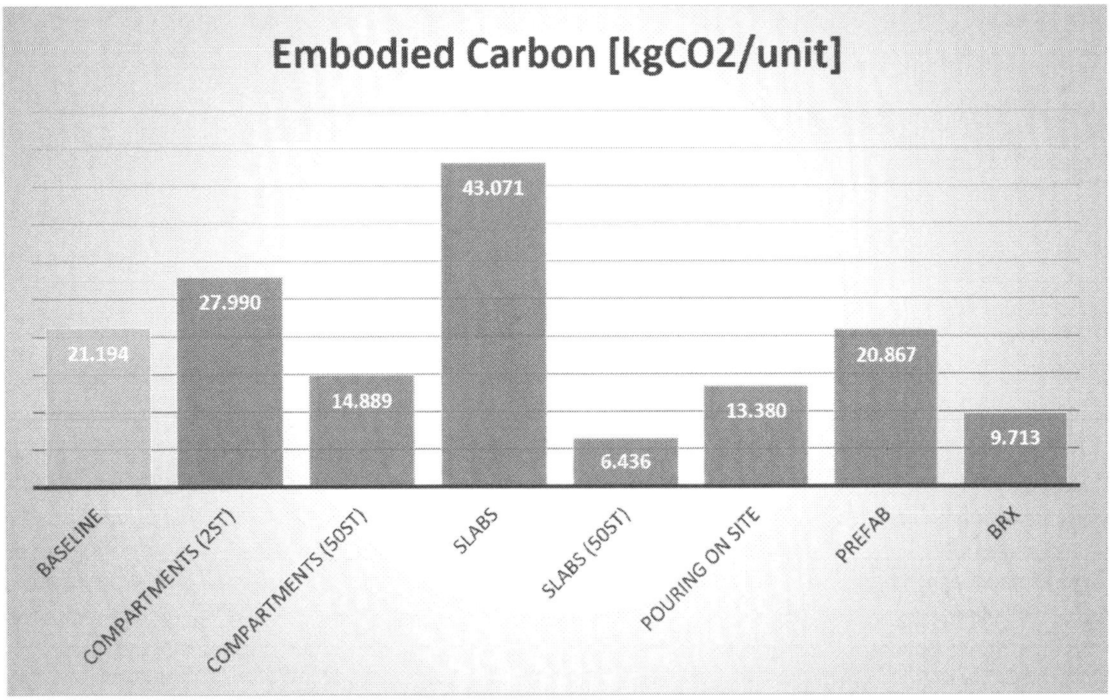


**Fig 7.** Embodied Carbon [$kgCO_2$/unit] per scenario.

The baseline scenario leads an embodied energy of 143.088 MJ/unit and 21.194 $kgCO_2$/unit, while scenario 1 , leads to an embodied energy of 284.086 MJ/unit and 27.980 $kgCO_2$/unit. This exceeds the baseline scenario by 99% on embodied energy and 32% on embodied carbon. However, upscaling this scenario to the full potential of extracting 50 apartments from 1 apartment building, leads to a saving of 66% on embodied energy and 30% on embodied carbon, compared to the baseline scenario.
Because of uncertainty in the occupancy rate of the machinery used, several sub-scenarios have been calculated (100%-75%-50% and 25% occupancy rate). Table 2 and 3 show the influence of the occupancy rate of the machinery for scenario 1 and scenario 1b.

**Table 2.** Influence of occupancy rate of machinery used on scenario 1.

| Occupancy rate | Embodied Energy [MJ/unit] | Embodied Carbon [$kgCO_2$/unit] | Percentage saved on EE compared to baseline. | Percentage saved on $ECO_2$ compared to baseline |
|---|---|---|---|---|
| **100%** | 284.086 | 27.980 | -99% | -32% |
| **75%** | 257.168 | 25.924 | -80% | -22% |
| **50%** | 230.230 | 23.858 | -61% | -13% |
| **25%** | 203.302 | 19.727 | -42% | 7% |

**Table 3.** Influence of occupancy rate of machinery used on scenario 1b.

| Occupancy rate | Embodied Energy [MJ/unit] | Embodied Carbon [$kgCO_2$/unit] | Percentage saved on EE compared to baseline. | Percentage saved on $ECO_2$ compared to baseline |
|---|---|---|---|---|
| **100%** | 48.580 | 14.889 | 66% | 30% |
| **75%** | 38.816 | 14.082 | 73% | 34% |
| **50%** | 29.053 | 13.275 | 80% | 37% |
| **25%** | 19.290 | 12.468 | 87% | 41% |

Table 2 and 3 demonstrate that the influence of the occupancy rate of machinery has a significant influence on the outcomes calculated for embodied energy and embodied carbon. The lower the occupancy rate, the higher the savings. The influence of the occupancy rate of scenario 2 is expected to be comparable to scenario 1 and 1b.

Results show that savings op to 70% on embodied energy and embodied carbon can be achieved under ideal circumstances. However, when a scenario involves fabricating new concrete, the results do not demonstrate large savings, ranging from -16% to 58% saving on embodied energy and 2% to 37% savings on embodied carbon compared to the baseline scenario. The reason for this is shown by analyzing the composition of a commonly used concrete mixture a "1,2,4 cement" (C20/25) in table 4.

**Table 4.** Analysis of a 1,2,4 concrete mixture, using the ICE database

| Component | Percentage | Embodied Energy [MJ/kg] | Embodied Carbon [$kgCO_2$/unit] |
|---|---|---|---|
| **CEM** | 14% | 0,786 | 0,136 |
| **Sand** | 29% | 0,002 | 0,001 |
| **Aggregates** | 57% | 0,047 | 0,003 |
| **Transport surcharge** | 57% | 0,018 | 0,001 |
| **Total** | **100%** | **0,853** | **0,141** |

Table 4 demonstrates that while only contributing 14% of the total amount of materials, cement accounts for more than 90% of the embodied energy and embodied carbon. Even if it is possible to use 100% coarse aggregates from granulated rubble, the effects on savings are small. Even more determinative for the outcome is the use of steel reinforcement in the concrete structure. According to the ICE database, the coefficient for using 97kg/m$^3$ reinforcement steel in a concrete structure will double the embodied energy and greatly increase (>30%) the embodied carbon of the concrete mixture. This conclusion also shows in the results when comparing scenario 3 , which includes 97kg/m$^3$ reinforcement steel, to scenario 5 , which uses no reinforcement steel. As expected, scenario 5 exceeds savings by > 50% on embodied energy and >30% on embodied carbon compared to scenario 3.

## 7. Discussion

The aim of this study is the analysis and comparison of the actual environmental impact of different methods of how concrete, from the existing building stock, not designed for reuse, can be (re)used in a more circular manner.

This study demonstrated that savings up to 70% on embodied energy and 70% on embodied carbon could be obtained by upscaling to the maximum capacity of the high apartment block of 50 apartments. Further improvements in savings can be achieved with an optimal occupancy rate of machinery. In scenario's that use new concrete, savings drop drastically, ranging from -16% to 58% saving on embodied energy and 2% to 37% savings on embodied carbon compared to the baseline scenario. This is explained by the disproportionate share on embodied energy and embodied carbon of cement and steel in a concrete mixture.

Although end of life re-use is outside the scope of this study, it should be noted that when making new concrete products, end of life re-use and designing for re-use are important factors to consider, for example based on Design for Disassembly indicators [10]. Preventing the use of new cement will yield significant savings in the second lifecycle of the concrete product.
Scenario 1 and 2 do not suffer from the impact of using new cement and reinforcement steel and therefore appear to be the best circular scenario when scaled up to their maximum potential. The main contributor to the environmental impact for these scenarios, is the use diesel by heavy machinery. Most of the heavy machinery can be seen as a fixed cost in these scenarios. Therefore optimization in occupancy rates of this machinery and upscaling is very effective in these scenarios.
When strictly looking at the savings on embodied energy and embodied carbon compared to the baseline scenario, scenario 1 and 2 score highest. However, there are a lot of practical problems and risks which are not shown in the measured outcome of embodied energy and embodied carbon. Scenarios 1 and 2 are high risk, high reward scenarios. Sawing and hoisting the housing units is a risky undertaking. A single mistake could render the entire housing unit worthless (e.g. cracks or even collapsing on itself during execution of the scenario). Another problem is assessing the quality and durability of the existing building. Especially the quality of the concrete of old buildings varies greatly and is probably not designed for a life span of more than 50 years. Another obstacle for applying these scenario are current building legislations/regulations. The older existing building stock is not up to date to current building legislation. And legislation for building methods like scenario 1 and 2 is not yet available. This means that close cooperation with a willing municipal administration is necessary. All these points require attention and expertise which is not commonly available.
Strengths of this study are the widely availability of data from real world execution of the demonstrators and the usage of a widely accepted database on carbon and energy in materials [5]. Another point of strength is the narrow focus of the study and clear demarcation of the scenarios. Outcomes are hereby not clouded by peripheral issues, but give clear insight on the environmental effect of the circular concrete scenarios.
A limitation of this study is the repeatability of the scenarios. Each new to be realize project and to be harvested building can differ drastically. So for each new project one must assess if that project fits within the set boundaries for this study.

## 8. Recommendation and conclusion

Several recommendations can be given for lowering the use of (new) cement in concrete. The simplest method is lowering the percentage of (portland cement) clinker in the cement mixture. EU regulations have to be expanded (EN-197-1) to allow this to happen. Another option that is being explored, is extracting reactive (unhydrated) cement from rubble concrete. There can be up to 50% of unhydrated cement in concrete rubble. If this is recoverable as a usable ingredient for new concrete, great savings on embodied energy and embodied carbon in recycled concrete can be made [11]. Promising research for alternative binders in concrete is being done as well. This ranges from adding additives which lower

the necessary clicker content, to using geopolymer concrete, to reworking fine concrete particles into a cement-like material.
Although scenario 1 and 2 score the highest savings on embodied energy and embodied carbon compared to baseline, these cannot be deemed as the most promising scenarios. The risks during execution are high and the shortcomings on quality of the existing building stock and current building regulations are significant obstacles to enable these scenarios as directly and large scale applicable.
The most promising scenarios with the highest feasibility of implementation in the built environment are scenario 3,4 and 5. The risks during execution of these scenarios are low and current building regulation can be met with relative ease. Furthermore, there is a lot of room for improvements. Optimizing manufacturing processes and material usage has directly significant influence on the environmental impact of these scenarios. The possibilities of end of life re-use scenarios are also significant factors in this.

## 9. Acknowledgement

The SUPERLOCAL / UIA Super Circular Estate has received funding from the European Fund for regional development in the framework of the Urban Actions Initiative under number UIA02/240. Michiel Ritzen has received funding from the Dutch Organisation for scientific research (NWO) grant number HBOPD.2018.02.025. The authors would like to thank all the partners in the SUPERLOCAL project for their contribution to this research.

## 10. References

[1] Collection, United Nations Treaty. (2016). *Paris Agreement*. (7.d). Paris.
[2] Rijksoverheid. (2019). *Klimaatakkoord*. Den Haag.
[3] Rijksoverheid. (2018). *Betonakkoord voor duurzame groei*. Nootdorp.
[4] SUPERLOCAL. (2019). SUPERLOCAL Super Circular Estate. Retrieved from https://www.superlocal.eu/
[5] Hammond and Jones. (2011b). Inventory of Carbon & Energy (ICE) Version 2.0. from University of Bath https://drive.google.com/file/d/1rYHA-klJTrq9h2XyuhZltJ703Ey0I_Cb/view
[6] Hammond and Jones. (2011a). *A BRSIA guide*.
[7] Bijleveld, Bergsma, van Lieshout. (2013). *Milieu-impact van betongebruik in de Nederlandse bouw. Status quo en toetsing van verbeteropties.*
[8] GMB. (2019). *Ketenanalyse Beton*.
[9] Zaad, J. (2019). *Sub report B: feasibility of circular concrete scenarios*
[10] Cottafava and Ritzen. (2020). Circularity indicator for the built environment: bridging the gap between embodiedimpacts and design aspects. *Resources, Conservation and Recycling*.
11] SmartCrusher, bv. (2020). Smartcrusher. Retrieved from https://www.slimbreker.nl/smartcrusher.html

IOP Conf. Series: Earth and Environmental Science **855** (2021) 012005 doi:10.1088/1755-1315/855/1/012005

# Developments to come to a circular construction economy; experiences in facilitating a local soil and sand depot.

**A G Entrop**[1,2]

[1] Saxion University of Applied Sciences, School of Business, Building & Technology, M.H. Tromplaan 28, 7513 AB Enschede, the Netherlands

**Abstract.** In the construction industry the ambition to come to a circular economy will, with its vast material usage, be a hard nut to crack. This paper describes a research project part of the H2020 Cityloops project, in which it was expected that soil and sand depots could be an interesting subject to study. Already for many years and in multiple municipalities, these depots help to overcome mismatches regarding quality and quantity in demand and request of soil and sand. With the ambition to close material loops the construction industry might experience these kinds of mismatches also for an increasing number of other resources in the nearby future. Having positive experiences with her soil and sand depot, the municipality of Apeldoorn was open to give insights in the flows at her open-air soil and sand depot. The conducted research helped making processes transparent and valuable insights were gained. However, considering the great variety in building materials, products and components, it currently seems possible to handle only a small number of other material flows in the same way soil and sand are being handled and stored at a depot.

## 1. Introduction

The urge to strive for closed material loops is felt in the construction industry. However, before the built environment can be regarded as circular, not only the material usage in buildings needs to be circular, but also the public space surrounding them needs to be established in a sustainable circular way. This is a technological as well as an organisational and a policy challenge. In the construction industry multiple principals, contractors, architects and suppliers of building materials try to reduce the need for new virgin materials, put effort in reclaiming used materials, products and building components, and reduce the amount of waste originating from their production processes.

As Magnusson, et al. [1] explain for houses, roads and other civil engineering purposes; construction also involves management of soil. Soil (as well as rock materials) from quarries or excavated from other construction sites is also used for e.g. filling, permeable layers, base layers, landscaping and as filler in concrete and asphalt. In terms of energy use and greenhouse gas (GHG) emissions, both quarrying and heavy transport give rise to negative environmental effects [1], that do not match the ambitions of a circular economy. The extraction of sand in open water, which is quite common in the Netherlands, also has a significant impact on the environment, e.g. [2, 3].

Before buildings and infrastructure can be designed for a particular location, research is needed regarding the bearing capacity of the underlying soil. Hence, the bearing capacity of the soil needs to able to resist the load. Research is also needed on the presence of any toxic substances endangering the health of contracted builders and future building or infrastructure users. In the ground no contaminations should be present in harmful quantities. If the proper quantity or quality of soil isn't present at the

[2] Author to whom any correspondence should be addressed.

construction site, transport is needed. From a cost and environmental perspective again, the transportation of soil needs preferably to be reduced to a minimum. However, in many projects soil and sand of a better quality or in another quantity than directly available, are needed to match the aforementioned requirements. If sand is going to be used as an ingredient, also specific quality standards need to be matched, e.g. [4].

Hence, in the construction industry sand is an important resource [5]. Multiple entrepreneurial activities are offered to make sure sand of a certain quality is available at a specific location in proper quantities. These activities are often concentrated around so called soil and sand depots (in Dutch: grondbanken), where soil and sand are being collected, processed, inspected, labelled, and even can be put on display. In the Dutch municipality of Apeldoorn one already has positive experiences in reusing soil and sand locally through an open air soil and sand depot, where quantities and qualities of soil offered and requested by actors in the construction industry are registered by the municipality. It is experienced that much of the soil offered to the depot of Apeldoorn is often of a better quality than expected on beforehand.

Therefore, after assessing the quality of this important resource in shaping public spaces, the soil leaving the depot can comply with the requirements of multiple projects. We would like to extend this principle to other resources needed in (re-)developing and maintaining public spaces, so that material loops can be closed. Attention will be paid to the processes involving the soil and sand depot of the municipality of Apeldoorn. Data was collected on how much soil with which quality was being handled. These processes may form a blue print to close circles for other materials in the construction industry.

Considering that sand is an important resource to the construction industry and social and political developments are pushing the construction industry to move towards a circular economy, e.g. [2, 3], the research described in this paper focuses on the possibility to facilitate circular processes in that same industry by positioning an existing soil and sand depot as an example for other resources. Therefore, the main research question is: how can the quest for a circular construction industry benefit from the experiences gained in facilitating a local soil and sand depot?

To come to an answer to this question, Section 2 describes the theoretical background regarding the role of sand in construction processes. In section 3 the field of practice will be entered by discussing how sand is being handled before, on and after a soil and sand depot. What processes to match quantities and qualities take place in order to fill in demands? Section 4 shows the results by means of data of a local soil and sand depot in Apeldoorn. These data will be analysed in section 5 followed by a section with the discussions, after which conclusions will be drawn and recommendations are provided.

## 2. Theoretical background

Multiple sand typologies and qualities can be distinguished. A basic definition of sand can be found in the Glossary of Term in Soil Science [6], in which sand is defined as *a soil particle between 0.05 and 2.0 mm in diameter*. This same glossary also mentions five soil separates by the following names: *very coarse sand, coarse sand, medium sand, fine sand, or very fine sand.* Sand is as a matter of fact a specific soil textural class with soil being defined as: *the unconsolidated material on the immediate surface of the earth that serves as a natural medium for the growth of land plants* [6].

As already addressed sand is directly and indirectly an important resource, when it comes to shaping the built environment. Its direct use consists for example in offering soil for proper heights in landscaping and solid foundations for new buildings. Regarding indirect use one can think of the role of sand in the production of bricks, e.g. [7], and concrete. Global geotechnical criteria for some different forms of earthworks are concisely brought together by Kataguiri, et al. [8] in table 1.

The environmental impact of the construction industry is more than significant. Magnusson, et al. [9] identified environmental and economic benefits of reusing excavated soil and rock. They distinguished the following three situations:

- reusing soil and rock on-site, where the availability of space is an important boundary condition;
- reusing soil and rock in other projects, where coordination and joint planning are major issues;
- recycling soil and rock at a facility, where high investment costs and location optimization could be important barriers.

**Table 1**. List of main geotechnical criteria for reuse of excavation material in earthworks [8].

| Earthworks | Geotechnical criteria |
|---|---|
| Trench backfill | Cohesion and friction angle (shear strength)<br>Swelling and loss of strength by wetting<br>Compaction degree (dry density and moisture content)<br>Grain size distribution |
| Walls with reinforced soil | Cohesion and friction angle (shear strength)<br>Swelling and loss of strength by wetting |
| Paving layers | Swelling<br>Penetration resistance (California Bearing Ratio)<br>Grain size distribution |
| Vegetation cover replacement | Organic content<br>Clay content<br>Cation exchange capacity<br>Permeability coefficient<br>Acidity (pH) |
| Drainage | Grain size distribution<br>Hydraulic conductivity<br>Loss of strength by wetting |

Considering that large volumes and, therefore, weights of soil and sand are being transported when preparing construction sites and constructing buildings, the fuel and other energy use and emissions have been more than once subject of investigation. Devi and Palaniappan [10] for example studied the situation in India for the excavation and transport of soil. In their study, the fuel use, energy use and emissions were related to operational parameters and technical parameters (see table 2). Operational parameters are for example soil type, dumping distance and depth of excavation. The technical aspect considers the equipment used, like the excavators and trucks.

**Table 2.** The diesel and primary energy use of the excavation and transport of soil and rock [10].

| | | **Case 1** Soft dis-integrated rock | **Case 2** Sandy clay & black cotton | **Case 3** Loose soil & weather rock | **Case 4** Loose soil & weather rock | **Case 5** Weather & hard rock |
|---|---|---|---|---|---|---|
| Excavation of soil only | Diesel use (l/m$^3$) | 0.33 | 0.37 | 0.37 | 0.33 | 2.09 |
| | Primary energy use (MJ/m$^3$) | 14 | 16 | 16 | 14 | 89 |
| Excavation & transport of soil to dumping site | Diesel use (l/m$^3$) | 0.45 | 0.52 | 0.55 | 0.52 | 3.2 |
| | Primary energy use (MJ/m$^3$) | 19 | 22 | 23 | 22 | 135 |
| Excavation, transport of soil & levelling at dumping site | Diesel use (l/m$^3$) | 0.51 | - | - | - | - |
| | Primary energy use (MJ/m$^3$) | 22 | - | - | - | - |

## 3. Entering the field of practice

The last thing one wants is that land cannot safely be used due to polluted soil and sand. The Netherlands is a densely populated country. Approximately 17.5 million people live on a surface of 41,500 km$^2$, which puts quite some pressure on the land use. The estimation is that at around 250,000 sites the soil is strongly polluted [11]. At its peak, in 2004, 650,000 sites were thought to be seriously contaminated. After investigations, the soil at most of them appeared to be not seriously contaminated. At locations with severe soil contamination 30,000 sites were remediated, meaning that contaminants were removed or isolated. Governmental bodies have agreed on remediating and/or controlling the remaining 1,400 locations with urgent soil contamination not later than 2020 [12].

An AP04 inspection, also called batch inspection, is used to determine the ecological hygienic quality of soil. If one wants to use or reuse soil, this inspection is compulsory. The quality of the soil is then assessed [13], which starts with a preliminary historical investigation on the batch to check out if one needs to be suspicious about the soil's quality. When one has suspicion about the soil's quality, the location will be visited and the total batch will be defined. A single batch may not exceed 10,000 $m^3$. Per batch, one hundred grips of 0.5 m deep must be taken. If a meter is to be excavated, fifty holes of two times 0,5 m need to be drilled per batch. The various soil layers need to be distinguished, sand and clay, for example, have to be examined apart. If the preliminary investigation or field inspection shows that a small part of the batch is suspected of contamination, this will be investigated separately. In this way costs can be saved by preventing the entire batch from being remediated. The samples are examined in a laboratory, where it is determined whether the soil is clean, slightly polluted or heavily polluted [14].

The soil that does not need to be cleaned or disposed can be transported directly to another project or indirectly via the soil and sand depot. Three main classifications can be distinguished when it comes to assessing the quality of soil at the soil and sand depot, namely [15]:

- AW2000: AW stands for Background Value (In Dutch: AchtergrondWaarde). The AW2000 values for soil are determined on the basis of contents of substances, such as those occurring in the soil of natural and agricultural land that are not affected by local sources of pollution. Soil that meets the background value is sustainably suitable for any type of soil use and is assessed as clean or unpolluted soil;
- Residential: when soil does not comply with the Background Value, but does not surpass a locally established maximum value of non-original substances, the soil can still be used without remediation for residential purposes;
- Industrial: when soil does not comply with the Background Value and cannot be used for residential purposes, but does not surpass a locally established maximum value of non-original substances, the soil can be used without remediation for industrial purposes.

For the municipality of Apeldoorn, a so-called soil quality map shows the diffuse chemical quality for soil layers from 0.0 to 0.5 m below ground level (topsoil), from 0.5 to 1.0 m below ground level (intermediate layer; only for the city centre) and from 0.5 to 1.0 to 2.0 m below ground level (substrate). The map has been established on basis of the presence of the following substances: arsenic (As), cadmium (Cd), chromium (Cr), copper (Cu), mercury (Hg), lead (Pb), nickel (Ni), zinc (Zn), polycyclic aromatic hydrocarbons (PAH), extractable non-volatile organohalogen compounds ($EO_X$) and mineral oil. In time, the soil quality map may, in line with standard NEN-5740, be updated for new substances, namely: barium, cobalt, molybdenum and polychlorinated biphenyls (PCBs) [16].

When composing the soil quality map, the municipality of Apeldoorn was divided into seven sub-areas. Within these sub-areas, the chemical soil quality is expected to be consistent due to similarities in usage characteristics. The municipality of Apeldoorn uses the 90 percentile value (P90) for the urban area and the 80 percentile value (P80) for the rural area. If this value is below the national background value (AW2000), the municipality will apply the area-specific soil quality class (agriculture / nature, maximum value for housing or maximum value for industry). The applicable background levels (P90 or residential or agriculture / nature) based on standard soil (10% organic matter and 25% lutum) are stated for the topsoil and topsoil in table 3 [16].

**Table 3.** Top soil (0.0-0.5 m below ground level) and substrate (0.5-2.0 m below ground level) based on 25% lutum and 10% organic matter [16].

| Zone | Parameter (in mg / kg dry matter) | | | | | | | | | | | | | |
|---|---|---|---|---|---|---|---|---|---|---|---|---|---|---|
| | As | Ba | Cd | Cr | Co | Cu | Hg | Pb | Mo | Ni | Zn | PAK | oil | PCB |
| B3 | 27.0 | - | 1.2 | 62.0 | - | 54.0 | 0.8 | 210.0 | - | 39.0 | 200.0 | 6.8 | 190.0 | - |
| O3 | 20.0 | - | 0.6 | 55.0 | - | 40.0 | 0.20 | 50.0 | - | 35.0 | 140.0 | 1.5 | 190.0 | - |

## 4. Results

The soil and sand depot is located at the north-east site of the built-up area of the municipality of Apeldoorn. It is an open air facility, where soil is being stored as shown in figures 1 and 2. It has already been in use for many years. During these years data have been collected on what quantity and quality of sand was offered or taken by who and when. This section presents the collected data for the time period 2011 to 2016. Although some data of 2008, 2009, 2010 and 2017 is available, it was unfortunately not possible to retrieve complete lists of transactions for these years. By conducting interviews it was possible to work out the different transport paths soil can follow within the municipality of Apeldoorn. This scheme showing the important role of the AP04 inspection process is presented in appendix A.

**Figure 1.** Sand stored at the soil and sand depot of the municipality of Apeldoorn (by courtesy of Robert Rouwenhorst).

**Figure 2.** Sand site of Apeldoorn as can be viewed from south to north at Google Maps (by courtesy of Aerodata International Surveys).

Figure 3 shows that annually between 11,332 and 22,756 $m^3$ of soil passes through the depot. A truck load consists normally of at maximum 20 $m^3$ soil and weighs in that case approximately 34 Mg. The amount of soil being offered and taken logically strongly depends on which civil engineering works and building projects are undertaken in the municipality. Figure 4 shows the large differences between the project related batch sizes of soil. In 2013 5,275 $m^3$ of industrial classified soil was transferred for one project via the depot. In 2015 the smallest batch for one single project, of only 19 $m^3$, was transferred.

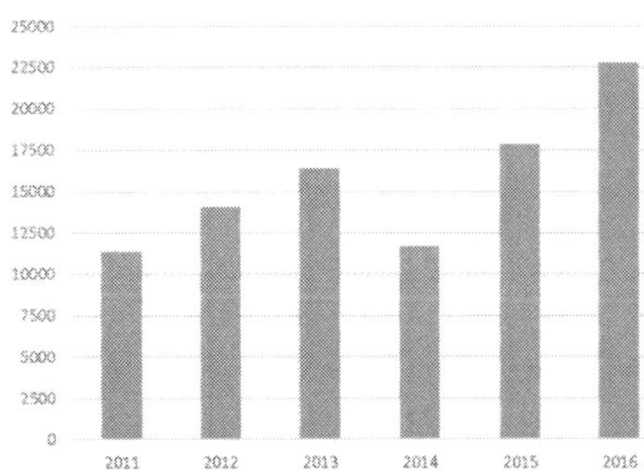


**Figure 3.** Amount of soil in $m^3$ offered by the soil and sand depot

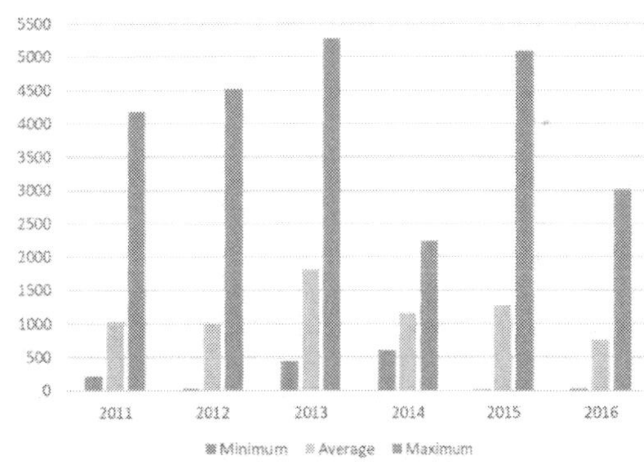


**Figure 4.** Indicative batch sizes in $m^3$ soil per project

**5. Analysis**

In a former section the existence of different quality labels for soil was explained. In this section that framework is used to analyse the flows of soil at the depot of the municipality. In October 2019 it was mentioned in a presentation of the municipality that of the around 200 batches of soil transferred between 2008 and 2017 none of them was more polluted than expected on the basis of the soil quality map.

By studying the data on the batches from 2011 till 2016, the author was able to compose figures 5 and 6. The data studied clearly shows that most of the incoming soil was rated as having a proper quality for industrial purposes (see figure 5), namely between 60.5% in 2015 and 94.5% in 2013 for the annual incoming volume of soil. However, after assessing the soil's quality properly, these values were lowered to between 1.8% in 2015 and 46.5% in 2012 for the annually outgoing volume of soil (see figure 6). The assessed soil of which the quality was not clearly documented, is classified in figure 6 as 'unknown'. Setting the worst case scenario as a basis by adding this 'unknown' quality category to the figures just mentioned, the maximum percentage rose to 71.2% in 2013 for the annually outgoing volume of soil. Based on this data the researchers have no substantiated reason to doubt the claim made in the presentation: the quality of the soil after analysis is higher than expected.

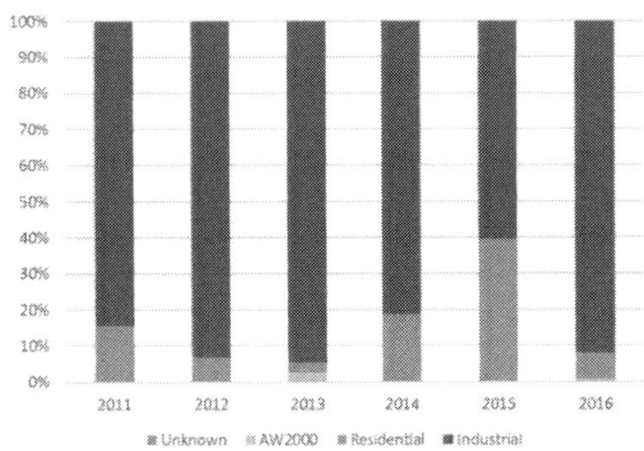


**Figure 5.** Estimated quality of incoming soil at the soil and sand depot

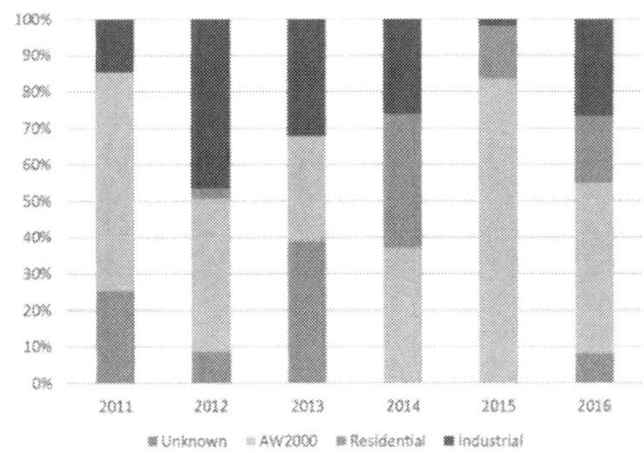


**Figure 6.** Assessed quality of outgoing soil at the soil and sand depot

## 6. Discussion

This study on the soil and sand depot of the municipality of Apeldoorn provided some interesting insights, that are in the noble opinion of the author worth sharing.

Firstly, this case shows the challenge to properly collect and store data related to incoming and outgoing resource flows. Although it seems relatively easy and straightforward to store the quantitative and qualitative information, it is without a 24-7 supervisor at the spot not easy to be unambiguously about what information exactly connects with which batches of soil in projects and at the depot. Volumes are rounded, soil from different sources could be mixed without notifications and a truck load might pass by unnoticed. A link between the physical product, soil, and the data providing information about its quality is not easily established and maintained, as could be the case for a tagged window frame for example.

Secondly, this case study shows the importance to check assumptions about the quality of resources at building sites. Standards are available on the handling of soil and soil quality can be objectively assessed by measuring the presence of multiple substances. To come to a circular economy, the construction industry needs also to be able to objectively assess the quality of other resources needed or materials coming available from for civil engineering works and constructions works. In Apeldoorn a positive effect is experienced about what opportunities can exist when assessing the quality of resources. Your resource might be more valuable than on forehand expected, resulting in a value increase for which little to no investments were to be made.

Thirdly, considering the large fuel consumption of equipment involved in handling soil and sand, a first step always needs to be if construction objectives can be met without relocating soil and sand at all. When soil and sand need to be relocated, then the distance between origin and destination needs preferably to be as small as possible and direct, by means of; without a stop at a depot. The challenge of finding an optimum between travel distances with direct replacing and shorter distances with indirect replacing applies in principle for every new building project without a closed soil balance.

## 7. Conclusion

Considering the data studied and unfolded in this paper, it can be concluded that the soil and sand depot in Apeldoorn fulfills its role with verve. The depot offers temporarily space for soil and sand in times that supply and demand are not well aligned. Due to preserved estimations of the soil's quality for the Apeldoorn's underground on forehand, the quality assessments on the volumes actually being transferred show smaller levels of pollution than expected. Given the fact that the batches did not undergo cleaning or purification processes to improve the soil's quality, the term upcycling does not apply. The added value of the soil and sand depot can be found in the space it has to offer to store and to combine batches, to facilitate AP04 inspection procedures and the immediate availability of this important resource.

However, currently most civil engineering works and buildings are mainly being composed of linear end-to-life components, products and materials. The main research question in this paper relates to a future in which the economy needs to be circular, namely: how can the quest for a circular construction industry benefit from the experiences gained in facilitating a local soil and sand depot? In landscaping and construction projects soil and sand are important resources that seem not easily to wear out or experience degradation. In that sense soil and sand are, compared to many other resources for constructions, relatively simple raw materials. The analysis shows that it is already rather complex to properly record the quantity and quality of soil and sand. A quick-scan on the historical background of a land plot only provides limited insights in soil quality. This should alarm us about the efforts needed to be undertaken, when it comes to investigating and administrating the historical background, former usage and current state of all other circular resources needed in construction.

It can be concluded that when the principles are applied currently in use for soil and sand, these processes will be time-consuming, prone to errors and costly. Logistic systems necessary to have all components, products and materials in place at the site in time are complex. Being a local depot, a soil and sand depot is able to operate as a node in the logistic system of the soil and sand market where supply and demand can converge. However, there will be a need for a new kind of depot for a circular construction industry in a new sort of logistic system. A system that is supported by automated sensors and a proper digital infrastructure linked to databases. The challenge will be to develop this logistic system manageably for all circular resources, because not every individual material, product or component can be thoroughly examined and tested by hand in detail. The specific principle regarding soil testing, where samples are taken from a maximum sized batch by procedures well established, does seem well applicable for multiple other already used building materials.

## References

[1] Magnusson S, Johansson M, Frosth S and Lundberg K 2019 Coordinating soil and rock material in urban construction - Scenario analysis of material flows and greenhouse gas emissions. *J. Clean Prod.* **241** 118236

[2] Jong M F de, Borsje B W, Baptist M J, Wal J T van der, Lindeboom, H J and Hoekstra P 2016 Ecosystem-based design rules for marine sand extraction sites. *Ecol. Eng.* **87** 271–280

[3] Veraart F, 2019 Land or lakes: gravel excavation in Dutch spatial and resources policies through the lens of sustainability developments, 1950–2015. *Land Use Policy* **82** 367-374.

[4] Tavira J, Jiménez J R, Ayuso J, Sierra M J and Ledesma E F 2018 Functional and structural parameters of a paved road section constructed with mixed recycled aggregates from non-selected construction and demolition waste with excavation soil. *Constr Build Mater.* **164** 57–69.

[5] Lai L W C, Chaua K W and Lorne F T 2016 The rise and fall of the sand monopoly in colonial Hong Kong. *Ecol Econom* **128** 106–116.

[6] Canada Dept. of Agriculture 1976 *Glossary of terms in soil science* (Ottawa: Research Branch)

[7] Guzlena S, Sakale G, Certoks S and Grase L 2019 Sand size particle amount influence on the full brick quality and technical properties. *Constr Build Mater* **220** 102–109

[8] Kataguiri K, Boscov M E G, Teixeira C E and Angulo S C 2019 Characterization flowchart for assessing the potential reuse of excavation soils in Sao Paulo city. *J. Clean Prod.* **240** 118-215

[9] Magnusson S, Lundberg K, Svedberg B and Knutsson S 2015 Sustainable management of excavated soil and rock in urban areas; a literature review. *J. Clean Prod.* **93** 18-25

[10] Devi L P and Palaniappan S 2017 A study on energy use for excavation and transport of soil during building construction. *J. Clean Prod.* **164** 543-556
[11] MilieuCentraal 2020 *Bodemvervuiling* (retrieved from https://www.milieucentraal.nl)
[12] Ministerie van Infrastructuur en Waterstaat 2020 *Atlas Leefomgeving* (retrieved from https://www.atlasleefomgeving.nl)
[13] Ministerie van VROM 2007 *Besluit bodemkwaliteit*
[14] Van Dijk Geo- en Milieutechniek b.v. 2020 *Partijkeuring grond* (retrieved from https://vandijktech.nl/milieu/partijkeuring-grond)
[15] Ministerie van VROM 1986 *Wet bodembescherming*
[16] Roording M and Teusink M 2016 *Verkennend bodemonderzoek en verkennend asbestonderzoek.* (Zelhem: Buro Antares)

**Acknowledgments**

The authors would like to express their gratitude to the public servants of the Municipality of Apeldoorn for providing valuable insights and data. This project has received funding from the European Union's Horizon 2020 research and innovation programme under grant agreement No. 821033.

## Appendix: detailed process scheme for soil/sand at the municipality of Apeldoorn

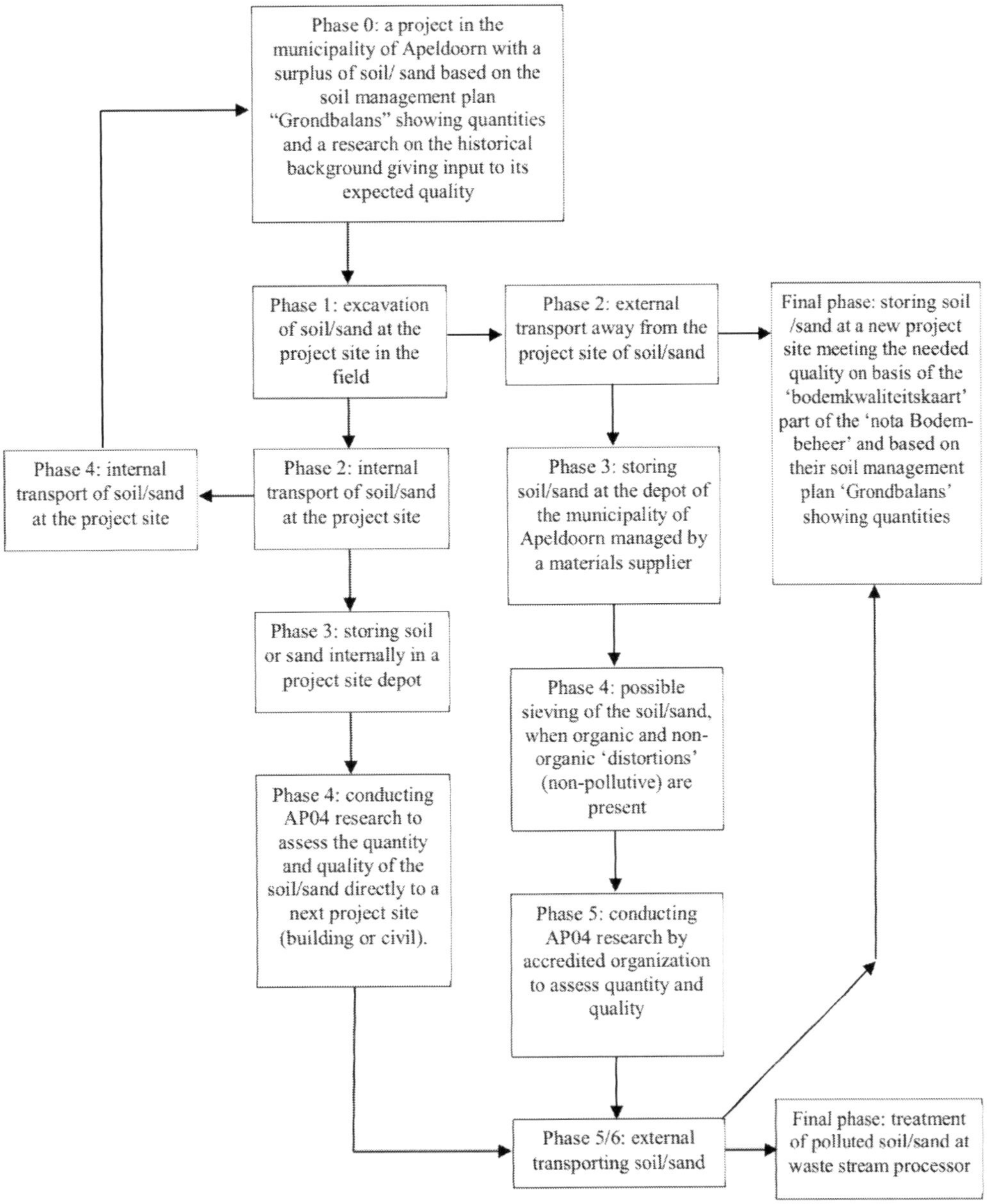

# Green public procurement: learnings from Pesaro city, Italy

**Usha Iyer-Raniga** [1,2] **and Margherita Finamore**[2,3]

[1] School of Property, Construction and Project Management, RMIT University, Melbourne VIC 3000, Australia
[2] United Nation One Planet Network Sustainable Buildings and Construction Programme, France, Paris
[3] Comune di Pesaro/Pesaro Municipality, RUP Servizio Lavori Pubblici/Public Works Department Piazza del Popolo, 1-61121 PU, Italy

**Abstract.** The development of mandatory green public procurement requirements is a significant and tangible contribution to undertaking planned, sustainable deconstruction (rather than demolition) approach to increase recycling materials, creating new businesses and new green jobs. Research undertaken of a procurement process for Pesaro, Italy demonstrates the application of an assessment approach and method based on holistic evaluation and interpretation of multiple sustainability aspects of buildings and their parts. Taking a holistic life cycle approach, this case study provides a clear evidence of how to foster 'circular' alternatives compared to linear, business as usual options in the management of public procurement. A competitive procedure in the procurement process was undertaken resulting in materials either being reused or taken to a recycling centre. The reuse of the materials during the new construction phase also included monitoring the quantity of all materials, preventing any waste and encouraging optimised reuse. This study shows that it is possible to drive the market towards a circular thinking and a circular construction site management is a critical requirement for transitioning towards a circular built environment. By ensuring that tendering processes support and develop sustainable recovery of waste materials, this study shows that it is possible to provide materials second and third life.

## 1. Introduction

The process of building and construction has an impact on the planet. Buildings and construction were attributable to 39% of energy and process-related CO2 emissions in 2019, of which 11% is accountable to the manufacture of building materials such as steel, cement and glass [1]. The latest IEA report (2019) [2] supporting the move to a decarbonised world posits that the transport, and the buildings and construction sector are key contributors to material use. Material use is directly linked to prevailing consumption and production practices. The process of procurement can influence the use of products and materials in buildings and the amount of waste produced. The amount of waste produced in a linear process generally does not get reused, recycled or repurposed into other useful products or materials. It usually ends up in landfill or is incinerated, and therefore, wasted. On the other hand, in a circular process, there are opportunities for nurturing second and third life of materials and products, therefore, such products and materials can remain in the value chain for reuse again after end of life. The intent with such an approach to circularity is to ensure that virgin materials are not used giving the planet an opportunity to regenerate.

Public procurement is a process by which public authorities, usually governments, can purchase works, goods or services from companies [3]. Green public procurement (GPP), as defined by the EU is a process where public authorities seek to 'green' or reduce the environmental impact of their procurement processes. In other words, GPP is the key to enhance resilience of the environment, if undertaken appropriately. Europe (EU) GPP provides support for public procurement through technical reports and tender criteria but leaves it to the public authorities to decide what or how they wish to include the requirements in their tender documents [4].

*1.1. Resource use and the situation in Europe*

Housing, nutrition and mobility together represent more than 82% of the total material footprint [5]. As reported in the "Global Construction 2030" the turnover in the construction sector is expected to increase by 85% to $15.5 trillion worldwide by 2030, mainly due to the action of three countries: China, US and India accounting for 57% of all global growth [6]. The rapidly growing construction sector is currently among the world's largest producers of waste: every year, 1.3 billion tons of construction and demolition waste is generated worldwide and half of it comes from construction [7][8]. As indicated in the IEA report [2] resource use through material efficiency is a step towards ensuring CO2 mitigation options, thus reducing emissions and supporting clean energy transitions. Resource use across the life cycle of buildings and construction, therefore, needs to be a key focus. In a clean technology scenario to 2060, where material efficiency is the focus (as opposed to the business as usual), materials efficiency contributes to approximately 30% of the combined reduction of commonly used materials in the sector: steel, cement and aluminium. In such a scenario, material demands would be reduced by 24% for steel, 15% for cement and 17% for aluminium [2]. In 2018, the first *Circularity Gap Report* published during the World Annual Forum in Davos stated that our global economy is only 9.1% circular [9]; with 4 Gigatons of cycled inputs, and 58 Gigatons outputs coming from extracted resources [10]. Out of the materials not cycled, the majority is lost beyond recovery — either dispersed in the form of emissions or unrecoverable waste.

Construction and demolition (C&D) waste is one of the largest waste streams in the EU (accounting for between 25-30% of all waste generated) [11]. Concrete, bricks, gypsum, wood, glass, metals and other common construction materials can all be recycled, saving space in landfills, and reducing the amount of virgin materials excavated for building activities. Demolition itself refers to a process resulting in waste and therefore of a lesser value. Instead, deconstruction refers to a process where it may be reused and reconstructed again, therefore, the value can either be retained or enhanced. This paper therefore, avoids the use of the word 'demolition', unless as part of standards; using the word 'deconstruction' instead to convey recovery options.

The EU Circular Economy Action Plan [12] highlights the importance of recovering valuable resources in the construction and deconstruction sector, and an EU-wide mandatory target for recycling 70% of construction and deconstruction waste in member states by 2020 has already been set. However, recovery varies greatly across the Union, from less than 10% to over 90%, and much work is still needed on the ground to adequately identify, separate, and recover valuable materials [11], so they are not lost forever, but may be reused or repurposed again. Consequently, there is a crucial need for new circular solutions, especially in the building sector and public procurement regulation can play a significant role in mainstreaming to drive circularity practices.

The building sector in Europe takes a strategic approach in comparison to most countries globally. Around 4 out of every 10 houses in Europe were built before 1960, a time when building practices were poor by today's standards [13]. The priority is to sustain and preserve what is already built, and in case renovation or deconstruction is needed, the idea is to proceed using GPP to enhance circular processes, where materials can re-enter the construction sector and be reused appropriately.

*1.2. Aim of the paper*

This study demonstrates how local authorities can contribute to the achievement of ambitious goals of materials recovery using GPP in the tendering process of public building works. It proposes actions to keep resources in the economy and retain the value of these resources, which will contribute towards delivery of a sustainable, low carbon, resource efficient and competitive economy in the building sector. By taking this approach, the procurement practice promotes sustainability, in accordance with national policies and priorities and can help the public sector construction firms take a more holistic approach to sustainability.

A holistic approach and the integration of the circular economy principles in public procurement is the key to meet the ambitious individual member states and the EU goals on circularity. Taking a holistic life cycle approach, this study explores how local authorities can contribute to the achievement of the SDGs using GPP in the tendering process of public building works. It proposes actions to keep resources in the economy and retain the value of these resources, which will contribute towards delivery of a sustainable, low carbon, resource efficient and competitive economy in the building sector. The results may be applied in the private sector as well.

This paper commences with an underlying approach to this research using the case study method. This is followed by an exploration of the literature on green procurement, focusing particularly on public procurement. The following section focuses on the importance of procurement, especially in the EU. The situation in Italy is presented next, with a description of the case study of Pesaro City. The underlying intent of the reuse of the building and its design rationale are then explained. Results and discussions follow next with the concluding section and recommendations at the end of the paper.

## 2. Research method and approach

This paper uses the case study approach. Case studies are a useful method in the built environment discipline to explore changes in practice or for policy making. Case studies are usually selected if they have something different to offer from the norm [14]. The case study method can offer a richness and in-depth understanding if appropriately selected and analysed.

In this paper, case study method has been used to analyse how changes in planning for procurement can support GPP and an overall intent to move to a circular economy. The case study focuses on the conversion of an old building in Pesaro City, Italy to a new school building whilst also considering urban regeneration. In doing so, several circular economy principals are followed: reuse of an existing building and site, reuse of building materials, reduction in waste, avoiding the use of new materials and supporting economic outcomes, all while regenerating a precinct. This is explained further in section 6.

## 3. Procurement in the literature

A systematic literature review on green procurement by Badi and Murtagh [15] indicates that engagement with the unique characteristics of the construction industry, where stakeholders get together to complete a project for a period of time, end to end perspective or design through to end of life of the project itself, a focus on the goals of sustainability, closing gaps in guiding the industry to meet and deliver sustainability outcomes, use of insights from other perspectives such as innovation and collaboration, and inclusion of critical stances beyond the typical profit margins are important. The role of green procurement and its contribution to circular economy has also been supported by Avsec and Kaucic [16].

Kuttinen and le Roux [17] have developed a guide for the Ministry of Environment, Finland to procure a low carbon building by considering the building's environmental impact over the life cycle, setting targets for carbon footprint, selecting appropriate design, material and building contracts to provide low carbon outcomes, comparing tenders, including low-carbon targets in the contracts and making use of experiences to document and share the learnings. Van Geet et al [18] state that product-service systems have a great potential to decouple consumption from economic growth. Sustainability or green benefits are not inherent in a procurement system; it needs to be specifically factored into or designed and developed to deliver sustainability outcomes. Till the recent past, the building sector has not really entertained service systems in procurement, but it is time product-service systems were considered as part of the procurement practices. Governments play a key role in helping businesses to develop new and innovative procurement models by adopting or adapting procurement processes to include tenders.

Ruparthna and Hewage's [19] paper presents a review of traditional and emerging procurement practices in the construction industry and call for a detailed knowledge of construction procurement as an area to be further investigated in greater detail. In a similar context with the building and construction sector in Malaysia, Bohari and Xia [20] discuss the importance of providing practical guidelines to the stakeholders in the sector. They suggest the importance of policy and guidelines, green practices, and environmental assessment. The role of green practices such as design, selection of construction materials, site management and waste management should be planned from concept design through to completion. The use of environmental assessment tools such as green building assessment schemes, life cycle assessment and the like, provide clear benchmarks for assessing the project, and attendant products and services. Research by Bohari et al [21] also supports Bohari and Xia's [22] findings, stating that green procurement is not widely understood or practiced in the Malaysian construction industry. There are, however, some guidelines on green procurement and green criteria involved in tender processes. A similar finding has also been endorsed by Bidin et al [23] and Rais et al [24]. Empirical research undertaken in Malaysia highlighted significant factors: perception, organizational management, regulatory environment, logistical, contractual, technical, strategic and financial aspects [25]. In neighbouring Indonesia, a similar finding was also observed with green procurement [26].

Focusing on the building and construction sector in China; Shen, Zhang and Long [27] examined the knowledge of green procurement in the country, the challenges and opportunities. The real estate developers do not know much about procurement and few, if any, have adopted a green procurement strategy. In a related paper [28] the marketing benefits, market pressure and internal pressure within organisation as the main reasons for behaviour related to green procurement are noted. The reasons to adopt green procurement are typically: policy pressure, marketing benefits and business benefits. In the Hong Kong context, Wong et al [29] state that the top three significant factors to enhance green procurement are mandatory environmental regulations, client requirements for tendering, and government requirements. In the UAE, a study by Al Nuaimi and Khan [30] showed that innovation capability of an organisation has an impact on implementation of green procurement. Employee commitment to change also influences green outcomes. Another study, also focusing on the UAE showed that the adoption of green construction factors including green purchasing by construction firms leads to improved sustainability performance [31].

In an international study [32] examining strategies for promoting green building technologies in the green building sector, an empirical approach using questionnaires with green building experts was undertaken. They identified government policy and regulation as one of the top four most important strategies to support the promotion of green buildings. Lozano and Witjes [33] discuss the importance of circular economy through collaboration between public procurement and businesses.

In terms of procurement policies in local governments, a study in Africa [34] showed that only the City of Cape Town and eThekwini municipalities had any green procurement strategies in their tendering practices or supply chain management. A similar study focusing on GPP in China [35] demonstrated that Chinese government procurement officials' knowledge of GPP is related to training, which in turn impacts the implementation policies and performance. Simply being aware of GPP does not produce improved GPP performance. Research by Liu et al [36] showed that in the case of local governments, improving GPP performance is positively linked to awareness of guidelines, coverage of GPP policy, and negatively with low administrative level.

In another study focusing on Valencia, Spain [37] it was found that the infrastructure, rather than the building and construction sector tended to employ environmental criteria for public procurement. The study highlighted the importance of GPP in local governments. In a previous study of sustainable procurement in Italy, Testa et al [38] also agreed that public green tenders support GPP. Good GPP performance needs support from local, national and regional initiatives.

Sonnichsen and Clement [39] state that eco-labels, standards, life cycle assessments and life cycle costing are all important considerations for circular public procurement, particularly when risk, timeliness and cost for public institutions are considered on a life cycle basis. A more recent paper by Lazaroiu et al [40] specifically focusing on green procurement focuses on how GPP drives the circular economy. It also focuses on the GPP of construction and building materials, environmental and supply chain management measures in GPP, procurement of sustainable innovation, environmental policy objectives of GPP with respect to energy, pollution, carbon footprint and climate change, GPP policies for production and use of sustainable goods and services and GPP as an inherent aspect of sustainability. All these elements may be linked together to ensure procurement policies and implementation practices support meeting all the factors to transition and retain a decarbonised world. The inclusion of GPP in the EU public procurement standards is an important step in meeting environmental targets [41].

## 4. Circularity and the importance of procurement

The Sustainable Development Goals (SDGs) defined by the United Nations 2030 Agenda for Sustainable Development came into effect after the Paris Agreement on Climate Change [42], focusing on ensuring triple bottom line (environment, economy and society) needs are supported. They also stress the importance of procurement in some of the goals, such as SDG 12 on sustainable consumption and procurement. GPP can therefore, play a key role in achieving the SDGs and can support countries and their governments in achieving targets associated with National Determined Contributions (NDCs) or National Adaptation Plans (NAPs). Circular approaches to buildings impact positively on materials, energy, waste, biodiversity, health and well-being, human culture and society. Therefore, local authorities have a crucial role in enabling the process towards circularity addressing public buildings. Additionally, circular buildings may produce multiple forms of value among which are material efficiency and attendant emissions.

As the world population increases and economies rely progressively more on outside resources to meet their demand for energy, water and food among others, communities are under enormous pressure to find these resources and accommodate waste and emissions. Within the current linear production and consumption economic model, only a small share of waste produced is reused, recycled or traded as secondary materials. Considering finite resource flows, economies will no longer be able to rely on these linear production and consumption models. A circular economy is a viable alternative to this model. It aims to retain more high value materials in the economy, increase the resilience of companies and protect economies to external shocks, incentivise innovation and support local labour markets.

As indicated in the previous section, public procurement can play a key role in transitioning to a circular economy. Including circular 'principles' into procurement practices can help the building sector take a more holistic approach to sustainability, from the first stages of a procurement to the end of product life, while also achieving potential savings. Due to the increased attention and the promotion and use of GPP, public authorities can influence the market by stimulating the industry to develop green technologies that cover the entire cycle of life of the products: from production to supply, and forms of transport, use, and reuse (reuse in a linear process is disposal), explored further below.

### *4.1. Circular public procurement in the EU*

The construction and refurbishment of buildings in an energy and resource efficient way is an important policy objective for Europe. The recast *Energy Performance of Buildings Directive* [43], the *Renewable Energy Directive* [44] and the *Energy Efficiency Directive* [45] together set out requirements for buildings to contribute towards ambitious EU targets for energy efficiency and renewable energy generation by 2020. Moreover, these policy instruments also require the public sector, through the procurement of refurbishment and new-build projects, to lead the way in delivering efficiency improvements and in deploying cleaner forms of energy generation. The *Roadmap to a Resource-Efficient Europe* [46] highlighted the significant impact of construction on natural resources. This is further emphasized in the Commission's Communications *Towards a Circular Economy* [47] and *Resource Efficiency Opportunities in the Building Sector* [48] which outline future policy initiatives to address construction and demolition waste.

The *EU Action Plan for the Circular Economy* [12] has established a concrete and ambitious programme of action which will help to 'close the loop' of product lifecycles. It proposes actions to keep resources in the economy and retain the value of these resources, which will contribute towards delivery of a sustainable, low carbon, resource efficient and competitive economy. This plan recognises public procurement as a key driver in the transition towards the circular economy, and it sets out several actions which the European Commission will take to facilitate the integration of circular economy principles in GPP.

The EU GPP criteria have been developed to facilitate the inclusion of green requirements in public tender documents. While the adopted EU GPP criteria aim to reach a good balance between environmental performance, cost considerations, market availability and ease of verification, procuring authorities may choose, according to their needs and ambition level, to include all or only certain requirements in their tender documents. The basic concept of GPP relies on having clear, verifiable, justifiable and ambitious environmental criteria for products and services, based on a life-cycle approach and scientific evidence base. In the communication *Public procurement for a better environment* [49] the Commission recommended the creation of a process for setting common GPP criteria.

The criteria used by Member States of the EU needs to be similar, to avoid a distortion of a single market and maintain healthy EU-wide competition. Having common criteria reduces considerably, the administrative burden for economic operators and for public administrations implementing GPP. It also supports trade across borders. Common GPP criteria are of a particular benefit to companies operating in more than a member state as well as SMEs (whose capacity to master differing procurement procedures is limited).

Since 2008, the Commission has developed more than 20 common GPP criteria [50]. The priority sectors for implementing GPP were selected through a multi-criteria analysis including; scope for environmental improvement; public expenditure; potential impact on suppliers; potential for setting an example to private or corporate consumers; political sensitivity; existence of relevant and easy-to-use criteria; market availability and economic efficiency. These criteria are regularly updated, and are based on data from evidence base, existing ecolabel criteria and information collected from stakeholders of industry, civil society and EU member states. The evidence base uses available scientific information and data, adopts a life-cycle approach, and engages stakeholders who meet to discuss issues and develop consensus. Criteria are continuously monitored and improved.

Thus, procurement can support circular economy, which in turn, helps drive sustainable development. By promoting a rethinking of business models supporting designing more durable and recyclable products, reusing materials in the production cycle and fostering a more responsible consumption, the circular economy approach can be an interesting implementation vehicle for SDG 12, pledging for more sustainable and responsible consumption and production patterns. Moreover, it is also equally relevant for the achievement of SDGs 6 (water), 7 (energy), 11 (sustainable cities and communities), 13 (climate action) and 15 (life on land) [51].

### *4.2. GPP in Italy*

The Ministry of the Environment in Italy has published guidance on their website for GPP in Italy [52]. One of the most important documents in the field is the National Action Plan on Green Public Procurement (GPP NAP). The GPP NAP sets rules for public procurement in Italy and provides a definition of Italy's national targets for GPP. In the context of EPDs the following aspects mentioned in the GPP NAP are of particular importance:

- the concept of minimum requirements within GPP
- instruments for the identification of environmental criteria
- focus areas within GPP

Minimum energy performance standards apply to public buildings and are set at national level based on a common EU methodology. In Italy, from 1 January 2019 due to the Ministerial Decree of 26 June 2015 of the Ministry of Economic Development, "Minimum Requirements"[53], all new buildings occupied and owned by public authorities must be "nearly zero-energy buildings" according to the definition of nZEB described in the EPBD 2010/31/EU on the energy performance of buildings [43]. The Energy Efficiency Directive also sets mandatory requirements regarding renovation of public buildings and purchase or new rental agreements meeting minimum energy-efficiency standards [54].

In addition, Italy has specific rules which create mandatory GPP standards for specific sectors or type of contracts. Minimum environmental criteria are technical indications mainly concerning environmental issues including the ethical and social aspects of the labour market. The criteria are developed to promote and ensure:

- efficiency and savings in the use of resources
- reduction in the use of hazardous substances
- quantitative reduction in waste products

Minimum Environmental Criteria (MEC) (CAM in Italian) established by the Plan for the environmental sustainability of consumption in the public administration sector and are adopted by Decree of the Minister of the Environment for the Protection of Land and Sea. Special MEC for the construction sector were established by the Decree of Ministry January 2017 and they have been made mandatory by art. 18 of Law 221/2015 and subsequently art. 34 of the procurement code - Legislative Decree 50/2016 (as amended by Legislative Decree 56/2017) [55], which sets down environmental and energy sustainability criteria and made their application compulsory for contracting authorities.

As well as adding value to environmental quality and compliance with social criteria, the application of the Minimum Environmental Criteria also meets the public administration's need to streamline its consumption and reduce its expenditure where possible. Now MEC regulations have been adopted for 17 categories of goods and services, while others are being defined. More specifically, tender notices published by contracting authorities are required to include at least the technical specifications and contractual clauses set down in MEC documents.

MEC must also be taken into consideration for the purposes of drawing up tender documents that apply the criterion of the most economically advantageous bid, thus acting on the recommendation, already set down in the foreword to the MEC, that tender notices should include the awarding criteria defined in the MEC.

Furthermore art. 95 of the procurement code (paragraph 6) provides the precise indications in term of the qualitative, environmental and social aspects associated with work contracts such as:

- the reduction of energy consumption and needed resources of buildings or product production
- the use of EU Ecolabel products
- the cost of operation and maintenance, including energy consumption, pollutant emissions and overall costs, including costs for mitigation of climate change impacts, referring to the entire life cycle of the asset offered, to incentivize the efficient use of resources and a circular economy approach to protect the environment and ensure a fair employment;
- the compensation of greenhouse gas emissions associated with the firm's activities calculated using methods established based on Recommendation No 2013/179/EU to measure the life-cycle environmental performance of products and production.

The Ministry of Environment published the first monitoring report of 2019 GPP showing the achievement of the objectives set by the Action Plan for the sustainability of consumption in the public administration sector on $6^{th}$ March 2020. The report illustrates the progress of the process of adoption of GPP in 2018 and describes the level of application of the Minimum Environmental Criteria (CAM) in the construction sector as required by art. 34 of the Procurement Code (Legislative Decree no. 50/2016) [55].

The report is based on a questionnaire that required quantitative data on the application of CAM, and quantitative-qualitative date related to the training courses organized to improve its development and related to a monitoring system to support the CAM application. Based on different categories of goods, services and works, the report analyses the GPP tendering processes adopted by Municipalities, Park Authorities, Regions and Metropolitan Cities throughout Italy. The date provided by the report are not enough to have a clear picture of the Italian situation because it has chosen a sample of 1086 municipalities called "Comuni Ricicloni" and only 40,6% have responded to the questionnaire. With respect to the construction sector only 19,3% declared their willing to adopt CAM in the building sector [56].

The next section explores how the procurement criteria were applied successfully in one example.

## 5. The case of Pesaro City

The ambition of Pesaro City is to accelerate the transition towards a circular economy, enabling construction firms to use a new system beyond the current outdated 'take-make-dispose' model and contribute to mainstreaming the circular economy concept within and outside the city.

As the first step to enhance the transition towards a circular economy the Public Works Department planned a holistic green building roadmap to provide concrete examples in mainstreaming circular practices. In terms of knowledge dissemination, circular building projects can serve for further awareness raising and experience sharing. Additionally, communication of public data through city portals, including the discussion and open data in relation to indicators, is a powerful tool towards the engagement and motivation of related stakeholders including citizens.

In 2017, the City of Pesaro launched a pilot project with the aim of encouraging a holistic approach towards circular economy principles to ensure the building design responds to environmental protection, health and well-being and other social considerations. From the awareness of the relevance of the building sector towards the development of better cities, the Public Works Department designed an action plan on circularity in the building sector making the environmental criteria mandatory in the tendering process.

*5.1. Planning the process*

Instead of taking a traditional approach to developing works procurement, the city decided to adopt a more circular approach which would be to build a sustainable new school using the current asset. The site consisted of old army barracks built in 1950 and therefore quite energy inefficient in operation. The building no longer served any purpose and the decision was made to reuse the site for a school with the aim of redeveloping abandoned urban areas. The purpose of the change in use was to obtain not only an energy efficient but a circular building with low environmental impact and low maintenance. The use of renewable energy is now almost expected in every development and the intent was to enhance the use of sustainable building materials and technologies, save raw materials and minimise waste from the construction of the building.

To make GPP work, it is essential to know how to make the most of public procurement procedures. A GPP policy can, if it is not carefully implemented, falter on practical issues such as which procedure to use, what criteria to apply, and how to properly assess and verify environmental claims. So, one of the practical steps towards circular procurement is to consider strategically how technical elements can be integrated into existing procurement practices and systems.

Background research needs to be undertaken at the outset. The first step in this is the preparatory stage, very crucial to ensure desired outcomes. Planning and a thorough analysis are essential before launching a tender to achieve environmental goals. The background preparation can also help to ensure that procedures are run efficiently and obtain best value for money over the entire life-cycle of the procurement process. As part of the development of the procurement procedure is to take into practice circular principles within the public procurement for works. This needs to be aligned with the technical and process requirements in the call for tender. Taking a step back and investigating the broader needs first allows the adoption of a functional and performance-based approach. To shift the tendering process to circular models requires a change of thinking including design, its use phase and end of life. It also requires identifying circular businesses to the wider market. Procurement therefore needs to consider life cycles and performance-based approaches. In the next section, the design intent of the project is described.

The background work and preparation lead to the call for tender. The call for tender describes technical specifications and provides measurable requirements against which the tenders can be evaluated. Performance-based criteria describe the expected results and which outputs are mandatory. In the case of this project, two criteria were selected as mandatory: (a) obtaining a nZEB school and (b) providing the environmental energy certification using the LEED certification scheme, achieving at minimum, the LEED Gold certification.

Even if contracting authorities have an obligation to get the lowest price for everything they procure, identifying the most economically advantageous tender does not necessarily mean choosing the cheapest offer. The GPP supports in finding the solution, which meets the requirements identified, including environmental ones, in the most cost-effective way. Best value not only measures the cost of the works, but also considers factors such as quality, efficiency, effectiveness and fitness for purpose. Protection of the environment and the final users can be one of these factors, which can therefore act as an equal consideration amongst others for the award of the contract.

*5.2. The design response*

The intent of design of the new building was to optimise circularity. The analysis of not only educational needs but also citizens expectations for use was the central theme: to make the building more circular in terms of its functionality and make sure that it can be used by everyone all day, every day. To respond to this need, the entrance to the building is a big hall to encourage circulation and interaction with others, so people can, meet others and stay to chat. In order to optimize space, this big hall has been planned as a mixed-use space that can be used like a conference hall by teachers and citizens like. The design of the circulation pattern itself encourages people to sit down and use the circulation space. Photos are provided in Figures 1 and 2. A library for communal use and classrooms for use by the teachers and students also open to the outdoors.

**Figure 1.** The entrance hall at the ground floor, Source: M Finamore

**Figure 2.** The entrance hall at the first floor, Source: M Finamore

The environmental impacts can be improved by the tendering process to build a low energy concept school using good design to support not only passive heating and cooling technologies, but relevant active mechanisms as needed. Moreover, considering how green requirements will affect the procurement process for works and implementing them in line with the legal obligation, materials and methods of production and provision are considered when defining appropriate selection criteria. As a result, the tender includes technical specification on energy efficiency, product lifecycle, waste management, water efficiency with attention paid to internal comfort to create the best indoor environmental conditions for the students and teachers. These results are guaranteed by the environmental energy certification made mandatory in the tender notice and measured in accordance with the Minimum Environmental Criteria (MEC).

As the school is built in a region with a Mediterranean climate, the building envelope also needs to prevent heat losses and simultaneously defend against overheating due to rising temperatures in spring and in autumn as the consequences of climate change. A thorough study of heat gains and losses was carried out to optimise the thermal capacity of the building envelope, which has been achieved by using a green roof, automated solar shadings and the ventilated wall.

The ventilated wall, made of porcelain stoneware, reduces the maintenance costs of the facade and, with the green roof and solar shading supports minimization of energy demand. As a result, mechanical plants are simpler and less expensive in terms of work costs and maintenance. All these technical and functional features increase building adaptability, durability, and resilience in accordance with the circular economy principles. All these features were part of the requirements for tender so the construction firms had to respond to these criteria as part of the bidding process.

*5.3. Reuse of an existing asset to a new asset*

Under the 2014 procurement directives used in Italy, all contracts below the threshold of 1million Euros must be awarded on the basis of the most economically advantageous tender (MEAT), an approach which makes it possible for the procurer to award and compare factors beyond price such as quality and sustainability. This study considers the reuse of an existing asset, thereby applying principles of the circular economy beyond merely a new school building's construction. The idea was to transfer the environmental requirements into technical elements to reduce the environmental impact of the building from the construction phase to its use. These elements have been developed through the development of 11 technical requirements as mandatory criteria chosen to obtain the best tender offer. Figure 3 shows the building as it currently stands.

**Figure 3.** The new school building, Source: M Finamore

It was also mandatory to certify the low environmental impacts by LEED certification and obtain at least the LEED Gold status. Though nZEB has a very low environmental impact — as attested by the LEED environmental certification-mandatory in the tender — the result has been an energy-efficient building with low environmental impact achieving the Platinum LEED certification for the school building, a much better environmental outcome than Gold status. The tender aims not only to obtain a nZEB building, paying particular attention to energy saving and the production of energy from renewable sources but also to the requirements related to the environmental protection the building. To this end, it is equipped with an automation system that optimizes the performance of the building itself.

Relevance is given to thermo-hygrometric comfort, acoustic comfort, correct use of natural and artificial light, optimisation of the building envelope to maximise free benefits of the sun, appropriate use of water, choice of materials, use of appropriate technologies to maximize the energy performance with adjustable solar shading and green roof designed not only to insulate but also to mitigate water use and the heat island effect. Also bearing in mind, the Mediterranean climate the green roof not only supports passive design, it also reuses rainwater collected on site. The green roof is made of seedlings called *sedum* that requires very little water in the said climate. The photovoltaic panel plant on the central part of the roof provides energy for nearly all the electricity needs of the school. This is shown in Figure 4.

**Figure 4.** The green roof, the PV panels and the white membrane, Source: M Finamore

All these needs were transformed into technical criteria and tested by the certification process. In particular, for the deconstruction and waste management processes, the application of the principles of the circular economy was developed by monitoring all input and output materials in the construction site during the construction phase and monitoring the former deconstruction phase of three buildings replaced by the new school to encourage greater reuse of waste materials arising from the deconstruction process. The support of the environmental certification chosen by the contracting authority to which the design and construction of the building has been tested, was crucial to enable systematic monitoring and data collection. As part of the mandatory requirements of the call for tender, it was also mandatory to take all materials to the recycling centres and to submit the certifications providing both the quantities taken to the recycling centres and the quantity recycled.

The material management system in the construction site was organized by:

- selective in-situ storage of deconstructed waste with the obligation to take all the waste to the recycling centre rather than a landfill
- selective collection of all packaging and processing waste, as well as the possible management of the surplus new materials by quantity, separating the materials by their composition (brick, reinforced concrete, wood, plastic)

- classifying in-situ the materials according to the European Waste Catalogue (EWC) codes and transporting them to the appropriate recycling centres.

This practice is in line with the MEC that can be clearly included in the requirements of the call for tenders as GPP.

## 6. Results and discussions

During the process of this project, 556,630 kg of waste was created through the conversion of the old military barracks to a new school for the twenty first century. Out of this total waste, 546,378 kg was recycled and recovered, achieving a 98% recovered waste.

Due to the use of the LEED certification scheme, all material streams were reported and separated into seven categories to support reuse of those materials to include the supply chain as a whole; also, a requirement under the said certification scheme. This led to the following material streams being reported with their attendant proportion of reuse:

- 170201 Wood 100%
- 170101 Concrete 100%
- 150106 Mixed packaging 10% - 30%
- 170405 Iron and steel 100%
- 170107 Concrete / brick / ceramic 100%
- 150101 Paper and cardboard packaging 100%
- 170904 Demolition (deconstruction) waste 100%

In the building process, itself, various categories of stakeholders were involved such as:

- GSE (Energy Services Management)
- Green Building Council Italia (GBC Italia)
- Professional orders
- Construction sector companies
- Trade associations
- Recycling centers and their administration

In particular, two local recycling centres were involved in this project: (a) a centre for the recovery of inert materials from deconstruction, which, through the use of primary and secondary mills, are separated through sieves, allowing them to obtain a multiplicity of grain sizes for different uses in construction and road works, and (b) a material storage centre for re-use of recovery of paper and collection of iron and plastic.

In this example from Pesaro, the waste management of deconstruction and construction waste was measured. The measurements were both in terms of the quantity of materials taken to the recycling centres and the quantity of the recycled products used. Such type of monitoring supports the recycling centre's market in a pro-active manner, thus spending less on waste management processes (8 €/t for the recycling plant against 80-120 €/t for the public landfill).

This practice was possible as the construction companies were informed about monitoring already and therefore, they had no choice but to follow the instructions for a proper management of the deconstruction, reconstruction and waste generated by the processing and packaging of the materials. These specific requirements by the Municipality of Pesaro were included in the call for tender and in the contract specifications, otherwise such measurements would not have been possible. This inclusive tender practice can be even more rewarding if:

- such measures are included through the implementation of the CAM;
- the waste management process is activated during the construction phase to increase the quantities of the material to be taken to the recycling centres, to have a wider amount of materials to be recycled, thus encouraging a competitive construction market.

From a circular economy perspective, the administrative barrier related to the definition of 'waste' can be overcome: material from the deconstruction of a building is generally not dangerous or polluting material, but when it is taken to the recycling centres it must be treated in accordance with the legal definition with attendant administrative burdens that increases cost, which makes the product uncompetitive in comparison to that from virgin materials in the market.

To avoid confusion, mixed messages and streamline outcomes, it is therefore necessary:

- to set up a score to each of the technical criteria and balance the weighting between them to ensure that the process is transparent in the selection of the right firm to be awarded the contract
- that the recycled materials are economically competitive compared to a new product through easier administrative procedures and incentives;
- to establish a procedure to certify the quality of the recycled product in order to place it on the market competitively with new products. Even at the same costs, the market should choose recycled products in relation to CAM requirements due to the impact that new materials create on the environment
- government procurement can pave the way for private companies to support such practices as such companies are loathe to take risk.

## 7. Conclusion

This case study of Pesaro City demonstrates that involving all parties from design and raw materials suppliers to end users, service providers and recyclers, including the associated information flows are crucial to develop and implement circular economy principles. This cannot just simply happen, it needs to be planned and all stakeholders need to be involved in committing to a shared goal.

Implementing circular procurement means going beyond lowest price at purchase. At the award stage, the contracting authority can evaluate the quality and circularity aspects of the tenders using predetermined award criteria. For the transition to circular economy, GPP cannot be done at the last minute. Planning is essential for delivering environmental and social value while retaining economic benefits. Public authorities play a crucial role in implementing new business models, influencing the market towards sustainable materials and products and offer lessons learned to all the partners involved in the process. Public authorities can lead the way for industry to follow.

GPP can be seen as an important enabler for the development of a holistic approach for supply chain collaboration in closing and slowing resource loops in the construction sector. The involvement of supply chain partners from construction firms to suppliers including architects and the final users, need to be extended to the entire supply chain, during the entire lifetime of a building, so that GPP is retained during the operation phase of the building. In this sense, circular public buildings can offer examples for scaling up and mainstreaming such transitions to circular built environments. Therefore, it has the potential to be upscaled to become routine practice over time.

Recommendations arising from this study are as follows:

- At the outset, setting up the GPP is important as it enables holistic design approaches and supporting monitoring and reporting processes to guarantee the goal achievements from environmental perspective.
- Setting up clear guidelines and criteria are critical as it makes the process open and transparent to all parties, not just confined to the weighting criteria and can support lateral solutions. Combined with monitoring, transparency assists in benchmarking.
- Stakeholder engagement before setting up the GPP assists in ensuring that the intent is robust and the GPP criteria are not unrealistic.
- Involving all the parties in the process is a key factor to be successful and develop new opportunities in terms of new green jobs, enhance innovative energy transition solutions and give them a better understanding of the challenges presented by the circular economy and more responsible use of resources.
- Tracking/measurement of materials was built into the tender as in the case study described in this paper. This assisted in understanding the gains made as compared to a business as usual scenario and supports monitoring and reporting as well.
- Monitoring can also assist in increasing stringency over time. As knowledge and awareness in GPP becomes normal business practice, the strive towards best practice can also incorporate other considerations such as service approaches rather than products in buildings such as lighting service contracts, solar panel contracts, and the like.

- Definitions or clear understanding of terms should be provided so there is no 'waste' of materials. Instead the focus shifts to the many 'r's': recycle, repurpose, renew, regenerate and so on supporting holistic circular thinking in which the word *waste* is replaced by the word *resource*.

## 8. References

[1] Global Alliance for Buildings and Construction (GABC), International Energy Agency (IEA) and the United Nations Environment Programme (UNEP) 2019 *2019 Global status report for buildings and construction: Towards a zero-emission, efficient and resilient buildings and construction sector*, retrieved June 20 2020 from https://www.iea.org/reports/global-status-report-for-buildings-and-construction-2019

[2] IEA 2019 *Global status report for buildings and construction 2019*, IEA, Paris, retrieved November 4 2020 from https://www.iea.org/reports/global-status-report-for-buildings-and-construction-2019

[3] ICLEI 2017 *Public procurement for a circular economy: good practice and guidance.* European Commission and ARCTIK SPRL-Environmental Communications. European Union, retrieved June 25, 2020 from https://ec.europa.eu/environment/gpp/pdf/Public_procurement_circular_economy_brochure.pdf

[4] European Commission 2020 EU Green public procurement criteria, retrieved August 8 from https://ec.europa.eu/environment/gpp/eu_gpp_criteria_en.htm

[5] Institute for European Environmental Policy 2018 Backgrounder on sustainable Consumption retrieved November 27 2020 from https://ieep.eu/uploads/articles/attachments/b8d5e43d-7c50-42d5-944f-120dab41f816/DRAFT%20-%20Sustainable%20consumption%20-%2012%20Oct%202018.pdf?v=63706895555

[6] Global Construction Perspectives, Oxford Economics 2015 *Global construction market to grow $8 trillion by 2030: driven by China, US and India* retrieved November 21 from https://www.ice.org.uk/ICEDevelopmentWebPortal/media/Documents/News/ICE%20News/Global-Construction-press-release.pdf

[7] European Commission 2019 Eurostat Circular Economy retrieved 20th September https://ec.europa.eu/eurostat/web/circular-economy/publications

[8] Transparency Market Research 2018 *Construction Waste Market Global Industry Analysis, Size, Share, Growth, Trends, and Forecast 2017 – 2025* retrieved 20 November from https://www.transparencymarketresearch.com/construction-waste-market.html

[9] de Wit M Hoogzaad J Ramkumar S Friedl H Douma 2018 *The Circular Economy Gap Report* retrieved 10 October 2020 from https://www.legacy.circularity-gap.world/2018

[10] Haas, W., Krausmann, F. Wiedenhofer, D. and Heinz, M 2016 *How Circular is the Global Economy? A Sociometabolic Analysis.* In: Social Ecology: Society-nature Relations across Time and Space (Eds.) Haberl H, Fischer-Kowalski M, Krausmann F and Winiwarter V, (New York: Springer, Cham) DOI: 10.1007/978-3-319-33326-7_11

[11] EC 2019 Construction and Demolition Waste (CDW) retrieved 10 October 2020 from https://ec.europa.eu/environment/waste/construction_demolition.htm

[12] European Commission 2020 Circular Economy Action Plan retrieved 20 September 2020 from http://ec.europa.eu/environment/circular-economy/index_en.htm

[13] Artola I Rademaekers K Williams R Yearwood J 2016 Boosting Building Renovation: What potential and value for Europe? https://www.europarl.europa.eu/RegData/etudes/STUD/2016/587326/IPOL_STU%28201629587326_EN.pdf

[14] Rowley J 2002 Using case studies in research *Management Research News* **25** 116-27.

[15] Badi S and N Murtagh 2019 Green supply chain management in construction: A systematic literature review and future research agenda *Journal of Cleaner Production* **223** 312-22.

[16] Avsec S and B Kaucic 2017 *Green economy for sustainable environment: Slovenian perspective* 62-71, doi10.4467/25438700ŚM.17.069.7926

[17] Kuittinen M and S le Roux 2018 *Procurement criteria for low-carbon building Environment Guide 2017*, 50 Ministry of the Environment Department of the Built Environment, (Finland: Helsinki) ISBN: 978-952-11-4788-3

[18] van Geet C, Evans N, Jones M, Neudecker I, and L Turley 2015 *Using product-service systems to enhance Sustainable Public Procurement* United Nations Environment Programme, France

[19] Ruparathna R and K Hewage 2015 Review of contemporary construction procurement practices *Journal of Management in Engineering*, **31** 3 DOI 10.1061/(ASCE) ME.1943-5479.0000279

[20] Bohari A A M and B Xia. 2015 Developing green procurement framework for construction projects in Malaysia, *The 7th International Conference of SuDBE2015* Reading, UK

[21] Bohari A A M, Skitmore M, Xia B and M Teo 2017 Green oriented procurement for building projects: Preliminary findings from Malaysia *Journal of Cleaner Production*, **148** 690-700 http://dx.doi.org/10.1016/j.jclepro.2017.01.141

[22] Bohari A A M and B Xia 2015 Developing Green Procurement Framework for Construction Projects in Malaysia *The 7th International Conference of SuDBE2015*, Reading, UK

[23] Bidin Z A, Bohari A A M, Rais S L A, Saferi M M, and A Olanipekun 2020 Challenges and Drivers of Green Procurement among Construction Practitioners in Malaysia, International *Journal of Service Management and Sustainability* **5** 1, 149-75 DOI10.24191/ijsms.v5i1.9864

[24] Rais S L A, Bidin Z A, Bohari A A M, and M M Saferi 2018 The Possible Challenges of green procurement implementation *IOP Conf. Series Materials Science and Engineering* **429** 1-7 10.1088/1757-899X/429/1/012023

[25] Alqadami A T, Zawawi N A W A, Rahmawati Y, Alaloul W and A F Alshalif 2020 Key success factors of implementing green procurement in public construction projects in Malaysia *IOP Conf. Series: Earth and Environmental Science* **498** 1-9 DOI10.1088/1755-1315/498/1/012098

[26] Wibowo M A, N U Handayani and A Mustikasari 2018 Factors for implementing green supply chain management in the construction industry *Journal of Industrial Engineering and Management* **11**: 651-79 https://doi.org/10.3926/jiem.2637

[27] Shen L, Z Zhang and Z Long 2017 Significant barriers to green procurement in real estate development *Resources, Conservation and Recycling* **116** 160–68 http://dx.doi.org/10.1016/j.resconrec.2016.10.004

[28] Shen L, Z Zhang and X Zhang 2017 Key factors affecting green procurement in real estate development: a China study, *Journal of Cleaner Production* **153** 372-83 http://dx.doi.org/10.1016/j.jclepro.2016.02.021

[29] Wong J K W, J K S Chan and M J Wadu 2016 Facilitating effective green procurement in construction projects: An empirical study of the enablers *Journal of Cleaner Production* **135** 859-71 http://dx.doi.org/10.1016/j.jclepro.2016.07.001

[30] Al Nuaimi B K and M Khan 2019 Public-sector green procurement in the United Arab Emirates: Innovation capability and commitment to change *Journal of Cleaner Production* **233** 482-89 https://doi.org/10.1016/j.jclepro.2019.06.090

[31] Shurrab J, M Hussain and M Khan 2019 Green and sustainable practices in the construction industry - A confirmatory factor analysis approach *Engineering, Construction and Architectural Management* **26** 1063-86 10.1108/ECAM-02-2018-0056

[32] Chan A P C, A Darko and E E Ameyaw 2017 Strategies for Promoting Green Building Technologies Adoption in the Construction Industry—An International Study *Sustainability*, **9** 969 doi:10.3390/su9060969

[33] Lozano R and S Witjes 2016 Collaboration for *Circular Economy: Linking sustainable public procurement and business model* Utrecht University Faculty of Geosciences, Utrecht DOI: 10.13140/RG.2.2.36081.68969

[34] Agyepong, A O and G Nhamo 2017 Green procurement in South Africa: perspectives on legislative provisions in metropolitan municipalities *Environ Dev Sustain* **19** 2457–74 DOI 10.1007/s10668-016-9865-9
[35] Liu J, B Shi, J Xue and Q Wang 2019 Improving the green public procurement performance of Chinese local governments: From the perspective of officials' knowledge *Journal of Purchasing and Supply Management* **25** 13 https://doi.org/10.1016/j.pursup.2018.05.002
[36] Liu J, J Xue, L Yang and B Shi 2019 Enhancing green public procurement practices in local governments: Chinese evidence based on a new research framework *Journal of Cleaner Production* **211** 842-54 https://doi.org/10.1016/j.jclepro.2018.11.151
[37] Fuentes-Bargues J L, P S Ferrer-Gisbert, M C González-Cruz and M J Bastante-Ceca 2019 Green public procurement at a regional level. Case Study: The Valencia region of Spain *International Journal of Environmental Research and Public Health* **16** 2936 doi:10.3390/ijerph16162936
[38] Testa F, E Annunziata, F Iraldo and M Frey 2016 Drawbacks and opportunities of green public procurement: an effective tool for sustainable production *Journal of Cleaner Production* **112** 1893-900 http://dx.doi.org/10.1016/j.jclepro.2014.09.092
[39] Sonnichsen S D, and J Clement 2020 Review of green and sustainable public procurement: Towards circular public procurement *Journal of Cleaner Production* **245** 18 https://doi.org/10.1016/j.jclepro.2019.118901
[40] Lăzăroiu G, L Ionescu, C Uță, I Hurloiu, M Andronie and I Dijmărescu 2020 Environmentally responsible behavior and sustainability policy adoption in green public procurement *Sustainability* **12** 2110 doi:10.3390/su12052110
[41] Pouikli K 2020 Towards mandatory Green Public Procurement (GPP) requirements under the EU Green Deal: reconsidering the role of public procurement as an environmental policy tool, *ERA Forum* **23** https://doi.org/10.1007/s12027-020-00635-5
[42] United Nations 2015 *Adoption of the Paris Agreement* retrieved November 27 2020 from https://unfccc.int/resource/docs/2015/cop21/eng/l09r01.pdf
[43] European Commission 2010 *Directive 2010/31/EU of the European Parliament and of the Council of 19 May 2010 on the energy performance of buildings (recast)* retrieved November 2020 from https://eur-lex.europa.eu/legal-content/EN/TXT/?uri=celex%3A32010L0031
[44] European Commission 2009 *Directive 2009/28/EU of the European Parliament and of the Council of 23 April 2009 on the promotion of the use of energy from renewable sources* retrieved November 27 2020 from https://eur-lex.europa.eu/legal-content/EN/TXT/PDF/?uri=CELEX:32009L0028
[45] European Commission 2012 *Directive 2012/27/EU of the European Parliament and of the Council of 25 October 2012 on energy efficiency, amending Directives 2009/125/EC and 2010/30/EU and repealing Directives 2004/8/EC and 2006/32/EC*, (OJ L 315, 14.11.2012, p. 1) retrieved November 27 2020 from https://eur-lex.europa.eu/LexUriServ/LexUriServ.do?uri=OJ:L:2012:315:0001:0056:en:PDF
[46] European Commission 2011 *Communication COM (2011) 571 final Communication from the Commission to the European Parliament, the Council, the European Economic and Social Committee and the Committee of the Regions Roadmap to a Resource Efficient Europe* (Bruxelles) retrieved November 27 2020 from https://www.europarl.europa.eu/meetdocs/2009_2014/documents/com/com_com(2011)0571_/ com_com(2011)0571_en.pdf
[47] European Commission 2014 *Communication COM (2014) 398 final Communication from the Commission to the European Parliament, the Council, the European Economic and Social Committee and the Committee of the Regions Towards a Circular Economy: A Zero Waste Programme for Europe* (Bruxelles) retrieved November 27 2020 from https://ec.europa.eu/environment/circular-economy/pdf/circular-economy-communication.pdf

[48] European Commission 2014 *Communication COM (2014) 445 final Communication from the Commission to the European Parliament, the Council, the European Economic and Social Committee and the Committee of the Regions: Resource Efficiency Opportunities in the Building Sector* (Bruxelles) retrieved November 27 2020 from https://ec.europa.eu/environment/eussd/pdf/SustainableBuildingsCommunication.pdf

[49] European Commission 2008 *Communication COM (2008) 400 final/2 Communication from the Commission to the European Parliament, the Council, the European Economic and Social Committee and the Committee of the Regions: Public procurement for a better environment* (Bruxelles) retrieved November 27 2020 from https://ec.europa.eu/transparency/regdoc/rep/1/2008/EN/1-2008-400-EN-F2-1.Pdf

[50] European Commission 2016 *Commission Staff Working Document EU GPP Criteria for Office Building Design, Construction and Management retrieved November 27 2020* from https://ec.europa.eu/environment/gpp/pdf/swd_2016_180.pdf

[51] OECD Urban Studies 2020 *The Circular Economy in Cities and Regions: Synthesis Report* Paris: OECD Publishing retrieved November 27 2020 from https://doi.org/10.1787/10ac6ae4-en

[52] Ministry of the Environment 2020 *Minimum Environmental Criteria* retrieved November 27 2020 from https://www.minambiente.it/pagina/i-criteri-ambientali-minimi

[53] DECREE 26 June 2015 *Application of methodologies for calculating energy performance and definition of minimum requirements for buildings (15A05198) (GU Serie Generale n.162 of 15-3307-2015 - Suppl. Ordinario n. 39)* retrieved November 27 2020 from https://www.gazzettaufficiale.it/eli/id/2015/07/15/15A05198/sg

[54] European Commission 2012 *Directive 2012/27/EU on energy efficiency Annex III of the Directive defines the measures which must be taken by central government authorities, and which can be voluntarily adopted by other public authorities* retrieved November 27 2020 frohttps://eur-lex.europa.eu/LexUriServ/LexUriServ.do?uri=OJ:L:2012:315:0001:0056:en:PDF

[55] Legislative Decree 2016 *18 April 2016, n. 50 artt. 34 and 96 - Implementation of Directives 2014/23/EU, 2014/24/EU and 2014/25/EU on the award of concession contracts, public contracts and procurement procedures of entities operating in the water sector, energy, transport and postal services, as well as for the reorganization of the current regulations on public contracts relating to works, services and supplies (16G00062) (GU Serie Generalen.91 of 19-04-2016 - Suppl. Ordinario n. 10) (recast)* retrieved November 27 2020 from https://www.gazzettaufficiale.it/sommario/codici/contrattiPubblici

[56] CReIAMO PA 2020 *GPP Monitoring Report 2019* retrieved November 27 2020 from https://www.slideshare.net/ProgettoCReIAMOPA/creiamo-pa-report-sul-monitoraggio-gpp-19

IOP Conf. Series: Earth and Environmental Science **855** (2021) 012007 doi:10.1088/1755-1315/855/1/012007

# The environmental impact of prefabricated renovation systems: a worthy alternative to on-site renovation?

**Y Decorte[1,*], S De Meyer[1] and M Steeman[1]**

[1] Ghent University, Faculty of Engineering and Architecture, Department of Architecture and Urban Planning, Sint Pietersnieuwstraat 41 B4, 9000 Ghent, Belgium

**Abstract.** The conventional renovation practices, which are mainly characterized by time-consuming manual on-site techniques, partly contribute to the low renovation rate. Accordingly, a faster and more efficient approach is necessary. The implementation of prefabricated systems could offer a possible solution. These systems are increasingly being studied, but little is known about their environmental impact. Hence, this study investigates the environmental impact by means of Life Cycle Assessment (LCA) of two prefabricated façade renovation systems being a timber frame and a sandwich panel; and compares it to a well-known on-site technique, External Thermal Insulation Composite System (ETICS). First, reference designs are assembled. Subsequently, the impact of different life cycle stages is determined in order to clearly indicate differences between on-site and prefabricated systems. More specifically, the production, transport, replacement and end-of-life stage are assessed. In the end, the environmental impact is examined over time combining all stages. The results show that the prefabricated systems are not yet a worthy ecological opponent to ETICS. As of the production stage, the environmental impact appears to be higher. Optimising the reference systems through an extensive redesign could lead to more competitive or even favourable results in terms of environmental impact.

## 1. Introduction

### *1.1. Need to improve the energy performance of the existing building stock*

Renovation of the existing building stock is perceived by the European Green Deal as a cornerstone in the transition towards climate neutrality by 2050 [1]. In Europe, buildings contribute to 50 % of primary energy use and 40 % of all greenhouse gas emissions [2]. This strongly relates to the age of the buildings. The European Union contains over 220 million building units, which represent 85 % of the building stock, that already existed in 2001. Moreover, 85 - 95 % of these buildings will still be used in 2050 [1].

European climate and energy objectives have been prescribed to stimulate the required conversion. According to the 2050 long-term strategy, the greenhouse gas emissions should be reduced by 80 % compared to the 1990 levels and all buildings must become energy-neutral [3]. Similar targets apply to the housing stock in Flanders. Therefore, a vast 96.5 % of the existing dwellings need to be energetically upgraded in the next 30 years. This corresponds to 2.9 million dwellings in total, or over 95,000 dwellings per year. Moreover, this implies an annual renovation rate of 3 %, which represents 3 times the current rate [4].

The current way of renovating buildings is inadequate to address these large numbers and thus partly contributes to the low renovation rate. The renovation industry at present is mainly characterized by time-consuming manual on-site practices leading to a relatively high labour cost and a long duration of the renovation due to phased works and weather dependence. In addition, occupants typically need to

* Corresponding author: yanaika.decorte@ugent.be

relocate because of major nuisances. Consequently, there is a need for a faster and more efficient renovation method that outweighs these disadvantages. The implementation of prefabricated systems could offer a possible solution.

### *1.2. Prefabrication as a possible solution*

Prefabrication is an off-site production process that takes place in a specialized facility under controlled circumstances where all building materials and parts are assembled into building elements or even complete three-dimensional modules. In that way, the on-site phase is typically limited to the installation of the prefabricated elements. The systems combine thermal insulation, acoustic performance, fire resistance, air- and watertightness and finishing layers in one element. Furthermore, new windows, HVAC installations and renewable energy systems (e.g. PV panels, solar collectors) can be integrated in the façade and/or roof panels in advance [5].

In case of new build, prefabrication is already frequently applied. The associated advantages and disadvantages are increasingly being studied in recent years. Several researchers have addressed a positive impact in terms of time schedule, cost, quality control, health and safety, material waste and nuisances [6, 7, 8]. For example, Bikitsga et al. [6] report that the safety on site is expected to be higher compared to traditional methods, because the project duration is significantly reduced, fewer workers are generally required on site and scaffolding is often not necessary. In turn, Yuvaraj et at. [7] indicate that two-thirds of the construction firms that build in a modular and/or prefabricated way confirm a reduced project schedule up to 35 % compared to traditional construction methods.

Despite the advantages, this type of industry has not yet been accepted as a standard approach in the Belgian renovation market. The well-known on-site methods still take preference due to the lack of technical knowledge [5]. In response, several European research projects have been conducted to complement this knowledge. Examples are TES EnergyFaçade [9], MORE-CONNECT [10], AIM-ES [11], and INDU-ZERO [12]. The state-of-the-art of prefabricated renovation systems that have been developed in EU-funded projects is extensively documented by ProGETonE [13]. The environmental impact of the devised solutions is, however, not assessed in these projects.

### *1.3. Environmental impact of prefabricated renovation systems*

As mentioned, prefabrication can offer many benefits to upgrade the existing building stock in a fast way. The growing importance of sustainability, however, raises the question whether prefabricated renovation systems are also an ecologically valuable alternative to on-site renovation. As dwellings become more energy efficient due to energy reduction efforts, the more important the environmental impact of material use (i.e. embedded energy) becomes. It is thus important to look beyond the energy use phase and to take into account the full life cycle of a building, from raw material extraction to end-of-life treatment.

In literature, no consensus is found regarding the environmental impact of prefabricated solutions [14,15]. Environmental benefits are found related to the different life cycle stages. Teng et al. [14] indicate a carbon reduction of 15.6 % in embodied energy due to a more efficient material use through standardization and industrialization. In addition, an average construction waste reduction of 52 % is reported by Jaillon et al. [16]. On the other hand, additional transport flows are required to bring the produced materials to the assembly facility, but the required flows to the construction site can in turn be reduced [17].

Hence, this paper aims to systematically evaluate the environmental impact of prefabricated renovation systems per life cycle stage to identify both impact reductions and additions compared to on-site renovation techniques. In particular, the environmental impact of two prefabricated façade renovation systems is assessed, namely of a timber frame panel and a sandwich panel. The results are then compared to the environmental impact of the well-known on-site technique ETICS (External Thermal Insulation Composite System).

## 2. Reference renovation systems

To investigate important differences in the environmental impact between prefabricated and on-site renovation techniques, three reference renovation systems are elaborated: a prefabricated sandwich

(SW) and timber frame (TF) panel and an on-site External Thermal Insulation Composite System (ETICS). The different wall constructions are shown in Figure 1.

Each wall has an equal functional unit being 1 $m^2$ façade renovation strategy with a heat transfer coefficient or U-value of 0.23 $W/m^2K$ and a total lifespan of 60 years. In addition, the systems have an identical finishing in order to eliminate environmental discrepancies related to the applied finishing materials. On the other hand, the mounting mechanisms are different. Whereas the prefabricated systems are mounted to the existing wall via a double rail mechanism in galvanized steel, the on-site system is glued directly to the existing structure by means of an adhesive mortar combined with HDPE plugs. The prefabricated panels are therefore easier to demount, for example, to be integrally reused. Note that, in practice, an adaption layer (e.g. mineral wool) will be installed between the existing wall and the prefabricated panels to level out irregularities and avoid air gaps. The thickness of the layer must be at least 3 cm and depends on the cavity width and the irregularities of the façade [18]. This layer is not considered, but will create an additional environmental impact on the one hand and lower the U-value on the other hand. Moreover, only the newly added materials associated with the reference systems are included within the system boundaries. The choice of ETICS could, however, lead to a different end-of-life scenario regarding the existing wall (i.e. glued to the existing wall), but this falls outside the scope of this paper.

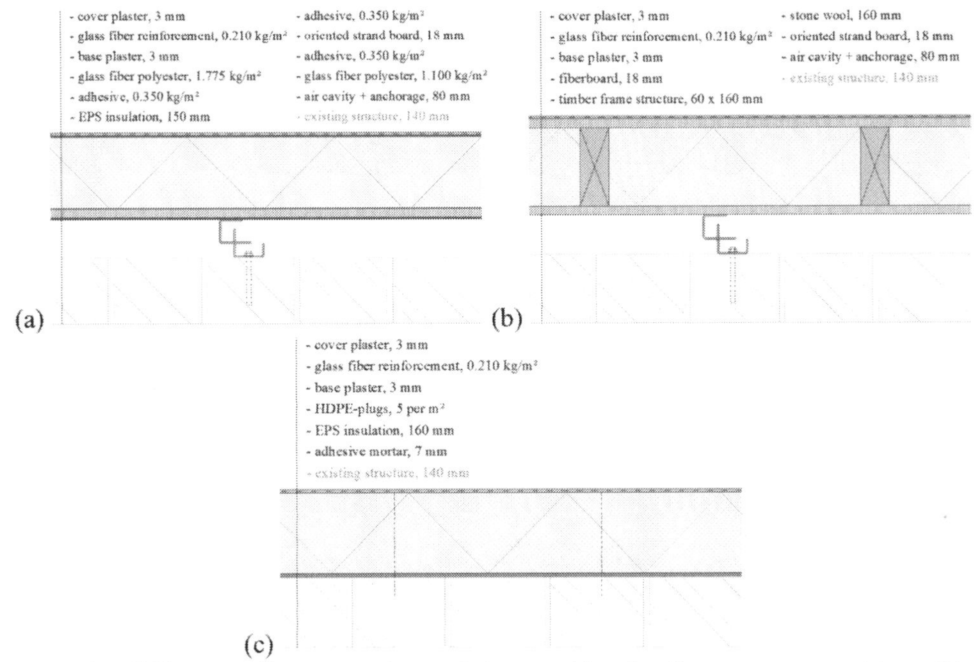


**Figure 1.** The different wall constructions of the considered reference systems: (a) prefabricated sandwich panel, (b) prefabricated timber frame panel, and (c) ETICS.

## 3. Methodology

The environmental impact is determined at element level by means of LCA according to ISO 14040 [19] and ISO 14044 [20]. To this end, the software SimaPro v9.1.0.7 is used. Environmental data are obtained from the Swiss Ecoinvent database v3.6 [20]. This widely accepted life cycle inventory database contains all required processes that apply as an average valid for all countries in the world ('GLO') and European countries specifically ('RER'). The cut-off modelling approach is selected, i.e. an upstream allocation method assigning the recycling activities and avoided virgin materials to the life cycle which will use the recycled materials as an input. Furthermore, the ReCiPe 2016 (H) method is applied to translate the inventoried data into environmental impacts [22]. The method considers 18 midpoint and 3 endpoint indicators. The results are converted into a single score, expressed in points.

The goal of this study is to gain insight into the relationship between the environmental impact of on-site and prefabricated renovation solutions. Therefore, the research concerns a cradle-to-grave analysis over two life cycles of 60 years each. However, the operational energy use stage is excluded. This phase is in fact considered equal for the three cases, since their U-value is equal. Hence, the impact of the production, transport, replacement and end-of-life stage is included. To clearly indicate the differences between the on-site and prefabricated reference systems, each phase will be elaborated on separately in the following section.

## 4. Results

### *4.1. Production stage*

The production stage includes the extraction of raw materials and the manufacturing processes with regard to constituting building materials, mounting mechanisms and material losses. All related environmental inputs and outputs are included in the generic material processes derived from the Ecoinvent database.

*4.1.1. Building materials*. Since the reference systems have a different composition, the environmental impact will differ. Figure 2 represents the environmental impact variation between the production stage of the three reference systems. Moreover, the impact to produce each constituting building material per reference system is shown. The on-site ETICS has the lowest environmental impact as the impact to produce the sandwich and timber frame panel (without mounting mechanism) is respectively 2.4 and 1.7 times higher. This is highly related to the simple composition of ETICS where only few constituting materials are required. The prefabricated systems, on the other hand, consist of multiple additional materials, namely adhesives, glass fibre polyester and wood, resulting in a higher impact. In turn, the impact of the finishing materials is identical, because these are assumed equal. The insulation materials have a quasi-similar impact.

Since ETICS is, based on its reference design, preferred over the prefabricated systems, simplifying the composition of the prefabricated systems and/or selecting materials with a lower environmental impact could lead to more competitive or even favourable results. An extensive redesign, however, falls outside the scope of this paper. Yet, the figure highlights the important contributors and necessary focal points for this redesign, e.g. insulation and OSB both for the timber frame and sandwich panel, and glass fibre polyester in case of the sandwich panel.

*4.1.2. Mounting mechanisms*. As can be seen on Figure 2, there is also a significant contribution of the mounting mechanisms, ranging from 30% (ETICS) to 38% (TF). Whereas adhesive mortar and HDPE plugs are used to mount the ETICS, a double rail anchorage system in galvanized steel is used in case of the prefabricated panels. The galvanized steel is produced via an oxygen converter steelmaking route (i.e. BOF) and thus mainly consists of virgin iron ore. The impact of the rail anchorage system is found 2.7 times greater than the ETICS mounting mechanism. This further increases the impact difference between ETICS and the prefabricated systems. In order to reduce the environmental impact of the prefabricated systems, three alternative mounting mechanisms are investigated.

The current anchorage system consists of two vertical rails per m² renovation panel composed of galvanized steel that are fixed to the existing wall and the prefabricated panel with 3.33 bolts per m² (GS rails). The first alternative is elaborated by reducing the required amount of steel. Therefore, pointwise fixed hooks in galvanized steel are used instead of full rails (GS hooks). It is assumed that 6.25 hooks and 25 screw per m² are required resulting in a mass reduction of 80%. The second alternative has the same design as the original rail mechanism, but here the material is changed to polyvinylchloride (PVC rails). PVC has a similar environmental impact per kilogram, but its density is almost six times smaller. The bolts however are still made of galvanized steel. The last alternative is to glue the prefabricated systems to the existing structure in a similar way as the ETICS (adhesive). The weight of the required materials is listed in Table 1 for the original mounting mechanism (GS rails) and the proposed alternatives (GS hooks, PVC rails and adhesive). Note that more detailed research is required to determine if these alternatives are feasible based on their load-bearing capacity.

Figure 2 shows that all proposed alternatives score significantly better than the original mounting mechanism. As mentioned above, this is related to the large amount of steel required in the initial mechanism. The variant with the steel hooks, in which the amount of steel is considerably reduced, yields an impact reduction of 27 %. A similar reduction is obtained by changing only the material of the original mechanism to PVC. Note, however, that steel typically has a longer lifespan than PVC. Although there seems to be no consensus in literature on the life span of PVC (30 to 60 years) and steel (30 to 100+ years) [23], a life span of 60 for PVC and 120 years for steel is considered in this study. For this reason, the alternative with the GS hooks is preferred over PVC rails. Lastly, using an adhesive mortar to mount the panels results in a more limited but still considerable reduction of 24 %. In addition, gluing disadvantages the circularity of the prefabricated panels and further complicates integral reuse of the panels. It can be concluded that special attention should be paid to the design of the mounting mechanism to minimize the environmental impact of prefabricated renovation solutions. A redesign of the mounting mechanisms alone, however, is not sufficient to obtain a competitive system with respect to ETICS in terms of material impacts.

**Table 1.** The weight of materials per mounting mechanism alternative.

| **Material** | **GS rails (ref.)** [kg/m²] | **GS hooks** [kg/m²] | **PVC rails** [kg/m²] | **Adhesive** [kg/m²] |
|---|---|---|---|---|
| Galvanized steel | 6.990 | 1.366 | 0.204 | 0.000 |
| PVC | 0.000 | 0.000 | 1.313 | 0.000 |
| Adhesive mortar | 0.000 | 0.000 | 0.000 | 3.000 |

*4.1.3. Material losses.* Throughout the life cycle of a product, a share of building material is lost due to transport, (incorrect) cutting, surpluses, breakage, theft and labour-related problems. This leads to an additional environmental impact since more material will need to be produced. The extent of these losses is, among other things, related to the construction method. The loss fractions in case of on-site techniques will typically be higher compared to prefabrication. On-site systems are in fact more susceptible to errors due to e.g. manual labour, careless handling, weather conditions, while in case of prefabrication, the controlled conditions in which the systems are produced will ensure that losses are kept to a minimum. For this reason, two loss fractions are distinguished: 10% for the on-site ETICS and 1% for the prefabricated systems. The loss fractions are based on the outer limits according to [24].

In Figure 2, the additional production impact due to material losses is marked in black. The environmental impact increases 10% for ETICS and 1% for the prefabricated systems, analogous to the corresponding fraction losses. However, the additional impact of ETICS does not compensate the initially higher impact of the production stage regarding the prefabricated systems. Note that these losses will not only have an additional impact on the production stage, but also on the transport, replacement and end-of-life stage, included in the sections below.

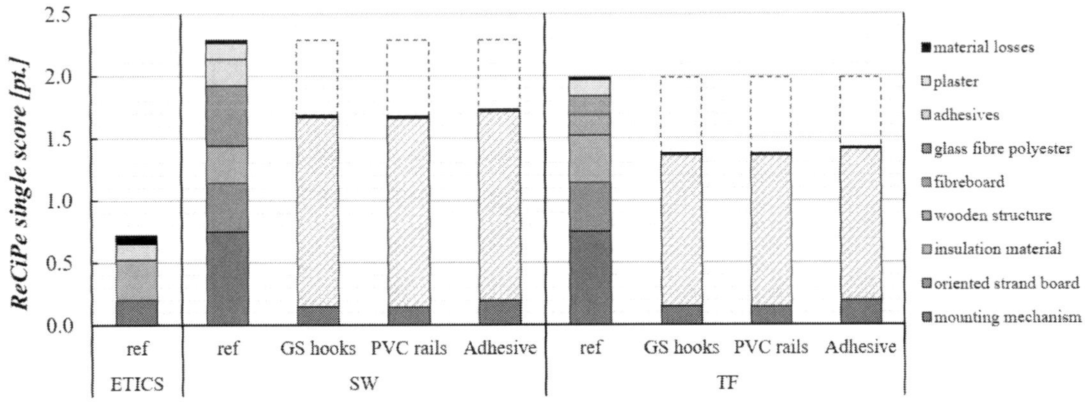


**Figure 2.** The environmental impact variation between the production stage of ETICS, sandwich (SW) and timber frame panel (TF). Per system, the left bar respectively indicates the impact of the reference systems (ref) per building material, while the other bars represent an alternative mounting mechanism: galvanized steel (GS) hooks, PVC rails and adhesive. The hatched bars refer to the total impact of the reference systems without consideration of the mounting mechanisms. The dashed line indicates the impact reduction due to the adjustment of the mounting mechanism.

### *4.2. Transport stage*

The required transport flows to deliver a finished product at the building site typically differ between prefabricated and on-site systems. An overview of the flows is provided in Figure 3. The first flow is identical for both systems, namely the transport from the material extraction site to the building material supplier (A - a). In case of on-site systems, the building materials are subsequently transported from the supplier to the construction site (B), whereas for prefabricated systems an extra stop is required. More specifically, the individual building materials are first transported to a specialized assembly facility (b) before being transported to the construction site (c). Moreover, an additional stop can also occur for both on-site and prefabricated systems. Either the finished building materials are transported directly from the supplier to the construction site or specialized facility, or they first pass by an intermediate distributor. For this study, it is assumed, however, that this additional stop does not occur in both cases.

The different assumptions for each transport flow in this study are as follows. For first transport flow (A-a), the predefined transport scenarios are applied that are already incorporated in the market processes of the Ecoinvent database. For the second flow (B - b), scenarios are acquired from Table 3 in [24]. This table gives an extensive overview of average transport distances and means of transport for twelve different product groups. Furthermore, an additional transport flow (c) is required for the prefabricated systems. It is assumed that the prefabricated panels are transported over a distance of 100 km by means of a heavy truck of the size class 16-32 metric tons gross vehicle weight and the emission class euro VI. These numbers are derived from a previous study [25]. Note that only one-way traffic is considered for each flow assuming that the vehicles are used for unrelated transport on their way back.

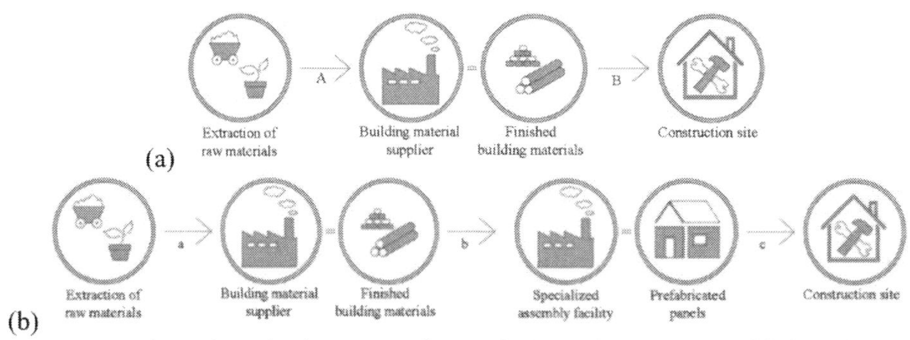


**Figure 3.** Overview of required transport flows of (a) on-site and (b) prefabricated systems.

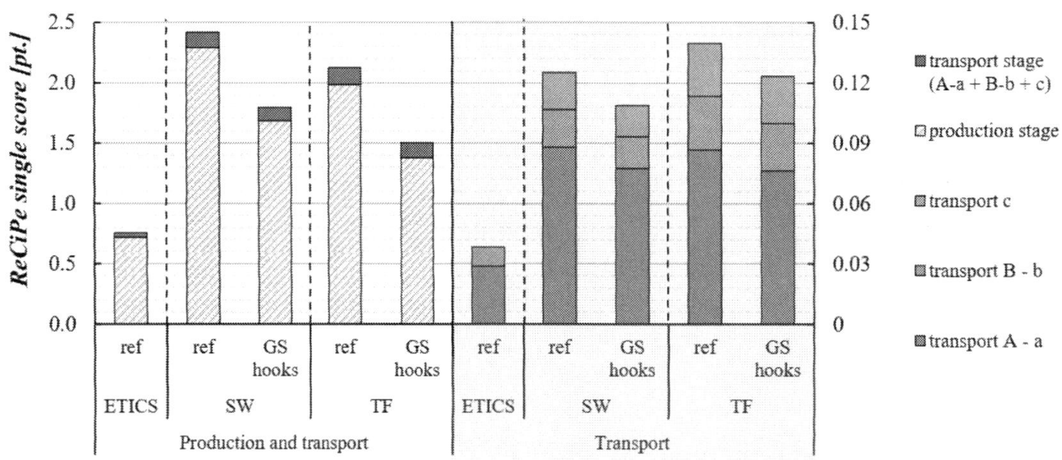


**Figure 4.** The environmental impact increase due to the inclusion of the transport stage of ETICS, sandwich (SW ref) and timber frame panel (TF ref) is shown on the left. In case of the prefabricated systems, the alternative mounting mechanism with hooks is also displayed (GS hooks). The hatched bars refer to the impact of the production stage of each system (cf. Figure 2). On the right, the contribution of each transport flow is depicted separately per system.

Figure 4 shows the environmental impact increase due to the inclusion of the transport stage on the left. The overall influence of transport seems to only represent a small part in the total environmental score, ranging from 5 % to 8 %. On the right, the impact share of the different flows is compared. The first transport flow (A – a) represents the largest share for each system ranging from 62% (TF) to 75% (ETICS). Furthermore, the ETICS system has the lowest impact score for each transport flow. On the one hand, this is because there is one transport flow less. On the other hand, this can be explained by the lower weight of the system in comparison to the prefabricated systems. This also clarifies why the impact of transport flow b and c of the timber frame system is higher than that of the sandwich panel. A redesign of the prefabricated reference panels could thus also reduce the impact of the transport stage.

### *4.3. Replacement stage*

Some materials or products will have a shorter lifespan than the assumed life cycle of 60 years. These materials or products will be replaced when they can no longer fulfil their function or because of aesthetic reasons. The lower the expected lifespan, the more the materials will need to be replaced. This results in a multiplication of the production and transport impact by the number of replacements. Moreover, each time a material has to be replaced, an end-of-life (EoL) impact will be allocated.

In this research, the finishing materials are the only materials that need to be replaced within the 60 years study period. It is assumed that the full plaster layer will be sent to landfill at year 20 and 40 and replaced by an identical plaster with the same production and transport impact as described above. As each reference system has the same finishing, the additional impact will be identical. The other constituting materials have a lifespan of 60 years, with the exception of steel for which a lifespan of 120 years is assumed. These materials will therefore not be replaced within a life cycle of 60 years. The influence of the replacement stage on the results is included in section 4.5.

### *4.4. End-of-life (EoL) stage*

Each material has a specific lifespan after which a certain EoL scenario will take place. A distinction is made between disposal (incineration and landfill), recycling and reuse. Incineration and landfill will have the highest impact on the environment as they lead to material loss, additional greenhouse gas emissions and the need to produce new materials from primary raw materials. Reuse, on the other hand, will result in the lowest impact. The environmental impact of recycling depends on the energy intensiveness of the recycling processes. Since a cut-off approach is followed, the burdens of recycling activities and benefits of avoided virgin materials are only considered when the recycled content is used as an input.

For this study, a specific scenario is described for each reference system in order to determine the environmental impact of the EoL stage. For ETICS, disassembly of the separate constituting materials is rather hard because all layers are glued. For this reason, reuse or recycling is considered impossible. Consequently, after 60 years it is assumed that the entire renovation package is disposed and sent to landfill. The EoL scenario of the sandwich panel (with hooks) follows the same reasoning. Analogously, the complete sandwich panel is disposed and sent to landfill. However, the galvanized steel hooks have a lifespan of 120 years and will therefore be reused till the end of the second life cycle. After 120 years, it is assumed that the steel is high-quality recycled. The impact of this recycling process is however not taken into account since a cut-off approach is followed. In that case, both the burdens and benefits of recycling are allocated to the life cycle that will use the recycled content. For the galvanized steel hooks used to mount the timber frame panel, an identical EoL approach is followed. Moreover, the timber frame structure consists of several dry connections making it feasible to disassemble the package till a certain extent. For this reason, only a few constituting materials are disposed and sent to landfill, namely the plaster layer and the fibreboard. The other constituting materials are assumed to be high-quality recycled; as a result, no burdens are assigned regarding the recycling process. The impact of these assumptions on the environmental impact over time is below explained in more detail.

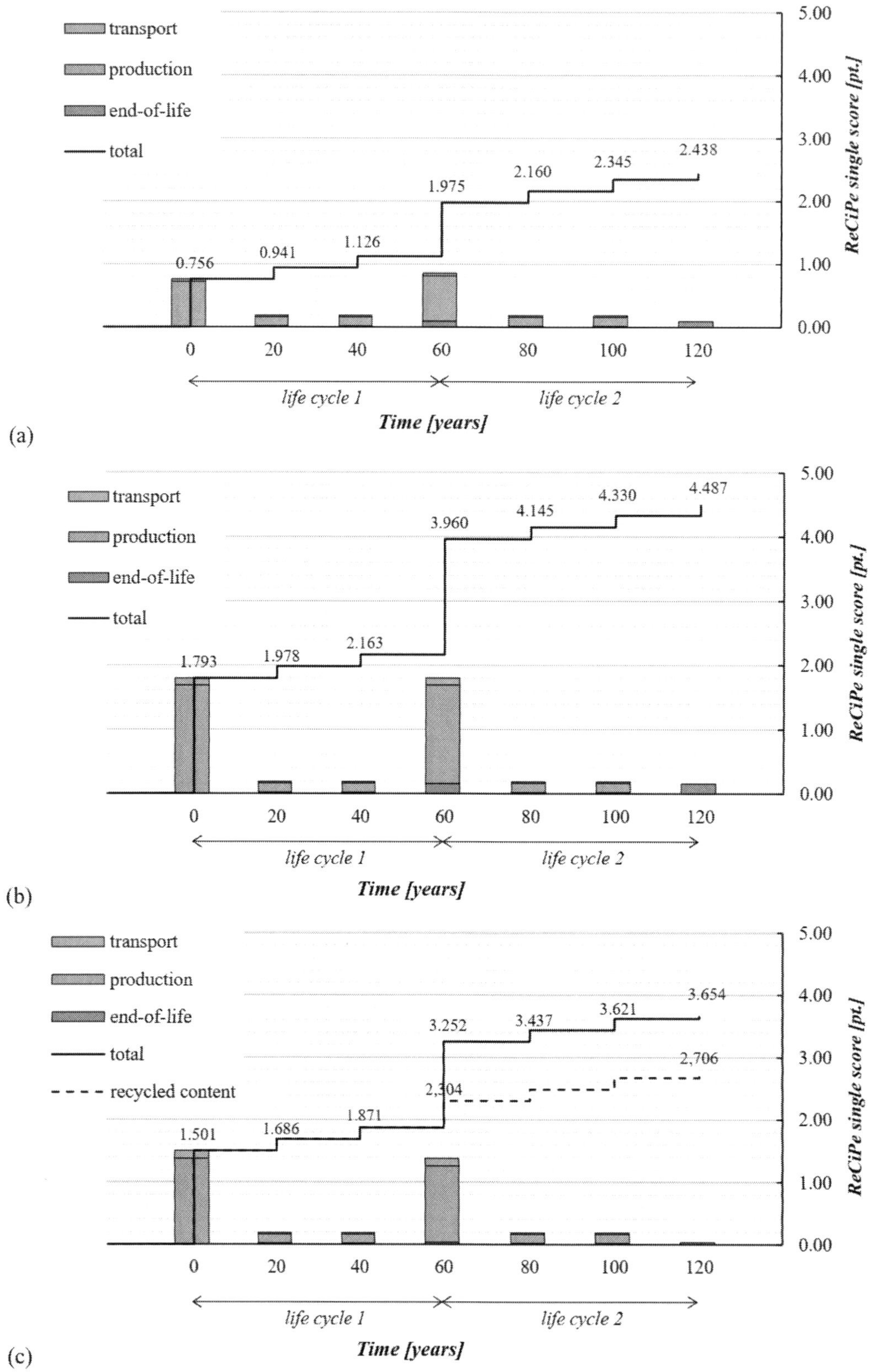


**Figure 5.** The evolution of the environmental impact over two life cycles of 60 years of (a) ETICS, (b) sandwich panel with hooks and (c) timber frame panel with hooks.

### *4.5. Environmental impact over time*

Figure 5 shows the environmental impact over time per reference system.

The first impact occurs in year zero. This is related to the impact of the production and transport stage as already elaborated in section 4.1 and 4.2.

Then, in year 20 and 40, the plaster layer is replaced for each system. This leads to an environmental impact with regard to the production (78%), transport (12%) and EOL (10%) stage. The impact is identical for each system (+0.185 pt/replacement).

After 60 years, the end of the first life cycle is reached, while the second life cycle starts. On the one hand, this causes a certain EoL impact. This impact is the highest for the sandwich panel (0.157 pt) followed by the ETICS (0.093 pt). The timber frame panel shows in turn the lowest impact (0.033pt). This is highly related to the large amount of materials that can be high-quality recycled as well as to the reuse of the steel hooks. On the other hand, the reference systems need to be produced and transported again leading to a similar impact as in year zero. However, as mentioned in section 4.4, the galvanized steel hooks will be reused in the second life cycle, leading to a lower production and transport impact of the prefabricated systems in comparison to year zero. Note that the recycled content from the first life cycle could be used to produce the timber frame panels for the second life cycle. This scenario is shown by the dashed line in Figure 5b. As a result, the environmental impact difference between the timber frame panel (2.304 pt) and ETICS (1.975 pt) is rather limited. However, in that case, the burdens of recycling would need to be addressed to the second life cycle. This is not included and falls without the scope of this paper. Moreover, it is also assumed that the materials are recycled for 100%, which is often not the case. The dashed line therefore represents the most optimal case. The real value will lie somewhere in between both boundaries.

Finally, the second life cycle follows a similar time lapse as the first life cycle. After 120 years, ETICS still has the lowest environmental impact. The impact of the sandwich and timber frame panel (without the use of the recycled content) is respectively 0.84 and 0.50 times higher than ETICS.

## 5. Conclusion

The literature review has shown that prefabrication can offer many benefits for the building sector. The growing importance of sustainability, however, raised the question whether prefabricated renovation systems show also benefits from an ecological viewpoint as well. Hence, this paper aimed to evaluate the environmental impact of prefabricated renovation systems per life cycle stage (production, transport, replacement and EoL) in order to identify impact reductions and additions compared to on-site renovation techniques. Therefore, the environmental impact of a reference sandwich and timber frame panel was assessed and systematically compared to the impact of the on-site ETICS by means of LCA.

The environmental impact to produce the sandwich and timber frame panel appears to be 2.4 and 1.7 times higher than the ETICS. On the one hand, the prefabricated systems consist of additional constituting materials resulting in both a higher weight and impact. Consequently, simplifying the composition of the prefabricated systems and/or selecting materials with a lower environmental impact could lead to more competitive or even favourable results. For the redesign, it is recommended to focus on the most important contributors, namely the insulation and OSB layer in case of the timber frame and sandwich panel, and the glass fibre polyester layer for the sandwich panel. It should however be noted that in this study only two types of prefabricated façade systems were examined, and variations on these types also exist. On the other hand, there is a significant contribution of the mounting mechanism. For the prefabricated reference systems a double rail anchorage system was assumed, leading to an impact which was 2.7 times greater than the adhesive mortar and HDPE plugs typically used to mount the ETICS. In addition, a redesign of the rail mechanism was proposed, which leads to an impact reduction of 24-27%. Changing the rails by pointwise galvanized steel hooks proved to be the most interesting option. A redesign of the mounting mechanisms alone was, however, not sufficient to obtain a competitive system with respect to ETICS.

Furthermore, the environmental impact with regard to the transport stage also seemed to be higher for the prefabricated systems due to the extra transport flow which is required and the higher weight. The share of the transport stage compared to the production stage is, however, relatively small. A redesign of the prefabricated reference panels could also reduce the impact of the transport stage.

No differences were noticed between the replacement stage of the systems, since only the finishing materials had to be replaced within the 60 years life cycle, which were the same.

Finally, per reference system EoL scenarios were devised. The impact of the EoL stage seemed to be the lowest for the timber frame panel due to circularity of the system which allows high-quality recycling of the constituting materials and reuse of the steel hooks. It is however rather complex to make assumptions about future waste treatment processes based on the current processes, since many processes are not yet optimized. Moreover, there is currently no consensus on an appropriate EoL allocation approach for circular constructions. The aim of this study was to provide some insight in how to address the benefits of circularity from a practical viewpoint with respect to the prefabricated systems based on a more limited number of scenarios. It was not the intention to perform a detailed EoL scenario analysis.

After enumerating the results of all stages, the prefabricated systems were no worthy ecological opponent to ETICS at this point. The initial high impact of the production stage could not be compensated by other benefits. Optimizing the reference prefabricated systems through an extensive redesign would be the first crucial step which could lead to more competitive or even favourable results. In addition, the advantages of reversible connections are not yet fully assigned. It is expected that the individual material layers will then be easier to separate and that material fractions will not be contaminated, which will increase the potential for high-quality recycling. There is, however, currently no consensus on how to include circular aspects, yet. E.g. an unambiguous definition of the EoL allocation is required to quantify the environmental impact of these circular aspects. Hence, module D (i.e. benefits and burdens regarding reuse, recovery and recycling potential) is presently excluded from the LCA system boundaries.

## 6. References

[1] European Commission 2020 *A Renovation Wave for Europe - greening our buildings, creating jobs, improving lives* (Brussels: European Commission)

[2] European Union 2019 *Level(s): Taking action on the TOTAL impact of the construction sector* (Luxembourg: European Union)

[3] European Commission 2019 *Energy performance of buildings directive: Facts and figures* (Brussels: European Commission)

[4] Flemish government 2020 *Long-term strategy for the renovation of Flemish buildings* (Flemish government)

[5] Steeman M, Van Den Bossche N and Maroy K 2016 Hygrothermal behaviour of prefabricated façade elements for building renovation *Central European Symposium on Building Physics* (Dresden) pp 197-204

[6] Bikitsha L and Haupt T 2011 Impact of prefabrication on construction site health and safety: Perceptions of designers and contractors *ASOCSA: The 6th Built Environment Conference* (Johannesburg) pp 196-211

[7] Yuvaraj V and Dharman R 2016 Present application of resource management in prefabricated and in-situ construction *Civil Engineering and Urban Planning: An Int J.* **3** (2) 63-74

[8] Afzal M, Maqsood S and Yousaf S 2017 Performance evaluation of cost saving towards sustainability in traditional construction using prefabrication technique *Int. J. of Research in Engineering and Science* **5** (5) 73-79

[9] Lattke F, Larsen K, Ott S and Cronhjort Y 2010 *TES EnergyFaçade: Pefabricated Timber Based Building System for Improving the Energy Efficiency of the Building Envelope*

[10] Op 't Veld P, Carrabs M and van Oorschot Z 2019 *MORE-CONNECT: Development and advanced prefabrication of innovative, multifunctional building envelope elements for Modular Retrofitting and smart Connections* (Brussels: European Commission). Available from www.moreconnect.eu/

[11] Dubois S, Remy O and de Bouw M 2016 *Retrofitting with AIM-ES: Architectural Industrialised Multifunctional Envelope System* (Brussels: BBRI)

[12] INDU-ZERO [Internet]. Available from https://northsearegion.eu/indu-zero/

[13] d'Oca S, Opt 't Veld P and Tisov A 2017 *ProGETonE: Brief report on the state of the art of deep renovation to nZEB and pre-fab system in EU.* Available from https://www.progetone.eu/
[14] Teng Y, Li K, Pan W and Ng T 2018 Reducing building life cycle carbon emissions through prefabrication: Evidence from and gaps in empirical studies *Building and Environment* **132** 125-136
[15] Janssen A, Putzeys K, Debacker W, Geerken T, Allacker K and De Troyer F 2010 *Onderzoek naar mogelijke nieuwe bouwconcepten en het effect ervan op het gebruik van oppervlaktedelfstoffen* (Brussels: Flemish Planning Bureau for the Environment)
[16] Jaillon F, Poon C and Chiang Y 2009 Quantifying the waste reduction potential of using prefabrication in building construction in Hong Kong *Waste Management* **29** 309-320
[17] Lindblad F, Lehman K and Aggerstam E 2019 A Comparative Study of the Environmental Impact from Transportation of Prefabricated Building Elements Using Wood or Concrete *Int. J. of Engineering and Technology* **11** (3) 154-161
[18] Maroy K 2017 *Luchtrotaties in de adaptatielaag* (Gent: Ghent University)
[19] ISO 14040 2006 *Environmental management - Life cycle assessment - Principles and framework*
[20] ISO 14044 2006 *Environmental management - Life cycle assessment - Requirements and guidelines*
[21] Ecoinvent Centre 2019 *The ecoinvent Database* [Internet]. Available from https://www.ecoinvent.org/database/
[22] Huijbregts M, Steinmann Z, Elshout P, Stam G, Verones F, Vieira M, Hollander A, Zijp M and Van Zelm M 2016 *A harmonized life cycle impact assessment method at midpoint and endpoint level. Report I: Characterization* (Bilthoven: National Institute for Public Health and the Environment)
[23] OVAM 2018 Rapport: Technische levensduur van gebouwcomponenten (Mechelen: OVAM)
[24] OVAM 2018 *Environmental profile of building elements* (Mechelen: OVAM)
[25] Decorte Y, Steeman M, Krämer UB, Struck C, Lange K, Zander B and de Haan A 2020 Upscaling the housing renovation market through far-reaching industrialization *Proc. Beyond 2020: World Sustainable Built Environment Conference* 588 (3) 1.08

**Acknowledgments**
The described work in the paper is part of the Interreg project INDU-ZERO which is co-funded by the North Sea Region Program 2014-2020. In addition, the author, Y. Decorte, would like to acknowledge FWO and Ghent University for supporting and hosting the work she does in her role as SB PhD fellow at FWO (grant number: 1S11621N).

# Development of a circularity assessment method for facade systems

**Lisa Van Gulck[1,3], Eline Leenknecht[1], Emiel Debusseré[1], Jona Van Steenkiste[2], Marijke Steeman[1] and Nathan Van Den Bossche[1]**

[1] Ghent University, Faculty of Engineering and Architecture, Research group of Building Physics, Sint Pietersnieuwstraat 41 B4, 9000 Ghent, Belgium
[2] Bureau Bouwtechniek, Kammenstraat 18, 2000 Antwerp, Belgium
[3] Corresponding author: lisa.vangulck@ugent.be

Various authorities identify the circular building strategy as the best way to reduce the environmental impact of the building sector. The EURECA project aims to develop a circular facade system for the renovation of high-rise buildings. The circularity of the facade systems proposed within the project should be evaluated in an objective manner. Current circularity assessment methods exist, however, they appear unfit to evaluate facade systems at an early design stage. Based on the analysis of existing assessment methods a new circularity assessment method is developed. The developed method allows to measure the circularity on element level with a limited amount of required information, allowing early stage decision-making. The method considers the parameters recycling, environmental cost, expected service life, component dependency, layer dependency and flexibility for reuse. The method is tested on four facade renovation systems: standard ETICS, circular ETICS, ventilated facade with rigid insulation and ventilated facade with flexible insulation. The circularity of each system can be represented by a radar chart, giving the score per parameter, or by a single score. In addition to the circular aspect, the financial aspect is added in the evaluation of the facade systems by using the Pareto front method.

## 1. Introduction

### *1.1. Context*

Population growth and the rapidly changing demands of society require a continuous stream of new buildings. In Europe the building sector is responsible for half of all energy use, 40% of all greenhouse gas emissions, half of all raw material extraction and a third of all water use [1]. These astounding numbers show the need for a shift in our way of building. Various authorities identify the circular building strategy as the best way to move forward and reduce the environmental impact of the building sector. The circular building strategy is the translation of the circular economy to the building sector.

The recognized urgency to move towards a circular economy is positive, but also results in a certain level of confusion [2]. A study by Kircherr et al. shows there are over 114 definitions of the circular economy [3]. Most of these definitions refer to a combination of reduce, reuse and recycle activities. This broad comprehension of the circular economy illustrates the numerous aspects it entails, making it difficult to define what exactly a circular product is and how to know which aspects to focus on. Applied to the building sector this results in the question 'what criteria must a building meet in order to be considered a circular building'?

Different methods to measure the performance of a building on circularity exist. These assessment methods allow to objectively compare the circularity of different building solutions and facilitate the practical implementation of the circular building strategy. Within each assessment method different aspects of circularity are quantified; each method has a set of circular parameters on which the building solutions are scored.

### *1.2. EURECA project*

EURECA, or Effectively Upscaling the Renovation of Envelopes with a Circular Approach, is a two-year research project funded by the Flemish government under the Circular Flanders initiative [4]. The aim of the project is to develop a circular facade system that is scalable and allows existing high-rise buildings to be renovated quickly in a circular manner. Building types with a repetitive facade have the greatest potential for this project. In Flanders, many apartment buildings from the 1960s and 1970s are in need of a renovation. These apartment buildings and university buildings with a similar typology are interesting for the EURECA project. The facade renovation system will be tested on three case studies: two high-rise apartment buildings and a university building. The modular facade renovation system must be demountable and should have potential for future reuse in another building. Apart from the Building Physics research group of Ghent University, the project consortium consists of Algemene Bouw Maes nv, a general contractor, the multidisciplinary research and consultancy agency specialized in building techniques Bureau Bouwtechniek and the aluminum constructor De Witte.

A decisive factor to evaluate the possible facade renovation systems (different construction methods and accompanying materials) on their application potential is their degree of circularity. One system may use materials with a low environmental impact but which are connected irreversibly, while a second system might have reversible connections and independent layers but requires more materials. To make objective decisions regarding the circularity of the different facade systems, their circularity will be measured by a circularity assessment method.

### *1.3. Objectives*

Different circularity assessment methods exist, such as 'Material Circularity Indicator' [5], 'C-calc' [6], 'Label Circular Building' [7] and 'Building Circularity Index' [8]. However, when analysing these four assessment methods no method is completely appropriate to evaluate the circularity of the facade renovation systems for the EURECA project. The existing methods often evaluate circularity on building level, while for the facade renovation systems an evaluation on element level is necessary. Most methods require a lot of detailed information, e.g. specific material properties such as geographical origin. The assessment method for EURECA should facilitate early stage decision-making when only aspects such as construction method and general choices of materials for the facade system are known. Lastly, many assessment methods also evaluate project specific parameters such as the use of BIM-models and material passports. Again, these parameters are less relevant for the early stage decision-making regarding the facade systems.

The goal of this research is to develop a circularity assessment method that can be applied for the comparison of the different facade renovation systems of the EURECA project. This method will allow to make objective decisions regarding circularity on element level, without having the requirement of a lot of detailed information or the choice of a specific project. The assessment method is thought out for renovation systems, but will also contain interesting information for new build scenarios. Existing circularity assessment methods provide very useful information on which parameters and scoring systems can be applied when evaluating circularity. This information will serve as a basis for the developed method.

## 2. Methodology

### *2.1. Research steps*

In a first research step, four existing circularity assessment methods are briefly discussed, it is stated why the methods are not fully suited for the assessment of facade renovation systems at an early design stage, and which aspects (circular parameters and scoring systems) are useful for the development of the new circularity assessment method.

Using the insight and information gained from analysing existing methods, the new circularity assessment method is developed in a second research step. The circular parameters included in the developed method are described and their accompanying scoring system is worked out. The developed method is tested on four facade renovation systems: standard ETICS, circular ETICS, ventilated facade with rigid insulation boards and ventilated facade with flexible insulation. Subsequently, the method is analysed to identify the most objective way to present the results of the assessment. Should circularity be represented by a single score, should weighting factors be applied?

In reality the financial aspect plays a crucial role when making renovation decisions. Because circular building solutions are often considered (too) expensive their practical implementation can be limited and linear building solutions can be favoured. In a third and last research step the financial aspect is included in the evaluation of the facade renovation systems. This aspect is not part of the circular score, but in addition to the circular score, it is a criterion on which the evaluation of the facade systems is based. The aim is to identify facade renovation systems that have both a solid circular score and a financially interesting profile. A common method to find the optima between two criteria is the Pareto front [9,10].

*2.2. Financial aspect*
To include the financial aspect in the evaluation of the facade renovation systems, the Net Present Value (NPV) of each system is calculated. The NPV is the sum of the initial investment cost and the discounted future costs occurring over the life span of the building. For the calculations, a real discount rate of 1,8% is used [11] and a life span of thirty years is considered. The investment, replacement and maintenance phase are taken into account. Construction costs are based on the ASPEN price dataset, valid for the Belgian context [12]. This research looks at the micro-economic level, taking into account costs as paid by the end consumer (including taxes). For renovation and replacement works a VAT rate of 6% is used.

## 3. Research

*3.1. Existing circularity assessment methods*
Four existing circularity assessment methods are briefly discussed. It is analysed why the methods are not completely adequate for assessment of facade renovation systems at an early design stage and which aspects (circular parameters and scoring systems) are useful for the development of the new circularity assessment method.

*3.1.1. Material Circularity Indicator (MCI).* The Material Circularity Indicator (MCI) developed by the Ellen MacArthur Foundation measures the circularity of material flows of a product [5]. The parameters used to calculate the MCI relate to material properties such as the recycled and reused content, the recycling efficiency, fraction sent to reuse and fraction sent to recycling. A detailed bill of materials is required to compute the MCI.

As detailed insight into the material flows is necessary, the MCI is not appropriate for early stage decision-making. The method is suited for optimizing products, allowing companies to identify additional circular value of their products and materials. Furthermore, the MCI is not developed specifically for the building sector. For the assessment of facade systems a larger focus on construction techniques (connectors, layers,…) is desired.

*3.1.2. C-Calc.* C-Calc developed by Cenergie, a firm specialized in innovative sustainable energy in the building sector, is a tool that measures the circularity of buildings based on the materials used on or removed from site (50%), the flexibility of the building for future use or deconstruction (30%) and project management (20%) [13].

C-Calc is not completely adequate to assess facade renovation systems at an early stage because the tool evaluates on building level, contains parameters that require detailed knowledge about the applied materials (e.g. geographical origin) and parameters that relate to project management (e.g. material passports). For the eventual facade system used in the EURECA project it is assumed that no reused materials are used for the initial construction of the system. However, future reuse of the materials or modules of the system should be inherent to the developed facade system. Although the C-Calc parameter 'fraction of reused materials (in initial construction)' is decidedly a relevant parameter for the evaluation of the circularity of building elements, the parameter seems less relevant to distinguish the different facade systems of the EURECA project. Future evolutions of the developed method could entail the inclusion of this parameter.

Parameters used in C-Calc that are interesting for the assessment of the facade systems are: the environmental impact, the recycled content and recycling potential of materials, the type of connection and the use of independent layers.

*3.1.3. Label Circular Building.* Label Circular Building developed by Flemish Construction Confederation, VLISOG and BBRI aims to be a guideline for circular construction. Because it allows to calculate a score it can also be used to measure the circularity of buildings and building processes [7]. Currently only a Beta version of the tool exists. Label Circular Building consists of a quantitative (60%) and qualitative (40%) assessment. The four themes of the quantitative assessment are environmental impact (10%), build and design for change (20%), urban mining (20%) and transition (10%).

As illustrated by the name, Label Circular Building evaluates circularity on building level. The quantitative themes 'urban mining' and 'transition' are not suitable for evaluating facade renovation systems at an early stage because they relate to topics such as building and material information and synergy with the environment.

The dimensions of construction elements should be modular to enable reuse in other projects and should be manageable to ensure easy handling. Within the EURECA project there is a separate research on the 'optimal' dimensions of the facade system taking into account the easy (de)construction of the system and the possibility of reuse on different facades. As with the C-Calc parameter 'fraction of reused materials', parameters that relate to the dimensions of a system are surely relevant for the evaluation of the circularity of building elements but are less useful for the developed assessment method for the EURECA project .

Other parameters than those assessed in C-Calc that are fit for the assessment of the facade renovation systems are: layering according to life span, simple and fast connections and the use of a limited amount of materials.

*3.1.4. Building Circularity Index (BCI).* According to Alba Concepts circularity consists of two components: origin and application of materials. These two components are measured by the Material Index with the parameters 'origin of materials', 'disposal scenario', 'technical life span' and 'volume' and by the Detachability Index with the parameters 'type of connection' and 'accessibility of connection' [14,15].

Alba Concepts considers a building a hierarchical cluster of elements and elements a hierarchical cluster of products. They have developed the Building Circularity Index (BCI), for which it is necessary to determine the Product Circularity Index (PCI) and Element Circularity Index (ECI).

The method developed by Alba Concepts forms a strong starting point for the developed assessment method, but certain parameters that are evaluated by the other assessment methods, such as environmental impact and layering according to life span, are also valuable to take into account when assessing facade renovation systems.

*3.2. Circularity assessment method for facade renovation systems*

Using the information from the first research step the circularity assessment method for facade renovation systems at an early design stage is developed. The goal is to use parameters for which the input can be found easily from transparent and publicly accessible sources. In what follows, the different parameters for the developed method are described and their corresponding scoring systems worked out. For the assessment method the following rule applies: the higher the score, the less circular a solution.

*3.2.1. Parameters.*

*1. Recycling (%).* The recycled content and recyclability of a material are taken into account by the parameter 'recycling'. NIBE, an independent consultancy firm specialized in sustainable building, has developed the concept 'material reutilization'[16]. For different construction materials they have defined the material reutilization score (%), which takes the recycled content of the material into account for 1/3 of the score and the recyclability of the material for 2/3 of the score.

In the developed method the value for the parameter 'recycling' is defined for each individual material layer of the facade system according to: *100% – material reutilization value in NIBE.* For example, the reutilization value for stone wool is 7% according to NIBE. Therefore the value of the layer stone wool for the parameter 'recycling' is 93% (100-7).

*2. Environmental cost (€/m²).* The environmental impact of a facade system is represented by the parameter 'environmental cost'. The environmental cost is the amount of damage to the environment and/or people expressed in terms of the financial amount needed to avoid the potential harm or to resolve the harm incurred [17]. The Belgian life cycle assessment tool TOTEM is used calculate the environmental cost [18]. The analysed assessment methods C-Calc and Label Circular Building also use this tool to determine the parameter 'environmental impact'.

When calculating the environmental cost of a building element, TOTEM makes a distinction between the environmental cost due to the materials and the environmental cost due to the operational energy use (i.e. transmission losses through the building element). For the developed method only the environmental cost due to the materials is considered.

*3. Expected service life (years).* The parameter 'expected service life' accounts for the circular principles 'durable materials' (long life span) and 'layering according to life span'. The service life of construction materials can be found in 'Rapport Technische Levensduur van Gebouwcomponenten' by OVAM [19].

In the assessment method the 'expected service life' of each individual material layer of the facade system is defined: *100 - expected service life of a material.* The principle 'layering according to life span' is taken into account by introducing the following rule: the service life of a layer cannot be longer than that of the previous (more internal) layer. The determination of the service life of each layer should be done from the inside to the outside of the construction.

*4. Component dependency (%).* The more independent the components within a layer are, the easier future replacements, repairs or adaptions of one component can happen without influencing the other components. The parameter 'component dependency' takes this principle into account. Table 1 gives an overview of the scoring system for the parameter.

***Table 1.*** *Scoring system for parameter 'component dependency'.*

| Component dependency | Score (%) | Additional clarification |
|---|---|---|
| Completely independent | 0 | |
| Remove connectors to separate (material and connectors remain intact) | 10-20 | *How complex is the removal of the connector* |
| Sequential disassembly: disconnect all to disconnect one | 40-50 | *How many elements need to be disconnected (few large – many small)* |
| Destroy connectors to separate | 60-80 | *How labor-intensive is the destruction of the connector* |
| Completely dependent | 100 | |

As an example, bricks connected with a traditional cement mortar are inseparable and are scored 100, while bricks connected with lime mortar can be separated with a lot of labor and are scored 80. Rigid insulation boards with a tongue and groove joint are scored 40, while flexible insulation that is connected with plugs to the bearing layer is scored 0.

*5. Layer dependency (%).* While the parameter 'component dependency' takes into account the dependency of the components within a layer, the parameter 'layer dependency' represents the dependency between the layers. It takes into account the type of connection between the layer and the previous (more internal) layer. The scoring system for this parameter is based on the scoring system developed by Alba Concepts for the parameter 'type of connection' which is part of the Detachability Index [20]. Table 2 presents the scoring system for the parameter 'layer dependency'.

***Table 2.*** *Scoring system for parameter 'layer dependency' (based on Alba Concepts).*

| **Layer dependency** | | **Score (%)** |
|---|---|---|
| Dry connection | Dry connection | 0 |
| | Click connection | 0 |
| | Velcro connection | 0 |
| | Magnetic connection | 0 |
| Connection with added elements | Bolt and nut connection | 20 |
| | Spring connection | 20 |
| | Corner connection | 20 |
| | Screw connection | 20 |
| Direct integrated connection | Peg connection | 40 |
| | Nail connection | 40 |
| Soft chemical connection | Sealant connection | 80 |
| | Foam connection (PUR) | 80 |
| Hard chemical connection | Glue connection | 100 |
| | Cement-bound connection | 100 |
| | Chemical anchor | 100 |

*6. Flexibility for reuse (%).* The parameter 'flexibility for reuse' represents the possibility of a material layer to be adapted (altering of dimensions) for it to be reused in another building. Table 3 shows the scoring system for the parameter.

***Table 3.*** *Scoring system for parameter 'flexibility for reuse'.*

| **Flexibility for reuse** | **Score (%)** | **Additional clarification** |
|---|---|---|
| Easily adaptable | 0-20 | *How labor-intensive are actions* |
| Flexible because of small dimensions, but hard or impossible to adapt | 40 | *E.g. bricks, small tiles,...* |
| Adaptable by more complex/specialized actions | 50-60 | *How specialized are the used machines,...* |
| Reuse not possible | 100 | |

For instance, the dimensions of bricks or tiles cannot be easily changed, but these building blocks allow a certain degree of flexibility because of their small dimensions. Therefore they are attributed a relatively low score.

*3.2.2. Calculating the circular scores.* Per facade system, a score is defined for each parameter. The parameter 'environmental cost' is determined for the whole system at once; for the other parameters, first a score is established per material layer and subsequently these scores are added up to a total parameter score. To comply with the circular principles 'independent layers' and 'reversible connections' building solutions often require additional materials, which results in an additional material layer that must be taken into account for the total parameter score. Therefore, a distinction is made between parameters for which the material aspect is the most important (parameters: recycling, estimated service life, flexibility for reuse) and parameters which focus on the layers of the construction (parameters: component dependency, layer dependency). For the three parameters with focus on the material aspect, the total score for each parameter is obtained by multiplying the score of the material layer with the relative contribution of that material to the environmental cost of the facade system (this info is provided by the TOTEM tool). It is desirable that materials with a high environmental impact can be recycled, have a long service life and have potential for reuse. Therefore, it is fair that they have a higher contribution to the total score of these three parameters. For the two parameters that focus on the layers of a construction, a total score per parameter is obtained by simply adding up the scores of the materials layers.

When comparing the different facade systems, for each parameter an average score is defined by taking the average of the scores of all systems for that parameter. Next the score for each system on that parameter is expressed relatively to the average. This gives a relative score per parameter and converts all scores to the same unit. Finally, all the relative parameter scores of that system are added to a single score. The developed method does not set benchmarks, but is about comparing the circular scores of the facade renovation systems and choosing the most interesting system. The method allows to compare each parameter individually and to compare the single scores.

When assigning the scores to each parameter it is important to apply a holistic approach. For example, if a layer cannot be disassembled (parameter layer dependency), that layer and the previous layer are also not flexible for reuse or the components of the previous layers can also not be adjusted independently (parameter component dependency). The different parameters are interconnected and applying logical reasoning when assigning the scores is important.

*3.2.3. Application circularity assessment method to facade renovation systems.* The developed method is applied to four facade renovation systems: a standard and a circular variant of ETICS and a ventilated facade with rigid and with flexible insulation boards.

ETICS is a common renovation strategy that has proven to be an interesting choice from the environmental viewpoint because of its relatively simple construction [21]. However, due to its plaster finishing and the glued fixation of the insulation, ETICS is not a circular construction method. A more circular variant can be obtained by replacing the mortar connection with a mechanical connection of PVC-profiles [22]. The circular variant of ETICS does still have a plaster finishing. For the ventilated facade the variant with flexible insulation requires a more extensive supporting structure than the variant with rigid insulation boards. Figure 1 and 2 show drawings of the facade systems.

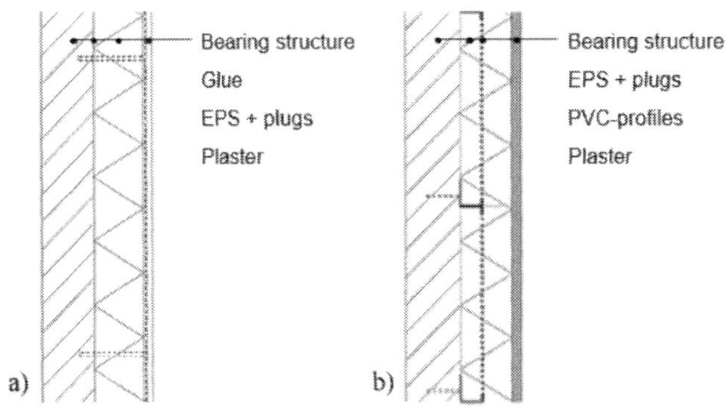


***Figure 1.*** *Drawings facade systems: a) ETICS standard; b) ETICS circular.*

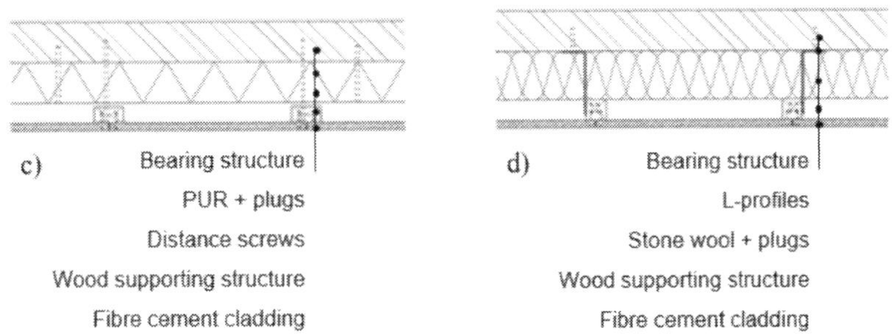


***Figure 2.*** *Drawings facade systems: c) ventilated facade rigid insulation; d)ventilated facade flexible insulation.*

Table 4 shows the development of the circular score per parameter for each facade renovation system. The different systems have the same U-value. For each system the different material layers are shown and their relative contribution to the environmental cost is stated. The connectors for each layer are mentioned; if one layer is connected with different types of connectors to the previous one, the most 'negative' connector counts. The material impact of the connectors is included in the environmental cost. If a relative score is highlighted green the system scores better than average on that parameter. If the relative score is highlighted red, the system scores worse than average on that parameter.

For the parameter 'component dependency' plaster is assigned the maximum score because although to a certain extent it is possible to repair plaster locally, this repair will always remain visible. EPS also gets the maximum score for this parameter since the plaster layer cannot be removed without damaging the EPS or other parts of the plaster and it is not realistic that EPS boards will be replaced or adapted independently.

The circular variant of ETICS has a higher environmental cost than the standard variant (PVC profiles have a higher material impact than glue) but has a better score on the parameter 'layer dependency'. The mechanical connection of the PVC profiles is a reversible connection while the glue connection is not. The biggest difference between the variants of the ventilated facade is in the parameter 'environmental cost'. The ventilated facade with flexible insulation requires a more extensive supporting structure (and connections), resulting in a higher material impact.

The ventilated facade systems have an additional material layer compared to the ETICS systems. This enables the ventilated facade systems to comply with the circular principles 'independent layers' and 'reversible connections', and, as a result, to have a better score than ETICS for the parameters 'component dependency', 'layer dependency' and 'flexibility for reuse'.

If the relative scores per parameter are added to a single score per facade renovation system (without weighting factors), the ventilated facade variants have a better circular single score than the ETICS variants and the ventilated facade system with hard insulation boards has the best circular single score.

The scores of the facade renovation systems on the parameter 'recycling' are relatively high. The values defined in NIBE are most likely on the conservative side. If more specific material choices have been made, more correct information on the recycled content and recycling potential of the material can be found in the product-specific EPDs.

### *3.3 Representation of circularity*

For each facade system Table 4 gives the score per parameter and a single score by adding the relative scores of each parameter. This single score makes it easy to compare the different systems and choose the most circular one. It is also possible to use weighting factors following the logic that some parameters are more important regarding circularity than others. The four analysed assessment methods also allow to express circularity by a single score (or by a label in the case of C-Calc). For C-Calc and Label Circular Building the relative share of the themes to the total score is mentioned. However, it seems rather subjective to decide which parameters contribute more to the circularity of a building than others. For example, for the developed circularity assessment method the parameters that evaluate a more general sense of sustainability (e.g. parameters recycling, environmental cost and expected service life)

IOP Conf. Series: Earth and Environmental Science **855** (2021) 012008 doi:10.1088/1755-1315/855/1/012008

***Table 4.*** *Circularity assessment method applied to four facade renovation systems.*

| **ETICS stand** | *Contribution environmental cost* | **Recycling** | *Relative* | **Environmental cost** | *Relative* | **Expected service life** | *Relative* | **Component dependency** | *Relative* | **Layer dependency** | *Relative* | **Flexibility for reuse** | *Relative* | **Total** |
|---|---|---|---|---|---|---|---|---|---|---|---|---|---|---|
| EPS 15cm | 0,63 | 97 | | | | 37 | | 100 | | 100 | | 100 | | |
| Plaster | 0,37 | 100 | | 4,37 | | 67 | | 100 | | 100 | | 100 | | |
| | | 98,11 | 1,00 | 4,37 | 0,67 | 48,10 | 0,95 | 200 | 1,82 | 200 | 1,82 | 100 | 1,46 | 7,71 |

*Connectors*
Insulation: Glue + plugs

| **ETICS circ** | *Contribution environmental cost* | **Recycling** | *Relative* | **Environmental cost** | *Relative* | **Expected service life** | *Relative* | **Component dependency** | *Relative* | **Layer dependency** | *Relative* | **Flexibility for reuse** | *Relative* | **Total** |
|---|---|---|---|---|---|---|---|---|---|---|---|---|---|---|
| EPS 16cm | 0,66 | 97 | | | | 37 | | 100 | | 20 | | 100 | | |
| Plaster | 0,34 | 100 | | 5,47 | | 67 | | 100 | | 100 | | 100 | | |
| | | 98 | 1,00 | 5,47 | 0,84 | 47,2 | 0,93 | 200 | 1,82 | 120 | 1,09 | 100 | 1,46 | 7,14 |

*Connectors*
Insulation: PVC profiles + plugs

| **Ventilated hard** | *Contribution environmental cost* | **Recycling** | *Relative* | **Environmental cost** | *Relative* | **Expected service life** | *Relative* | **Component dependency** | *Relative* | **Layer dependency** | *Relative* | **Flexibility for reuse** | *Relative* | **Total** |
|---|---|---|---|---|---|---|---|---|---|---|---|---|---|---|
| PUR 9cm | 0,36 | 95 | | | | 37 | | 40 | | 20 | | 10 | | |
| Wood support structure | 0,07 | 100 | | | | 50 | | 0 | | 20 | | 20 | | |
| Fibre cement cladding | 0,57 | 100 | | 5,9 | | 63 | | 0 | | 20 | | 50 | | |
| | | 98,18 | 1,00 | 5,9 | 0,90 | 52,63 | 1,03 | 40 | 0,36 | 60 | 0,55 | 33,33 | 0,49 | 4,33 |

*Connectors*
Insulation: Plugs
Wood support structure: Distance screws
Cladding: Screws

| **Ventilated soft** | *Contribution environmental cost* | **Recycling** | *Relative* | **Environmental cost** | *Relative* | **Expected service life** | *Relative* | **Component dependency** | *Relative* | **Layer dependency** | *Relative* | **Flexibility for reuse** | *Relative* | **Total** |
|---|---|---|---|---|---|---|---|---|---|---|---|---|---|---|
| Stone wool 13cm | 0,16 | 93 | | | | 25 | | 0 | | 20 | | 0 | | |
| Wood support structure | 0,09 | 100 | | | | 50 | | 0 | | 20 | | 20 | | |
| Fibre cement cladding | 0,75 | 100 | | 10,41 | | 63 | | 0 | | 20 | | 50 | | |
| | | 98,89 | 1,01 | 10,41 | 1,59 | 55,77 | 1,10 | 0 | 0 | 60 | 0,55 | 39,32 | 0,58 | 4,82 |

*Connectors*
Insulation: Plugs
Wood support structure: L-profile + screws
Cladding: Screws

| | | | | | | | | | | | | | | |
|---|---|---|---|---|---|---|---|---|---|---|---|---|---|---|
| | AVERAGE | 98,30 | | 6,54 | | 50,92 | | 110 | | 110 | | 68,16 | | 6 |

could be given a lower weighting factor than the parameters focusing more specifically on circularity (e.g. parameters component dependency, layer dependency and flexibility for reuse). However, making the distinction between sustainability and circularity is very challenging since circularity is a part of sustainability and this results in an overlap between both topics [23].

Even without using weighting factors, the representation of the circularity of a construction element by a single score can push more nuanced information to the background. Therefore it could be argued that the relative scores per parameter should be displayed separately, for example with a radar chart. This allows to identify the parameters for which the facade system could be further optimized, without worsening the scores of the other parameters. Figure 3 shows a radar chart for the four facade systems. Each axis of the radar chart represents a different parameter. The closer the lines are to the centre of the chart, the more circular a solution.

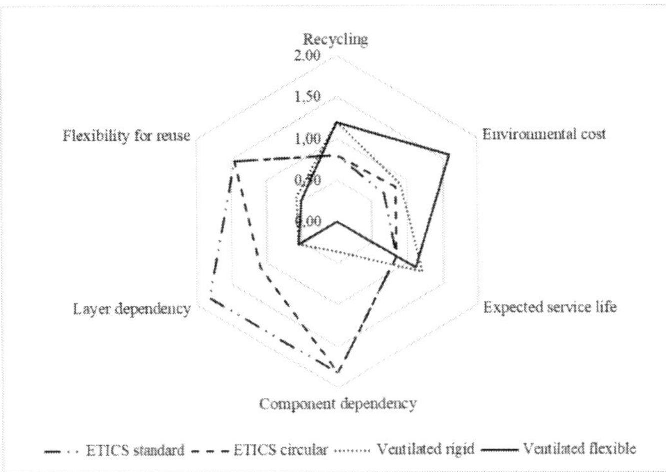


***Figure 3.*** *Radar chart for circularity of four facade renovation systems.*

### *3.4. Inclusion of the financial aspect*

In addition to the degree of circularity of the facade systems, the financial aspect will also play a decisive role when evaluating the different systems. To include the financial aspect in the evaluation the NPV of each facade system is calculated. To find the optima between the two criteria circular score and NPV the Pareto front method is used. The rule 'the lower, the better' applies for both criteria. Figure 4 shows a diagram with on each axis a different criterion. The circular score (single score as determined in Table 4) and NPV of the different facade renovation systems are plotted.

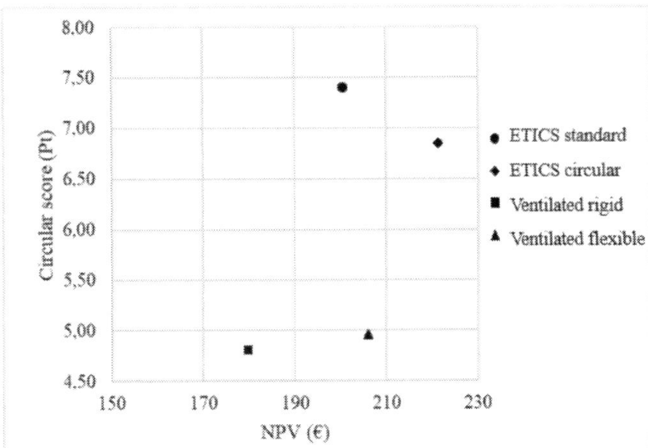


***Figure 4.*** *Multi-objective optimization circular score and NPV for four facade renovation systems.*

The most interesting facade system is clearly the ventilated facade with rigid insulation boards. It has the best circular score and the lowest NPV. The extensive maintenance of ETICS significantly increases its NPV. ETICS is neither financially nor circularly an interesting option. Only four facade systems are evaluated in this study. If the circular score and NPV of more systems is determined and plotted there will most likely not be one optimal solution, but rather an actual Pareto front.

## 4. Discussion and conclusion

The EURECA project aims to develop a circular facade system for the renovation of high-rise buildings. To objectively evaluate the circularity of the facade systems proposed within the project, a circularity assessment method must be used. Different circularity assessment methods exist but none are completely suited to evaluate facade renovation systems at an early design stage. Based on the evaluation of four existing assessment methods, this research develops a new circularity assessment method, suited to measure the circularity on element level with a limited amount of information necessary, allowing early-stage decisions.

The assessment method considers six parameters: recycling, environmental cost, expected service life, component dependency, layer dependency and flexibility for reuse. The method is tested on four existing facade renovation systems: standard ETICS, circular ETICS, ventilated facade with rigid insulation boards and ventilated facade with flexible insulation.

The ventilated facade systems have an additional material layer compared to the ETICS systems, which enables them to comply with the circular principles 'independent layers' and 'reversible connections'. As a result the ventilated facade systems have a better score than ETICS for the parameters 'component dependency', 'layer dependency' and 'flexibility for reuse'.

If per facade renovation system the relative scores of each parameter are added to a single score, without weighting, the ventilated facade with rigid insulation boards is the most circular system, standard ETICS the least. The use of weighting factors can be considered subjective and the representation of the circularity of a facade by a single score can erase nuances. A possible representation of the circularity of a facade system is through a radar chart, with each axis representing a circular parameter. This gives more background information but renders decision-making less straightforward because there is not one single score to take into account.

In reality, the financial aspect is a determining parameter when making renovation decisions. Therefore, the financial aspect is included in the evaluation of the different facade renovation systems by calculating their NPV. Using the Pareto front method, the ventilated facade with rigid insulation boards is circularly and financially the most interesting facade system. In this study the NPV was calculated as if the facade renovation system will stay on the same building for thirty years. It would have been more accurate to consider a scenario where the system is reused multiple times after a shorter time frame on the building, e.g. a scenario where the system is reused three times, each time for a time frame of about fifteen to twenty years on a building. However, the reuse of construction elements is currently no common practice and the prices related to deconstruction and remounting are often unknown.

The assessment method is developed for a renovation scenario. However, in hindsight many parameters and their corresponding scoring systems used in the developed assessment method are independent from the scenario 'renovation' or 'new build'. The developed circularity assessment method can also be used for the new build scenario, but as opposed to the renovation scenario, the bearing structure is taken into account for the scoring of the parameters because this layer is new and therefore its properties can be influenced. The developed assessment method is also suitable for the evaluation of other building elements besides the facade.

**Acknowledgements**
This research was supported by the initiative Circular Flanders [OC-SO-2019 320]. Alongside, Lisa Van Gulck, a doctoral research fellow of the Research Foundation – Flanders (FWO), acknowledges the support of the FWO [grant number: 1S00221N].

## References

[1] European Commission 2019 *Levels Report: Taking Action on the TOTAL Impact of the Construction Sector* (Luxembourg: Publications Office of the European Union)
[2] Bronsvoort E and van Oppen C 2020. *Madaster circularity indicator:* https://www.madaster.com/en/latest-news/blog/madaster-circularity-indicator (accessed October 2020)
[3] Kirchherr J, Reike D and Hekkert M 2017 Conceptualizing the circular economy : an analysis of 114 definitions *Resour Conserv Recycl* **127** 221–32
[4] Circular Flanders 2019 *EURECA*: https://vlaanderen-circulair.be/en/cases/detail/eureca (accessed October 2020)
[5] Ellen MacArthur Foundation 2015 *Circularity Indicators - An Approach to Measuring Circularity* (Ellen MacArthur Foundation)
[6] Cenergie 2020 *C-Calc*: https://www.cenergie.be/nl/diensten/advies/c-calc (accessed October 2020)
[7] Vergauwen A 2020 *Circulair Gebouw – Leidraad Meetsysteem* (WTCB)
[8] Alba Concepts 2020 *Building Circularity Index*: https://albaconcepts.nl/building-circularity-index/#slide-4-informatie (accessed October 2020)
[9] Sierra-Pérez J, Boschmonart-Rives J and Gabarrell X 2016 Environmental assessment of façade-building systems and thermal insulation materials for different climatic conditions *J. Clean. Prod* **113** 102–13
[10] Ostermeyer Y, Wallbaum H and Reuter F 2013 Multidimensional Pareto optimization as an approach for site-specific building refurbishment solutions applicable for life cycle sustainability assessment *Int J Life Cycle Assess* **18** 1762–79
[11] Federal planbureau 2019 *Economische Vooruitzichten 2019-2024 – Versie van Februari 2019* (Brussels: Federaal Planbureau)
[12] ASPEN 2019 *ASPEN INDEX - Onderhoud, Ombouw, editie 53.* (East-Flanders: ASPEN)
[13] C-Calc 2020 *Webinar C-calc (10/06/20)*: https://www.youtube.com/watch?v=TkdeRExZ0qw&feature=youtu.be (accessed June 2020)
[14] Alba Concepts 2015 *Building Circularity Index:* https://albaconcepts.nl/building-circularity-index-nieuwsbericht/ (accessed November 2020)
[15] Alba Concepts 2015 *100% circulair in 2050, is jouw organisatie er klaar voor?* (Rosmalen: Alba Concepts)
[16] NIBE 2011 *Bron 2 Bron*: https://www.nibe.info/nl/b2b (accessed October 2020)
[17] OVAM 2020 *Environmental Profile of Building Elements - update 2020* (Mechelen: OVAM)
[18] TOTEM 2018 TOTEM: https://www.totem-building.be/ (accessed October 2020)
[19] Vandenbroucke M 2018 *Rapport Technische Levensduur van Gebouwcomponenten* (OVAM)
[20] Alba Concepts 2019 *Een meetmethode voor losmaakbaarheid* (Eindhoven: Alba Concepts)
[21] Wang W, Zmeureanu R and Rivard H 2004 Applying multi-objective genetic algorithms in green building design optimization *Build Environ.* **40** 1512-25
[22] Zolpan 2019 *Gamme Armaterm – Guide de Composants* (Zolpan)
[23] Rahla K M, Bragança L and Mateus R 2019 Obstacles and barriers for measuring buildings' circularity *IOP Conf. Ser.: Earth Environ. Sci.* **225** 012058

IOP Conf. Series: Earth and Environmental Science **855** (2021) 012009 doi:10.1088/1755-1315/855/1/012009

# Circularity Assessment using the UN 2030 Agenda Indicators

**P Huovila**[1, 3] **and U Iyer-Raniga**[2]

[1] Green Building Council Finland, Lapinrinne 3, 00100 Helsinki, Finland, Coordinator, OPN SBC Programme

[2] School of Property, Construction and Project Management, RMIT University, 124 Latrobe Street, Melbourne VIC 3000, Australia, Co-lead OPN SBC Programme

[3] pekka.huovila@figbc.fi

**Abstract.** Established in 2015, the One Planet Sustainable Buildings and Construction (SBC) Programme has studied the state of play for circular built environment in different regions: Africa, Asia, Europe, Latin America, Middle East, North America and Oceania. One aspect of this study has been assessment for circularity using the 2030 Agenda indicators. This paper describes how nature's resources are used in buildings to provide more comfortable and healthier spaces while generating less waste and emissions with the aim to change linear processes to circular, thereby increasing resource efficiency and reducing waste and emissions, while also providing local jobs.

A set of core indicators have been selected to assess the circularity at global and regional levels. It is intended that the indicators selected for SBC become a starting point for prioritising the built environment sector to embrace circularity principles and practices.

## 1. Introduction

Adopted in 2012 at the World Summit on Sustainable Development, the 10-Year Framework of Programmes on Sustainable Consumption and Production (10YFP) is a global commitment to accelerate the shift towards sustainable consumption and production in both developed and developing countries. Target 12.1 [1] of the 2030 Sustainable Development agenda calls for the implementation of the 10YFP. The One Planet Network (OPN) has formed to implement the commitment of the 10YFP.

Established in 2015, the One Planet Sustainable Buildings and Construction (SBC) Programme [2] of the One Planet Network contributes to accelerating the shift towards sustainable consumption and production (SDG 12) in the buildings and construction sector. The SBC programme aims to promote circularity principles and approaches to support countries in achieving their sustainability commitments and development objectives. The work in SBC is guided by the Paris Climate Agreement [3], UN 2030 Agenda for Sustainable Development [4] and the New Urban Agenda [5].

Circular economy can be the means to achieve the Paris Climate Agreement targets. It is mentioned in New Urban Agenda and several times in the UNEA4 resolution 'Innovative Pathways to Achieve Sustainable Consumption and Production' [6]. It is, however, not explicitly cited in the 2030 Agenda where it is split across several of the 17 Sustainable Development Goals (SDGs), their 169 Targets and associated indicators. The aim of this paper is to present findings of a global survey to identify SDG indicators for circular built environments. It is logical to use the SDGs as a yardstick for monitoring and tracking our transition to a circular built environment as it aligns with the Paris Agreement for mitigation and the SDGs for mitigation, adaptation and resilience to chart a forward direction towards a decarbonised future. The SDGs do not mention circular economy but the principles underpinning SDGs and circular economy are similar focusing on environment, economic and social impacts; to ensuring low resource impact and use, while taking account of equity and economic priorities such as jobs.

The paper commences with an understanding of existing indicators in the literature to measure and track circular built environment. This is followed by an explanation of the survey undertaken and its findings. These are mapped against the state of play of recent reports undertaken for the OPN SBC programme on circular built environments. Discussions are followed by conclusions.

## 2. Circularity indicators for the built environment

Ellen MacArthur Foundation (EMF) have identified five research themes to chart a direction for a circular economy. Among these are questions around enabling conditions for transition and tracking progress. To support circularity outcomes, EMF have designed *Circulytics* – a tool for measuring circularity but this is not focused specifically for the built environment [7]. The EU has developed policies and guidelines to support the building and construction industry to transition to a circular economy while also meeting the SDGs [8]. According to the Circularity Gap report [9], the circularity gap has been widening rather than narrowing. Countries are critical facilitators of the circular economy and have an important role to play in the movement towards circularity. While countries are different, there are common challenges and opportunities to be explored.

Nations such as Japan have also been grappling with CE through targeting waste, reducing landfill and taking an 'eco' approach to planning, design, construction and operation of the built environment since the start of the 21st century [10]. Developing countries such as China, have focused CE on a broader systemic policy, concentrating on the national government to include development planning and requiring the collaboration of various government agencies across various levels and its focus is on energy efficiency and solid waste management [10]. Yet, countries that are ahead of the pack in terms of CE and those putting policies in place are struggling with CE indicators as existing measures such as recycling rates, CO2 emissions, material flow analysis are not designed for closed-loop systems. While some of these may be okay at a local level or an individual building level, they do not provide a clear pathway for a system where waste from one cycle may become the resource for another.

The concept of a carbon footprint for measure has been used to compare sectors and countries. For example, Hertwich and Peters [11] calculate the greenhouse gas emissions for 73 nations and 14 world regions and find that the built environment consumes a significant proportion. Operation and maintenance of residences and mobility contribute to over a third (36%) of GHG emissions. Material flow analysis has been used by authors such as Krausmann et al [12] but this relies on the availability of complete data which is not always the case.

Authors such as Marvuglia et al [13] suggest emergy-based indicators to measure circularity. Emergy accounting allows accounting for ecosystem services and resources, and therefore, water and energy flows directly or indirectly affecting a product or service may be calculated. However, there are limitations. There are problems with interpretation of the data, communication is problematic, and uncertainties are hard to deal with this accounting approach. The complexity of the accounting processes itself is also difficult in this emerging area.

A report by the WBCSD and Climate-KIC [14] found there were many frameworks used by organisations to measure circularity. This creates competing and often conflicting indicators rather than a harmonious approach and difficulties in comparing 'apples with apples'. Considering the maturity of circular economy and the development of indicators, a sensible approach to developing indicators for circular built environment was to use the SDGs itself as the timeline for the SDGs is from 2015-30. The SDGs cover qualitative and quantitative indicators, and they also include environmental, economic and social indicators. Another reason for favouring the use of the SDGs was not to 'reinvent' indicators which may require data collection and analysis, but to also bring the importance of the buildings and construction sector in line with reporting on Paris Agreement. It also supports alignment with countries to consider the buildings and construction sector as part of their Nationally Determined Contributions (NDCs) and National Adaptation Plans (NAPs).

## 3. Survey on 2030 Agenda and circular built environment

The SBC programme, led by the Ministry of the Environment, Finland and co-led by RMIT University and UNEP, deployed a survey to study how circularity of the built environment is present in the 2030

Agenda targets and indicators. The global survey, that is still open, has had responses from over 100 experts from all continents. The split of responses is shown in Figure 1.

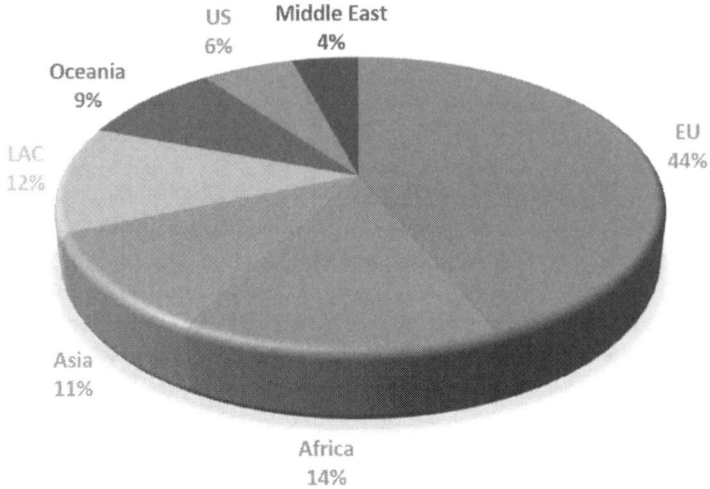


**Figure 1.** The survey sample of 119 respondents.

### *3.1. Sustainable Development Goals and circularity*

The first question in the survey deals with Sustainable Development Goals to assess which of the SDGs relate closest with circular built environment (CBE). The scale used is the following:

- 0 has no importance to CBE
- 1 may have some importance (local/regional specificity)
- 3 has minor importance to CBE
- 5 is important to CBE
- 9 is crucial, core CBE goal.

Figure 2 shows the percentages of the respondents judging the SDGs to be crucial for circular built environment. SDG 11 *Make cities and human settlements inclusive, safe, resilient and sustainable* and SDG 12 *Ensure sustainable consumption and production patterns* scoring almost 80 % followed by SDG 13 *Take urgent action to combat climate change and its impacts* and SDG 9 *Build resilient infrastructure, promote inclusive and sustainable industrialization and foster innovation* assessed by over 60 % of experts.

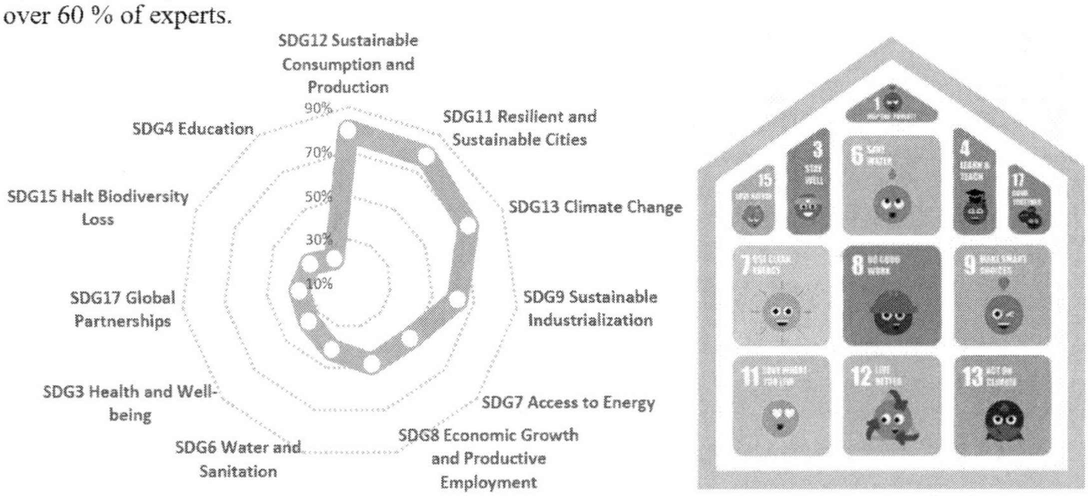


**Figure 2.** Circular Built Environment related Sustainable Development Goals. ©Ninni Westerholm emojis: https://www.oneplanetnetwork.org/sustainable-lifestyles-and-education/good-life-goals

IOP Conf. Series: Earth and Environmental Science **855** (2021) 012009 doi:10.1088/1755-1315/855/1/012009

The other relevant SDGs were judged to be SDG 7 *Ensure access to affordable, reliable, sustainable and modern energy for all*, SDG 8 *Promote sustained, inclusive and sustainable economic growth, full and productive employment and decent work for all* and SDG 6 *Ensure availability and sustainable management of water and sanitation for all* was acknowledged by over 40 % of the respondents.

SDG 3 *Ensure healthy lives and promote well-being for all at all ages*, SDG 17 *Strengthen the means of implementation and revitalize the global partnership for sustainable development*, SDG 15 *Protect, restore and promote sustainable use of terrestrial ecosystems, sustainably manage forests, combat desertification, halt and reverse land degradation and halt biodiversity loss* and SDG 4 *Ensure inclusive and equitable quality education and promote lifelong learning opportunities for all* were ranked as core indicators by over 20 % of the experts that completed the survey.

### *3.2. Circularity indicators in the 2030 Agenda*

The second question in the survey is about the 2030 Agenda indicators that can be used to assess the state to form a baseline and for setting the objectives that can be monitored. The scale was the same as for SDGs and the data was analysed following the same logic scoring the percentage of respondents evaluating our shortlisted 57 indicators to be core to circularity.

Some of the indicators relate with two or more targets which is the case of indicator 12.2.1/8.4.1 *Material footprint, material footprint per capita, and material footprint per GDP* addresses to target 12.2 *By 2030, achieve the sustainable management and efficient use of natural resources and target 8.4 Improve progressively, through 2030, global resource efficiency in consumption and production and endeavour to decouple economic growth from environmental degradation, in accordance with the 10-year framework of programmes on sustainable consumption and production, with developed countries taking the lead.*

The third indicator receiving almost 80% of the votes was 12.5.1 *National recycling rate, tons of material recycled* addressing target 12.5 *By 2030, substantially reduce waste generation through prevention, reduction, recycling and reuse*. The fourth indicator, that received over 70 % support was 11.c.1 *Proportion of financial support to the least developed countries that is allocated to the construction and retrofitting of sustainable, resilient and resource-efficient buildings utilizing local materials* addressing target 11.c *Support least developed countries, including through financial and technical assistance, in building sustainable and resilient buildings utilizing local materials*.

These four indicators were selected as global CBE core indicators in the SBC Programme. It should be noted, however, that in the annual review process [15], a better indicator is being sought for 11.c.1 dealing with local materials [16]. Ten secondary circularity indicators were selected in the SBC Programme based on the survey as shown in Figure 3.

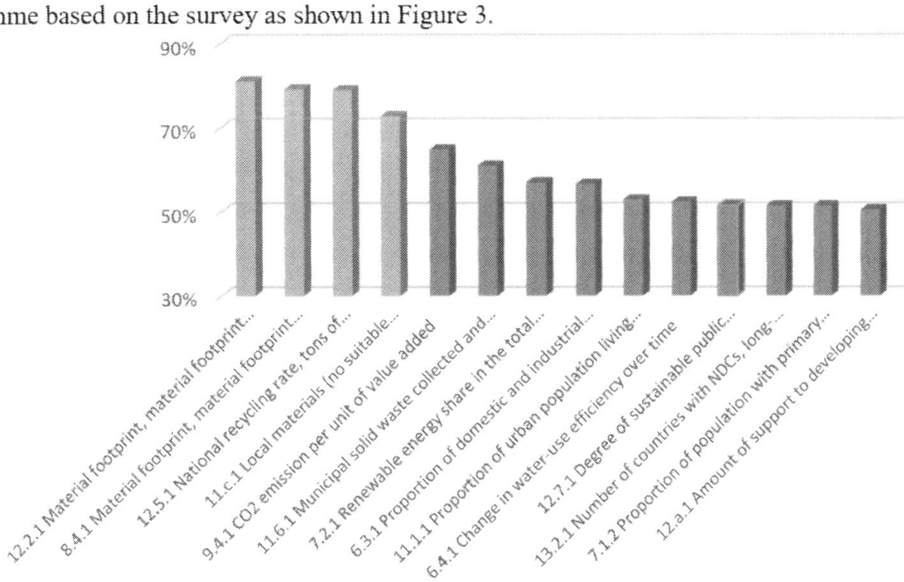


**Figure 3.** Key CBE indicators based on survey.

The ten secondary indicators and their respective targets are listed below:

- 9.4.1 *CO2 emission per unit of value added* addressing target 9.4 *By 2030, upgrade infrastructure and retrofit industries to make them sustainable, with increased resource-use efficiency and greater adoption of clean and environmentally sound technologies and industrial processes, with all countries taking action in accordance with their respective capabilities*
- 11.6.1 *Proportion of municipal solid waste collected and managed in controlled facilities out of total municipal waste generated, by cities* addressing target 11.6 *By 2030, reduce the adverse per capita environmental impact of cities, including by paying special attention to air quality and municipal and other waste management*
- 7.2.1 *Renewable energy share in the total final energy consumption* addressing target 7.2 *By 2030, increase substantially the share of renewable energy in the global energy mix*
- 6.3.1 *Proportion of domestic and industrial wastewater safely treated* addressing target 6.3 *By 2030, improve water quality by reducing pollution, eliminating dumping and minimizing release of hazardous chemicals and materials, halving the proportion of untreated wastewater and substantially increasing recycling and safe reuse globally*
- 11.1.1 *Proportion of urban population living in slums, informal settlements or inadequate housing* addressing target 11.1 *By 2030, ensure access for all to adequate, safe and affordable housing and basic services and upgrade slums*
- 6.4.1 *Change in water-use efficiency over time addressing target* 6.4 *By 2030, substantially increase water-use efficiency across all sectors and ensure sustainable withdrawals and supply of freshwater to address water scarcity and substantially reduce the number of people suffering from water scarcity*
- 12.7.1 *Degree of sustainable public procurement policies and action plan implementation* addressing target 12.7 *Promote public procurement practices that are sustainable, in accordance with national policies and priorities*
- 13.2.1 *Number of countries with nationally determined contributions, long-term strategies, national adaptation plans, strategies as reported in adaptation communications and national communications* addressing target 13.2 *Integrate climate change measures into national policies, strategies and planning*
- 7.1.2 *Proportion of population with primary reliance on clean fuels and technology* addressing target 7.1 *By 2030, ensure universal access to affordable, reliable and modern energy services*
- 12.a.1 *Installed renewable energy-generating capacity in developing countries (in watts per capita)* addressing target 12.a *Support developing countries to strengthen their scientific and technological capacity to move towards more sustainable patterns of consumption and production.*

The first two indicators 9.4.1 and 11.6.1 were scored by over 60 % of the respondents while the eight following were supported by at least 50 % of the respondents. They are listed in the order of priority. In case any indicator had equal scores, an average of the scores were used to determine their order.

These four core indicators and ten secondary indicators were selected as the global CBE set in the SBC Programme based on first 100 expert responses as shown in Figure 4. Since the conditions and priorities vary globally, other indicators can be applied regionally, nationally or locally. The SBC Programme has organized virtual workshops to examine in depth regional indicator sets in Africa, Latin America and Asia.

**Figure 4.** SDG indicators for the SBC programme shown in the SDG circle. ©Ashley King, Ninni Westerholm

## 4. State of play for circular built environment

The SBC programme prepared seven regional state of play for circular built environment reports from Europe [17], North America [18], Oceania [19], Middle East [20], Asia [21], Africa [22] and Latin America [23]. Then a global report [24] was compiled to summarize main findings and draw first conclusions.

### *4.1. Regional priorities*

Our regional reports show that the different countries and continents are in various states of transition towards the adoption of circular economies, but that Europe is leading, largely due to having been engaged in this transition process for at least a decade. Other regions are still grappling with considering what circularity actually means for them and proactively drafting supporting policies and programmes with the goal of transitioning to a low-carbon future.

Tensions with respect to design, operation and deconstruction exist between new developments and existing stock. The focus in Asia, Africa and the LAC regions has been based on a linear framework for supporting the needs of the built environment, be it residential, commercial or infrastructure. It is anticipated that the global report will enable governments to consider the development (and review as appropriate) of their National Adaptation Plans and Nationally Determined Contributions. Policy-makers need to consider the quantum and quality of construction and deconstruction waste to ensure that it does not end up in landfill, and that promote policies and programmes to support the second and third lives of materials, while ensuring healthier built environments.

### *4.2. Lifecycle considerations*

Buildings usually last for 80 to 100 years, or more. Therefore, whether it is conversion of existing building stock or planning new buildings, the lifecycles of buildings must be considered a priority. The regional reports have shown that, with rising populations and urbanisation, the regions in the Global South are going to see a phenomenal increase in the number of greenfield developments. These need to be considered carefully.

In the Northern hemisphere where the built environment is already mature, the focus will be more on sustainable redesign, rebuild and more efficient use of existing spaces with increased densities, as well as on using materials in their second life where possible. Material use will continue. This needs rethinking and replanning so that virgin material use is arrested if not eliminated. This will call for closing local feedback loops wherever possible.

## 5. Discussion and conclusion

The results of the survey highlighted clearly, the indicators that the SBC programme can use to guide and direct a pathway for circular built environments regardless of scale; at regional, national or local levels. The four core indicators are all focused on materiality: material footprints, recycling and use of local materials. This indicates that there is a clear agreement among the survey respondents that material use is critical to circular built environments, regardless of the scale. This is hardly surprising as materials form the basic building blocks of the built environment.

The secondary indicators broaden the scope from purely material use and impact to related issues such as emissions, waste management, energy use, water use, impact per head of population, procurement, enabling policies, access to and diffusion of renewable energy in the built environment. This clearly demonstrates that the secondary indicators need to work in tandem with other factors that support planning design and operation of the built environment.

The principles underpinning circularity and the SDGs are similar. The intent of both approaches are to support social, environmental and economic outcomes so that future generations of humans can sustain life of this planet. While the immediate priorities in the developed world and the developing world are not the same, nevertheless, the solutions devised must be developed through a sustainable, safe and resilient built environment accounting for a changing climate, reducing our consumption and finding innovative and creative ways to remanufacture production to reuse waste, use energy and water efficient practices to support biodiversity, health and well-being in our indoor and outdoor environments. A change from the current linear paradigm will lead to more circular ways of planning, designing and operating our built environment, which in the process will need new skills and retraining, new jobs to support economic growth and bring people out of poverty. It requires a collaborative and integrative effort across present apolitical boundaries. As developing countries are contemplating their green field developments, a clear direction that takes them on a circular path exists through SDG indicators 12.2.1 (8.4.1), 12.5.1 and 11.c.1. Other indicators mentioned as secondary indicators may be prioritised within a region, country or locality depending on the climate and the context.

Developing countries can fast track their direction to the future by leap frogging from the linear structures and processes that the developed countries have followed to date. Policies, guidelines and capacities for developing a circular paradigm are available and can be aligned with NDCs and NAPs. Using the relevant indicators also supports monitoring and tracking to ensure desired outcomes are achieved.

## 6. References

[1] https://sdgs.un.org/goals/goal12 *Accessed 23 November 2020.*

[2] https://www.oneplanetnetwork.org/sustainable-buildings-and-construction *Accessed 26 November 2020.*

[3] United Nations 2015 *Paris Agreement*, 27 p. *Retrieved 27 November from* https://unfccc.int/files/essential_background/convention/application/pdf/english_paris_agreement.pdf

[4] United Nations 2015 *Transforming our world: the 2030 Agenda for Sustainable Development*. A/RES/70/1, 35 p. *Retrieved 23 November 2020 from* https://www.un.org/ga/search/view_doc.asp?symbol=A/RES/70/1&Lang=E

[5] United Nations 2017 *New Urban Agenda*. A/RES/71/256, 66 p. *Retrieved 23 November 2020 from* https://unhabitat.org/sites/default/files/2019/05/nua-english.pdf

[6] United Nations 2020 *Global indicator framework for the Sustainable Development Goals and targets of the 2030 Agenda for Sustainable Development*. A/RES/71/313, 21 p. *Retrieved 23 November 2020 from* https://unstats.un.org/sdgs/indicators/Global%20Indicator%20Framework%20after%202020%20review_Eng.pdf

[7] Ellen MacArthur Foundation 2020 *Circulytics – measuring circularity. Retrieved November 26 2020 from* https://www.ellenmacarthurfoundation.org/resources/apply/circulytics-measuring-circularity

[8] European Commission 2018 *Communication from the Commission to the European Parliament, the Council, the European Economic and Social Committee and the Committee of the regions on a monitoring framework for the circular economy (COM 2018) 29 Final.*

[9] Circle Economy 2020 *The circularity gap report 2020, When circularity goes from bad to worse; The power of countries to change the game*. PACE

[10] Yong G, J Sarkis, S Ulgiati and P Zhang 2013 Measuring China's circular economy *Science* **339** 1526-27 American Association for the Advancement of Science DOI: 10.1126/science.1227059

[11] Hertwich E G and G P Peters 2009 Carbon footprint of nations: a global, trade-lined analysis *Environmental Science & Technology* **43** 6414-20 DOI:10.1021/es803496a

[12] Krausmann F, S Gingrich, N Eisenmenger, K H Erb, H Haberl and M Fischer-Kowalski 2009 Growth in global materials use, GDP and population during the 20th century. *Ecol. Econ.* 68, 2696–705

[13] Marvuglia A, R Santagata, B Rugani, E Benetto and S Ulgiati 2018 Emergy-based indicators to measure circularity: promises and problems *Energy Policy Journal* **21** 4 179-96 DOI: 10.24425/124510

[14] World Business Council for Sustainable Development (WBCSD) and EIT Climate-KIC 2018 Circular metrics landscape analysis *European Institute of Innovation and Technology (EIT)* under grant agreement No TC2018A_2.1.4-CirMet_P308-1A.

[15] https://unstats.un.org/sdgs/indicators/indicators-list/ *Accessed 30 November 2020*

[16] United Nations 2019 *Innovative pathways to achieve sustainable consumption and production*. UNEP/EA.4/L.2, 4 p. *Retrieved 23 November 2020 from* https://www.iso20400.org/innovative-pathways-to-achieve-sustainable-consumption-and-production/

[17] Westerholm N 2020 *State of play for circular built environment in Europe. A report compiling the regional state of play for circularity in the built environment in Europe across the member states of the European Union, Iceland, Norway, Serbia and the United Kingdom*. October 2020, UN OPN SBC Programme, 47 p. *Retrieved 24 November 2020 from* https://www.oneplanetnetwork.org/linear-circular-first-global-mapping-circularity-built-environment

[18] Keena N and Dyson A 2020 *State of play for circular built environment in North America. A report compiling the regional state of play for circularity in the built environment in North America across the United States of America*. October 2020, Yale Center for Ecoystems in Architecture, Yale University and UN OPN SBC Programme, 54 p. *Retrieved 24 November 2020 from* https://www.oneplanetnetwork.org/linear-circular-first-global-mapping-circularity-built-environment

[19] Iyer-Raniga U 2020 *State of play for circular built environment in Oceania. A report compiling the regional state of play for circularity in the built environment in Oceania across Australia and New Zealand*. October 2020, RMIT University and UN OPN SBC Programme, 59 p. *Retrieved 24 November 2020 from* https://www.oneplanetnetwork.org/linear-circular-first-global-mapping-circularity-built-environment

[20] Al-Alawi B, Cavallari G, Cindric H, Gargallo M, Hegazy N, Saqib S, Shaka H and Tibi G 2021 *State of Play for Circular Built Environment in Gulf Cooperation Council (GCC) countries. A report compiling the regional state of play for circularity in the built environment in Gulf Cooperation Council countries across Bahrain, Kuwait, Oman, Qatar, Saudi Arabia and the United Arab Emirates*. February 2021, ARUP and UN OPN SBC Programme, 31 p. *Retrieved 18 February 2021 from* https://www.oneplanetnetwork.org/linear-circular-first-global-mapping-circularity-built-environment

[21] Niazi Z, Singh A and Sen I 2020 *State of play for circular built environment in Asia. A report compiling the regional state of play for circularity in the built environment in Asia across China, India, Indonesia, Nepal and Pakistan*. October 2020, TARA, Development Alternatives and UN OPN SBC Programme, 69 p. *Retrieved 24 November 2020 from* https://www.oneplanetnetwork.org/linear-circular-first-global-mapping-circularity-built-environment

[22] Gibberd J 2020 *State of play for circular built environment in Africa. A report compiling the regional state of play for circularity in the built environment in Africa across Egypt, Ethiopia, Ghana, Kenya, Malawi, Nigeria, Rwanda, South Africa, Uganda and Zambia*. October 2020, Gauge and UN OPN SBC Programme, 37 p. *Retrieved 24 November 2020 from* https://www.oneplanetnetwork.org/linear-circular-first-global-mapping-circularity-built-environment

[23] Moreno P 2020 *State of play for circular built environment in Latin America and the Caribbean. A report compiling the regional state of play for circularity in the built environment in LAC across Argentina, Brazil, Chile, Cuba, Ecuador and Mexico*. October 2020, Ecosur and UN OPN SBC Programme, 34 p. *Retrieved 24 November 2020 from* https://www.oneplanetnetwork.org/linear-circular-first-global-mapping-circularity-built-environment

[24] Iyer-Raniga U and Huovila P 2021 *Global state of play for circular built environment. A report on the state of play on circularity in the built environment across Africa, Asia, Europe, Gulf Cooperation Council countries, Latin America and the Caribbean, North America and Oceania*. February 2021, UN OPN SBC Programme, 43 p. *Retrieved 18 February 2021 from* https://www.oneplanetnetwork.org/linear-circular-first-global-mapping-circularity-built-environment

# Facilitating circular site preparation; developing the instrument PrCiSi

**A G Entrop[1,3], L Hagen[1] and A M Kuiper[2]**

[1] Saxion University of Applied Sciences, School of Business, Building & Technology, M.H. Tromplaan 28, 7513 AB Enschede, the Netherlands
[2] Municipality of Olst-Wijhe, Raadhuisplein 1, 8131 BN Wijhe, The Netherlands

**Abstract.** The urge to strive for closed material loops is felt in the construction industry. However, before the built environment can be regarded as circular, not only the material usage in buildings, but also preparations taken on sites need to be circular. In this paper a new instrument is presented to help project leaders at municipalities and contractors to assess how circular their process of site preparation is. The instrument, by the name of PrCiSi, has been developed on basis of a literature study on sustainable material usage and interviews among stakeholders in this niche market. The instrument has been tested on the site preparation processes for the residential neighbourhoods Olstergaard and Noorder Koeslag in the municipality of Olst-Wijhe. Of course, there is still room for improvement and more test cases are welcomed, but three main subjects, namely materials, equipment and personnel, seem to form a proper basis. A large number of test cases might help us to further develop the scoring system. However, being convinced that we can cross a boundary here, we are happy to share with you the current status to inspire you to use and improve this instrument with us.

## 1. Introduction

The impact of the construction industry on the natural environment is severe, natural areas are changed into predominantly hard solid surfaces, the energy use in the built environment is high and the industry puts huge claims on materials. The Dutch construction industry for example claims 50% of the total resource usage [1]. Sustainable building has been the concept that aims to reduce the environmental impact caused by infrastructure and facilities throughout their lifecycle and create healthy structures, environment friendly, comfortable and productive built environment [2]. Although especially sustainable energy use has been in the centre of attention for quite a while, now material usage rapidly gains attention in a society that embraces circularity. The Dutch government started to broadly communicate their ambitions in 2016 with a programme on circularity for all industries. The ambitions are to have a 50% circular economy in 2030 and a 100% circular economy in 2050 [3].

Multiple construction companies, architects and suppliers of building materials try to reduce the need for new virgin materials, put effort in reclaiming used materials, products and building components, and reduce the amount of waste originating from their production processes. However, it seems that little is known about what can be done in the stages at site before actual buildings are being constructed.

The goal of this research project is to develop an instrument for those who are involved in the process of designing and executing site preparations to contribute to a circular economy by means of reducing the environmental impact of materials and energy used. The research was executed within the research

[3] Author to whom any correspondence should be addressed.

group Sustainable Environment of Saxion University of Applied Sciences in close collaboration with the Netherlands municipality of Olst-Wijhe [4, 5].

Coming sections will address the research setup, theoretical framework and the process of designing the instrument. After these sections, the instrument will of course be tested on its usability, before a section with the conclusion is offered at the end of this paper.

## 2. Research setup

As in many research projects, our research also started with conducting a literature study. Two topics form the main focus in this study, namely circularity in the construction industry and site preparation. Having access to the international oriented ScienceDirect database and more national oriented governmental and educational documents, helped us to find out what a circular economy could mean for the Dutch construction industry. The amount of available documents on the process of site preparation was unfortunately considerable smaller.

Having available the theoretical framework, we then resided to experts in the field. Nine interviews were conducted to put the finger on where circularity and site preparation can meet each other. These experts have multiple years of experience in planning or executing site preparation in the Netherlands. They work at municipalities, contractors or consultancy firms. Conducted interviews took on average one hour.

The collected information forms the foundation of the instrument. Considering that municipalities give shape to zoning plans and often take the initiative to plan a certain residential area, the instrument needs to be able to help those working at municipalities and for municipalities in designing and executing site preparations. For this reason, we thought it would be a good idea to develop a list of questions that helps to start the site preparation design process and analyse site preparation plans. The answers to the listed questions make it possible to assess how circular the site preparation ambitions are.

## 3. Theoretical framework

The theoretical framework consists mainly of two pillars that are addressed in the coming two subsections, namely circularity in the construction industry and site preparation.

### *3.1. Circularity in the construction industry*

From a national and chronological perspective it is relevant to explain that the Netherlands already adopted a vision on how to cope with waste in 1979. This was Lansink's Ladder that offered a waste hierarchy [6]. Incorporated into Netherlands law on environmental conservation, this scheme originally contained seven steps to cope with waste disposal:

1. Prevention: try to create as little waste as possible;
2. Consideration of the raw materials: use raw materials which do not harm the environment after product usage;
3. Product reuse: try to reuse a product in its original state as often as possible;
4. Material recycling: when the product cannot be used in its original form, then try to recycle its materials;
5. Combustion as a source of energy: when the product is assigned to be waste, then it can be incinerated in order to generate heat and electric power;
6. Combustion: less favourable is to burn the waste without the generation of heat and electric energy;
7. Landfill: the least preferable option for disposing products and materials.

In the Dutch construction industry, with its vast flow of construction and demolition waste, separating different material streams on site is already being common practice for multiple decades. Meanwhile, our understanding of the environmental impact of not only waste, but also new materials has been increasing. Internationally, the thrive for sustainable development considering our needs and what this implies for fulfilling the need of future generations, introduced in 1987 [7], is nowadays firmly grounded in seventeen sustainable development goals [8]. Thanks to the efforts of Braungart and McDonough, the traditional Cradle-to-Grave approach in linear economies started by 2002 to be replaced by a Cradle-to-Cradle approach with three main principles [9]:

1. Waste is food, because everything is a nutrient for something else;
2. Use current solar income: energy that can be renewed;
3. Celebrate diversity: species, cultural, and innovation diversity.

Since 2010, the Ellen MacArthur Foundation has continuously been inspiring others to re-think, re-design and build a positive future through the framework of a circular economy. Multiple large companies have joined this journey and the butterfly shaped model for a circular economy with biological and technical materials is world famous [10].

The quest in literature to define what circularity implies in the construction industry, might start with a reference to Kirchherr, Reike and Hekkert and their analysis of 114 definitions of a circular economy [11]. These researchers were able to distinguish two types of approaches in defining a circular economy: a R-approach and a system approach. The R-approach concentrates, in order of favourability, around the strategies: refuse, rethink, reduce, reuse, repair, refurbish, remanufacture, repurpose, recycle and recover. In the system approach three levels of circular economy systems, where the transition needs to take place, are regarded, namely: at a macro, meso and micro-level. The macro-level considers a whole economic industry and the micro-level applies for example to an individual company.

Based on their sources and the applied coding framework, Kirchherr et al [11] *define circular economy as an economic system that replaces the 'end-of-life' concept with reducing, alternatively reusing, recycling and recovering materials in production/distribution and consumption processes. It operates at the micro level (products, companies, consumers), meso level (eco-industrial parks) and macro level (city, region, nation and beyond), with the aim to accomplish sustainable development, thus simultaneously creating environmental quality, economic prosperity and social equity, to the benefit of current and future generations. It is enabled by novel business models and responsible consumers.*

Considering that our research focuses on how building sites can be prepared circularly in the Netherlands, the efforts of the Dutch government to define this concept cannot be ignored. In 2018 the national Transition Agenda Circular Building of the Netherlands offered the construction industry the following translated description [12]: circular building means to develop, use and reuse buildings, areas and infrastructure without unnecessary depleting natural sources, without wasting the environment and without affecting ecosystems. It is a way of building that can be justified economically and that contributes to human and animal welfare. Here and there, at present and for the future.

### *3.2. Site preparation*

If we are trying to come to a sustainable built environment and a circular construction economy, it makes sense to also have a closer look at the process of site preparation regarding input, throughput and output. In introducing a three step method for sustainable land use by the name of Trias Toponoma, Entrop and Brouwers [6] mention that the space occupied by constructions used by the present generation should ideally not jeopardize the needs of future generations. Locations which are used for buildings, will only rarely and in a time-consuming process be turned back into their original natural state. Therefore, it is necessary to reflect carefully on whether a certain location should be used for construction activities at all. When a certain land plot is being prepared for new construction activities, the process encompasses multiple stages. The activities that make up a traditional construction preparation phase are, according to Segeren and Hengeveld [13]:

- Clearing the site: this can include fencing off the location, removing and replacing vegetation and diverting existing infrastructure (e.g. cables and pipes);
- Executing groundwork: it can encompass raising or excavating the site, placing soil in a depot for later usage, dredging ditches, digging trenches at the location of future (construction) roads, soil compaction and soil improvement works;
- Applying drainage or other forms of dewatering by means of temporary systems;

- Installing sewerage and other facilities for waste water and storm water;
- Constructing civil engineering works as part of the water management, road and sewerage system, for example, weirs, culverts, bridges, tunnels, pumping stations, etc.;
- Applying pavements: temporary roads are laid out and will be replaced during the development phase by definitive pavements;
- Creating green and recreational facilities: interventions such as the creation of open water are also part of the preparation phase. Smaller green areas including the planting of (new) trees, shrubs, etc. make part of the development phase;
- Installing cables and pipes: main systems for the various utilities are installed and will be connected to the buildings during their development phase.

The activities mentioned above are reflected in the STABU specifications (specification system for residential and utility buildings). However, two additional phases can be distinguished [14]:

- Installing construction site facilities: for example the placement of trailers and sheds, the provision of equipment and personnel and organizing the layout of the work site;
- Providing support and executing demolition works for current structures at the site.

According to [15], activities that are part of the phase where construction site facilities are installed, are:

- Installing temporary infrastructure: this includes access to the construction site, the construction roads on the construction site and the layout for available storage space;
- Placing temporary accommodations: this means that temporary housing for construction site personnel and management is placed and that storage space is created for materials sensitive to weather influences;
- Installing temporary installations: these are installations for security and surveillance purposes, temporary utilities in the form of electricity, phone lines and data lines, (potable) water and sanitary installations.

Before any work can be carried out on-site, a number of investigations must be carried out, which can include [16]:

- Ammunition research if the site is located in a former military area;
- WION - KLIC report in which information is collected about the presence of (underground) utility lines on and directly outside the site layout. Because this information is not always entirely reliable and other cables and pipes may also be present, test trenches must be dug to check the data available;
- Soil contamination testing to determine the presence and level of pollution;
- Groundwater research to determine its quality and level;
- Inventory of adjacent plots: when there are other constructions or infrastructures in the immediate vicinity of the site, it should be investigated whether these could be damaged by the foreseen construction work. Plans are necessary on how to cope with any foreseen risks;
- Trees and shrubs that need to be preserved, are measured and recorded in plans;
- To prevent the project from coming to a standstill when archaeological findings are made, a historical investigation beforehand is advised. Historical research is compulsory, when the site is located in an old town district;
- A risk inventory is drawn up, because unexpected things can always occur. This is used to estimate what might be in the underground and how to respond to it.

This part of the literature and document study made it possible to work out a flow chart of activities involved in preparing a site before constructing buildings, which is shown in figure 1.

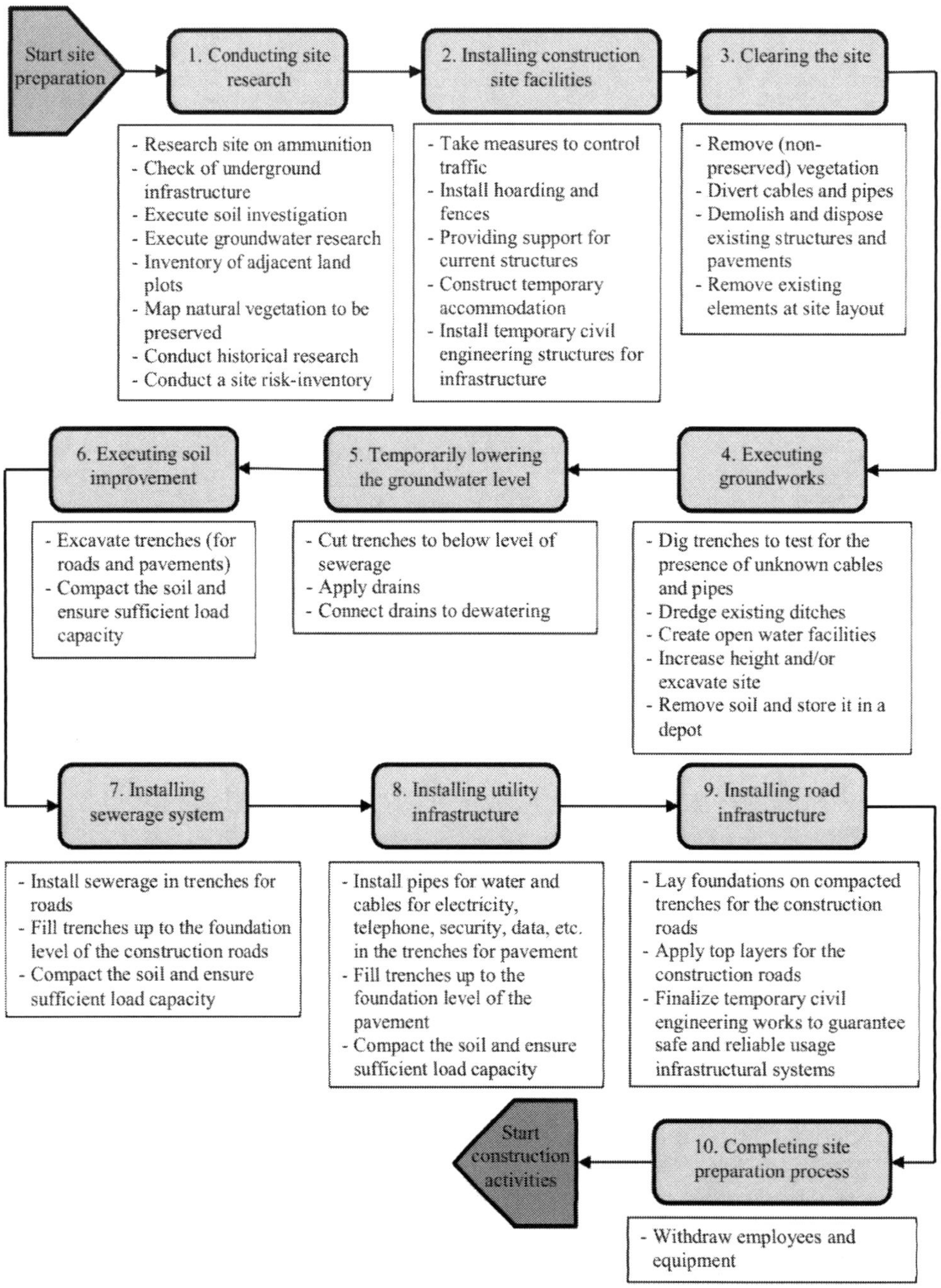


**Figure 1.** Flow chart of activities to prepare a site for (residential) construction activities [13-17]

To determine how circular the site preparation actually is, the measurement methods of circular constructing are transferred to site preparation. The input of a site preparation process consists, in line with [18] of land, materials, energy, and work. Materials need to be obtained, processed, transported, maintained and reused with the lowest environmental impact possible. Energy use needs to be reduced, to come from renewable sources and to be available on location in time as efficient and as effective as possible. The work involved in site preparation is done by employees and equipment (work and energy), which are deployed to execute using materials the activities on the site (land). Since the site itself is different for every project, the parameters that can be influenced and will be regarded are materials, equipment, and employees. Therefore, circular site preparation can be defined as follows: in circular site preparation, the site is seen as a system where materials, equipment, and employees are input influencing the level of circularity being subject to A) the R-approach: refuse, rethink, reduce, reuse, repair, refurbish, remanufacture, repurpose, recycle and recover and to B) the three step triad: reduce, renewable and efficient.

## 4. Designing the instrument

In the early days of sustainable building, developing and contracting entrepreneurs in the Dutch construction industry were asked to invest in sustainable measures with the help of long lists. These lists summed up all kinds of possible options to make urban plans and buildings (slightly) more sustainable. These lists were the Nationale Pakketen Duurzaam Bouwen (loosely translated in English: National Packages Sustainable Building). Currently, in these early stages of coming to a circular economy, the authors think that asking questions more or less in line with the setup of these former National Packages can help in making involved organisations more consciousness about the impact of their plans. Therefore, we decided to collect and structure questions to be asked to project developers at municipalities and at contracted or at to be contracted organisations in the preparation phase. The name of the instrument in which questions, answers and a rating system are brought together is PrCiSi, being a contraction of Preparation Circularity on Site.

### *4.1. Circular measures derived from interviewing experts*

As already was addressed in the section on the research method, a total of nine respondents were interviewed. The interview format used can be found in the appendix. Although four out of the nine respondents experience in the preparation phase barriers to circularity in forms of money and time, multiple opportunities were addressed. An important starting point is a process in which the site is properly analysed and an inventory is made of available materials. An online databank of locally available materials and products could help in the process of stimulating reuse. Multiple respondents have a scheme to accompany their vision of coming to a more circular site preparation phase, e.g. the Ladder of Lansink [19], metabolic scheme of Gladek [20], and a four step method, consisting of:

1. Reusing the materials available to close materials loops;
2. Preventing waste by using materials as economically as possible;
3. Using biobased solutions for the materials needed;
4. Construct flexible, adaptable and/or demountable, so that a building does not need to be demolished, but it can be adapted to future needs.

Regarding specific measures, individual respondents offered among others the following suggestions:

- The development of construction roads consisting solely of rubble foundations;
- A day for local residents to harvest or collect materials and products on site;
- Design constructions for the site that can later easily be disassembled.

### *4.2. Formulating questions for the instrument*

The input from literature and experts made us distinguish three categories of measures needed to come to a more circular economy in the preparation phase of building projects, namely materials, equipment and personnel. In the first version of our instrument the category materials consists of 46 questions, while in the categories equipment and personnel respectively 35 and 16 questions are being asked. Due to the high adoption rate of Microsoft Office, it was decided to construct the instrument in Excel.

IOP Conf. Series: Earth and Environmental Science **855** (2021) 012010 doi:10.1088/1755-1315/855/1/012010

### *4.3. Setting the scores in the instrument*

With almost one hundred questions in multiple categories, we felt the urge to make sure an easy overview can be maintained (see figure 2). Therefore, a first column indicates by using a mark and the colours green and red, if a certain question in the second column already has been answered or not. The answers to the questions can be selected from a drop down menu in the third column. Each individual answer corresponds with a numerical value in a fourth column. For this value the typical Dutch rating system is being used, which means that a 10 indicates an excellent score high in the R-approach with a focus on reducing the need for materials and a 1 stands for the worst possible outcome low in the R-approach. Depending on the specific question and the result of an accompanying literature study on that topic, it was defined what matches with which score exactly. This means that for materials 28, for equipment 12 and for personnel 7 different answer models were developed. In column five and six additional questions can be raised, explanations can be provided and proof needs to be given to substantiate the selected answer in the third column, before the score is taken into account for the overall category score. The user of the instrument or, so you wish, assessor will end up with three scores between 1 to 10, namely one score for materials, one for equipment and one for personnel. In this first version of PrCiSi every question in each individual category has the same impact in the overall category score. In the very last column an indicator shows at a glance the difference between sufficient and insufficient scores.

| | |
|---|---|
| Materials | 6.4 |
| Equipment | 4.0 |
| Personnel | 4.3 |

| | |
|---|---|
| Project: | Noorder Koeslag |
| Assesed by: | Anonymous |
| Date: | 25/08/2020 |

| Question completed yes/no | Question | Materials | Answer | Score | Additional question | Arguments, proof and substantiation | Score counted | Indicator insufficient score |
|---|---|---|---|---|---|---|---|---|
| | 1 | **Necessity of the need for** | | | | | | |
| v | 1.1 | Are crawl spaces going to be allowed or applied? | Yes | 1.00 | Why are crawl spaces applied or not? | *Unknowen* | 1.00 | insufficient |
| v | 1.2 | Is the soil at the project site contaminated? | No | 10.00 | What contamination is present at the project | *Unknowen* | 10.00 | |
| v | 1.3 | Is it possible to leave contaminated soil? | There is no contaminated soil | 10.00 | How can contaminated soil not have to be removed? | *Unknown* | 10.00 | |
| v | 1.4 | How is contaminated soil remediated? | There is no contaminated soil | 10.00 | | | 10.00 | |
| v | 1.5 | Is pipework going to be installed below groundwater level? | Yes | 1.00 | At what level should the piping be installed? | *Groundwaterlevel is lowered approximately 1.1* | 1.00 | insufficient |
| v | 1.6 | How will it be prevented that groundwater temporarily needs to be lowered? | Temporary lowering of groundwater cannot be prevented, source dewatering is used | 1.00 | | | 1.00 | insufficient |
| v | 1.7 | How is the material use for piping reduced or prevented? | No material reduction has been applied to the pipework | 1.00 | | | 1.00 | insufficient |
| v | 1.8 | How is the material use for the construction roads reduced or | No reduction in the material use of construction roads has been applied | 1.00 | | | 1.00 | insufficient |

**Figure 2.** Screenshot providing an impression of the PrCiSi instrument

## 5. Testing the instrument

In the municipality of Olst-Wijhe (with approximately 18,000 inhabitants) PrCiSi was tested by assessing two sites. The first site with a surface of 33,500 m$^2$, for 90 dwellings at maximum, was recently prepared and is known by the name of Noorder Koeslag. The second site still needs to be prepared for construction purposes and is called Olstergaard. This site covers an area of 35,000 m$^2$ and offers around 71 separate land plots to the housing association and for (collective) private commissioning.

The commissioned organisation that conducted the site preparation activities at Noorder Koeslag, was asked to make use of the instrument. This contractor experienced no severe problems in answering the questions except for rating the durability of the materials used for cables, pipes, civil construction works and the road infrastructure. When preparing the site, excess materials were made available for reuse. These materials were not reused on site, but did find a new purpose within the province of Overijssel. Materials needed for the roads in this project were also found within the province, but from a transport point of view it would have been better to collect them locally. The scores for Noorder

Koeslag were, as assessed by the contractor, a 6.66 for materials, a 3.7 for equipment and a 4.3 for personnel (see figure 2). These scores are the results of a great number of questions that could not be answered positively, because 43% of the questions in the category material were rated as insufficient, as well as 67% in the category equipment and a staggering 69% in the category personnel. Although little irrefutable proof to substantiate the answers was provided by the contractor, it was a positive experience for the researchers to find out that the contractor was willing to answer all questions, associated the questions asked to circularity and considered himself to be able to complete the instrument.

The test for Olstergaard could not fully be completed, because no specific contractor had yet been commissioned at the time the first version of PrCiSi became available. Because no contractor was commissioned, questions regarding the transport of materials, equipment and personnel could not be answered. Due to fluctuating ground water levels in the area, uncertainties exist in estimating how much ground water needs to be extracted and which method could be applied. Regarding material use the respondent expressed his doubts on reusing materials, when it comes to pipes, but also cables, their quality needs namely to assured. The scores for the categories equipment and personnel could not yet be assessed. The category materials scored a 8,1. This fairly high, but questionable score was possible, due to the relatively small material usage. In other words, not all infrastructure was installed yet, nor were infrastructural works designed or planned yet. The only possible solution here is to wait, before more information becomes available. All things considered Olstergaard was in fact not yet assessable.

## 6. Conclusion

In this paper the research process to come to a practical instrument to assess the rate of circularity of a site preparation process was explained. Although the authors are aware that among the readers of this paper questions might exist on how questions and answers in the tool are exactly formulated, conciseness is a great asset and only a Dutch version of the tool was developed and tested.

In this first version of PrCiSi every question in each of the individual categories has the same impact in the overall category score. However, testing the instrument and new insights might steer us in other directions, where the impact of individual questions is differentiated. With more sites to be tested, future research can also clarify more profoundly what can be regarded as insufficient, sufficient, good, excellent levels of circularity. Hence, this can set the pass mark for site preparation projects.

The instrument was originally developed to facilitate the sustainable property development process in Olstergaard. However, at the time a first (Dutch) version of the tool came available, not all necessary data could be provided. It is, therefore, recommended to revisit this case to re-explore the strong and weak points of the instrument and, of course, of the preparation phase for Olstergaard.

The authors are aware that the instrument still needs further development. Considering that granting high scores (marked 8 to 10) already for nowadays projects might not be the best incentive to stimulate developments towards a circular economy in 2050. Thus regarding verification, the instrument operates as expected, but the current way of computing might result in scores that overshoot the actual level of circularity. Further research is necessary to validate this instrument and the underlying model.

However, we are confident that providing information about the development of this instrument helps in raising circular awareness among those that are currently involved in site preparation. As Mulhall & Braungart [8] could phrase it: this first version of PrCiSi might make the current situation less bad.

## References

[1] Bruijn T de, Bults J, Engelsman L, Entrop B, Smit M, Straatman J and Vrielink R 2019 *Circulair bouwen; een transitieagenda voor Overijssel*

[2] Whole Building Design Guide Sustainable Committee 2014 *Sustainable*

[3] Ministry of Infrastructure and Environment and the Ministry of Economic Affairs 2016 *Nederland circulair in 2050; Rijksbreed programma Circulaire Economie*

[4] Hagen L 2020 *Komen tot een circulaire bouwrijpfase; handvatten voor een nieuwe ontwikkeling in het circulair bouwen middels de casus Olstergaard*

[5] Rikkert N, Jansen D and Morsink R 2020 *Het circulair bouwrijp maken van een (woning)bouwplan*
[6] Entrop A G and Brouwers H J H 2009 *J. Build. Apprais.* **5**, 293
[7] World Commission on Environment and Development 1987 *Our common future* (Oxford: Oxford University Press)
[8] United Nations 2015 *Transforming our world: the 2030 Agenda for Sustainable Development*
[9] Mulhall D and Braungart M 2010 *Cradle to Cradle criteria for the Built Environment* (Nunspeet: Duurzaam Gebouwd / CEO Media BV)
[10] Ellen MacArthur Foundation 2013 *Towards the circular economy; economic and business rationale for an accelerated transition*
[11] Kircherr J, Reike D and Hekkert M 2017 *Resour. Conserv. Recycl.* **127**, 221
[12] Nelissen E, Griendt B van de, Oppen C van, Pallada I, Wiedenhoff J, Waal J van der, Quist J, Engelsman L, Schaafsma M, Dreumel M van, Terwisscha van Scheltinga P, Broere P, Fraanje P, Mars P van der, Hoof S van and Bögl T 2018 *Circulaire Bouweconomie; transitie-agenda circulaire economie*
[13] Segeren W and Hengeveld H 1984 *Bouwrijp maken van terreinen* (Deventer: Kluwer Technische Boeken B.V.)
[14] Ketenstandaard Bouw en Techniek 2020 *STABU besteksystematiek voor de woning- en utiliteitsbouw*
[15] Flapper H 2005 *Jellema 12A Uitvoeren-Techniek* (Zutphen: ThiemeMeulenhoff)
[16] Centraal Bureau Bouwbegeleiding 2018 *Controleplan grondwerk bouwrijp maken*
[17] Pijpers I and Woude D van 2004 Jellema 1 Bouwnijverheid (Zutphen: ThiemeMeulenhoff)
[18] Bras-Klapwijk R M, Heijungs R and Mourik P van 2003 *Levenscyclusanalyse voor onderzoekers, ontwerpers en beleidsmakers* (Delft: Delft Academic Press)
[19] Lansink A 2017 *Challenging changes; connecting waste hierarchy and circular economy* (Nijmegen: LEA)
[20] Gladek E 2017 *The Seven Pillars of the Circular Economy*

**Acknowledgments**
The authors would like to express their gratitude to the Netherlands Province of Overijssel for providing financial support of the present research (Impuls Circulair Bouwen). We also would like to thank the respondents for their anonymous help.

IOP Conf. Series: Earth and Environmental Science **855** (2021) 012010 doi:10.1088/1755-1315/855/1/012010

**Appendix: interview format**

In July 2020, nine respondents were interviewed on circular site preparation. This appendix shows the interview format used.

| | | | |
|---|---|---|---|
| Name respondent | *< titles, initials, first name and family name respondent >* | | |
| Email address | *< email address respondent >* | | |
| Job description | *< respondent's job description >* | | |
| Name organisation | *< name of the organisation, where the respondent works >* | | |
| Work experience | *< relevant work experience of the respondent >* | | |
| Subject interview | Circular site preparation | | |
| Start time | *<##:##>* | End time | *<##:##>* |
| Date interview | *<##-##-2020>* | Date feedback report | *<##-##-2020>* |
| Location interview | *<due to covid19 regulations online meetings were held>* | | |

*A. Introduction*
As an introduction the first few questions were always related to completing the table with the personal and job characteristics, e.g.:

1. Which organisation are you working for?
2. How would you describe your job at this organisation?
3. What is your work experience regarding the preparation of sites?

*B. Definitions*

1. What does building circularly mean to you?
2. What does site preparation mean to you?

*C. Tender phase*

1. What are types of clients your organisation is working for?
2. Do these clients specifically ask for circularity or do you bring this subject up?
3. What could be improvements in the tender phase to execute the site preparation in a circular way?

*D. Circular site preparation*

D.1. Clearing the site

1. How can existing structures be handled at a site in the most circular way?
2. How can (non-preserved) vegetation be removed in the most circular way?
3. How can cables and pipes be diverted in the most circular way?

D.2. Executing groundworks

4. How can groundworks be executed in the most circular way?
5. How can contaminated soil be remediated in the most circular way?

D.3. Installing sewerage system and utility infrastructure

6. How can the groundwater be temporarily lowered in the most circular way?
7. How can the sewerage system be installed in the most circular way?
8. How can the utility infrastructure be installed in the most circular way?

D.4. Installing road infrastructure

9. How can the road infrastructure be installed in the most circular way?

D.5. Additional activities

10. Are there additional activities during the site preparation that could be executed in a circular way? If so, how could this be done in the most circular way according to you?

D.6. Withdrawing employees and equipment

11. How could equipment be deployed and withdrawn in the most circular way?
12. How could employees be deployed and withdrawn in the most circular way?

*E. General*

1. What are important laws and regulations that apply to site preparation? Do these laws and regulations have, in your opinion, a positive or negative impact on circularity?
2. How can municipalities facilitate the circular measures mentioned by you in section D?
3. When focusing on the maintenance of equipment, are there types of equipment we should not be using anymore during site preparation?
4. When focusing on the maintenance of materials, are there types of materials we should not be using anymore during site preparation?

*F. Completion*

1. What can in your opinion be interesting characteristics of and opportunities for a site to prepare it in a circular way?
2. What are in your opinion possible barriers to a site to prepare it in a circular way?
3. Do you have a reference site preparation project that offers information on quantities and materialization that can be used for this research project?
4. Do you have maybe any additional comments that are worth sharing in relation to this circular site preparation research project?

IOP Conf. Series: Earth and Environmental Science **855** (2021) 012011 doi:10.1088/1755-1315/855/1/012011

# Structural color coatings for high performance BIPV

R Habets[1,2], Z Vroon[1,2,3], B Erich,[1] N Meulendijks[1,2], D Mann[1,2] and P Buskens[1,2,4]

[1] The Netherlands Organisation for Applied Scientific Research (TNO), High Tech Campus 25, 5656AE Eindhoven, The Netherlands.
[2] Brightlands Materials Center, Urmonderbaan 22, 6167RD Geleen, The Netherlands.
[3] Zuyd University of Applied Sciences, Nieuw Eyckholt 300, 6400AN Heerlen, The Netherlands.
[4] Hasselt University (UHasselt), Institute for Materials Research (IMO), DESINe group, Martelarenlaan 42, 3500 Hasselt, Belgium

**Abstract.** Building integrated photovoltaics (BIPV) offer aesthetics and freedom of design for architects and home owners. This can accelerate implementation and free up new spaces for solar energy harvesting at building level, which is a necessary step towards a climate neutral built environment. Colored solar panels with high conversion efficiency and low cost price are an important development for large scale market penetration of BIPV. Here we report a solution processed structural color coating for solar panels and solar collectors. We show that virtually any color can be prepared, that the desired coating stack can be designed using optical calculations and that the exact color can be produced via a low cost solution process. Furthermore, we show that the light transmission for the colored glass plates is still very high, exceeding commonly used absorbing colors and enables very high solar cell efficiency. The colored PV panels have been tested in real environment and via accelerated lifetime testing for 3 years without any performance decline or degradation.

In the context of the Paris Agreement, and the European, national and regional objectives and associated legislation derived from it, the transition to a climate-neutral built environment is one of the main objectives [1]. In addition to reducing the total operating costs in buildings, on-site sustainable energy generation is key to realize net zero energy consumption. Here solar energy harvesting by photovoltaics is one of the most promising technologies to realize a self-sustainable built environment. Key aspects for architects, construction companies and home owners for widespread implementation of solar panels in buildings are freedom of design and aesthetics. Therefore building integrated photovoltaics (BIPV), where the solar panels are directly integrated in a building's façade, are an essential tool to offer architects a possibility to use essentially every part of the building surface to generate energy without impacting the aesthetics of the building [2]. A key aspect here is color. At the moment, most solar panels are usually dark blue or black in color. To offer differently colored solar panels is important for the acceptance and broad market introduction of BIPV, which is necessary to convince architects and consumers to integrate solar panels into building shells. There are several existing techniques on the market to realize colored BIPV products. One example is screen printing patterns of ceramic paints on a solar panels cover glass [3]. These materials derive their color from light absorption, which greatly reduces the electricity yield of the applied solar panels. Moreover, the applied colors are relatively dull. An option to reduce the impact of the front sheet coloration on the solar cell efficiency is by using an interference filter coating, which uses selective light reflection to create coloration [4]. The principle of selective reflection is based on nature. Beetles use this principle, for example, to achieve a metallic green color (Fig. 1). This leads to more sparkling (metallic) colors, and at the same time to much less loss of light entering the panel [5].

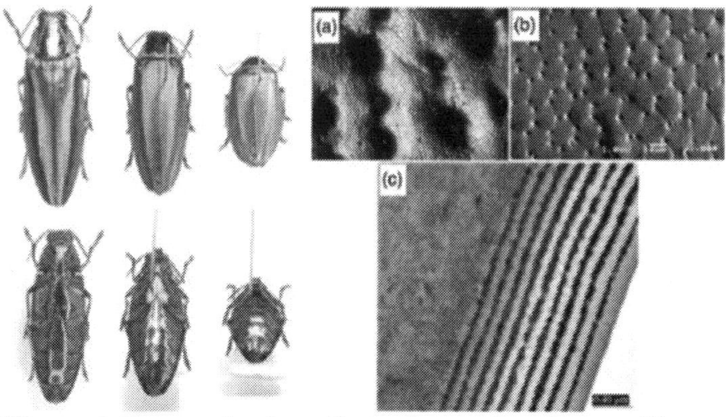

**Figure 1** Structural color effect using an interference filter as a functional coating, based on green beetles in nature (reprinted with permission from [4]).

To achieve selective reflection, so-called interference filters will be used as a functional coating. Interference filters are multi-layer coatings, which alternately have a high and low refractive index, and are all between 30 and 200 nm thick. By properly setting layer thickness and refractive indices, all possible colors can be made with one and the same coating liquid, including architecturally relevant colors such as terracotta and light gray. The concept of using interference filter coatings for coloration of PV and solar collector panels has been reported by Schüler et al. [5-7] and is currently on the market sold by SwissINSO [8,9].

Another important aspect for colored PV is cost price and embodied energy. Currently multi-layer interference coatings for coloration of solar panels are produced by magnetron sputtering. This technique is commonly used in the glass industry, e.g. for production of energy-efficient window coatings. But the technique is cost and energy intensive, which increases the consumer costs of the colored PV and the embodied energy. Therefore a cheaper, less energy demanding production process is necessary to improve the energy balance and cost price of colored BIPV.

Sol-gel processing is a proven technology to produce highly damage resistant optical coatings for outdoor applications. DSM sells sol-gel derived coating processes for the production of anti-reflective coatings onto PV panels [10]. Schott and Prince Optics produce all kind of optical coating products for outdoor applications that are produced by sol-gel/dip-coating technique [11]. Estimated production costs for 50-100 nm single layer sol-gel coatings on a large scale are 3 €/m$^2$ depending on the material costs and the production volume. The two main benefits of sol-gel/dip-coat processing is that very homogeneous coating systems in the range of 20-200 nm can be obtained. It is a technique that is used to produce very precisely optical lenses. For colored coatings it is necessary that homogeneous layers can be produced. Also the refractive index can be easily adjusted by mixing of components in the coating solutions. The second benefit is that for prototypes the scale up to sizes up to 1.0 x 2.0 m$^2$ can be achieved relatively easy with low investment costs [12]. The sol-gel/dip-coating technique is a process that can be described in three main steps [12].

- Pre preparation. In the first step coatings solutions are prepared by mixing alkoxides with solvents and the substrates are cleaned.
- Coat process. In the second step a cleaned glass substrate is placed in a dip-coat machine and brought in the coating solution. The substrate is taken with an adjusted withdrawal speed from the coating solution.
- Cure process. After the coating step the coating is dried and cured.

For design purposes of the multi-layer interference stacks the thickness dependent optical constants and with them the transmission and reflection of multi-layer stacks can be calculated [6,13]. Via reverse engineering a required coating stack with material and layer thicknesses can be calculated to reach a desired color. For a proof of concept, we selected three commonly used colors, i.e. yellow green, tomato red and signal blue. We simulated 5 different coating stacks made of 3-5 alternating layers of $SiO_2$ and $TiO_2$, which resulted in reflection and transmission spectra shown in Figure 2. With these coating stacks the three selected colors could be achieved.

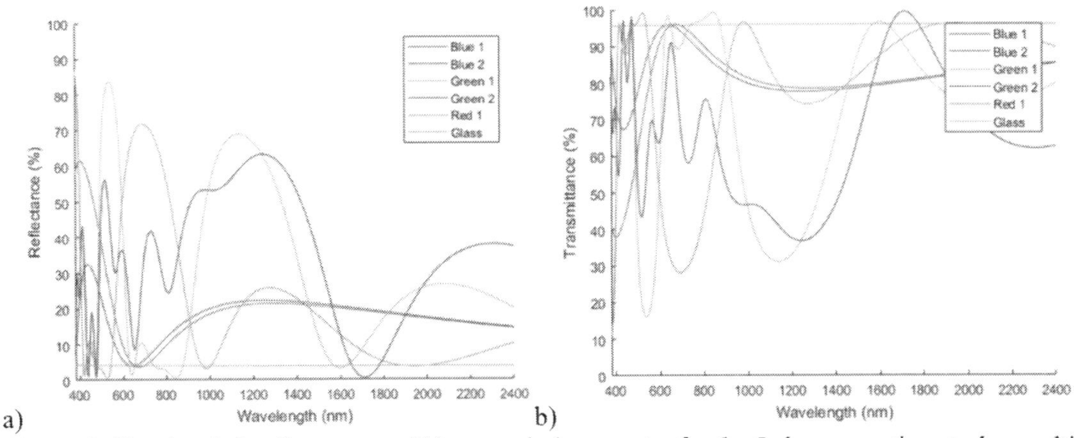


**Figure 2.** Simulated a) reflectance and b) transmission spectra for the 5 chosen coating stacks resulting in blue, green and red colored glass.

For experimental validation we selected two coating stacks representative for a blue and red color respectively. We used dip coating to prepare the alternating $SiO_2$ and $TiO_2$ coating stacks at the desired thicknesses resulting in the colored coatings predicted via simulations. Hereby we were able to coat 25 x 40 $cm^2$ glass plates with a red and a blue colored coating (Figure 3a,d). From Figure 3a and b it is clear that the blue sample is blue and the red sample is clearly red. The red sample is close to the terracotta color. In Figure 3b,e and Figure 3c,f the transmission and reflection data of the blue and red sample are given in the wavelength range 300-2400 nm. The reflection and transmission have been measured on five spots on the sample (for both samples), the curves are overlapping each other totally. This is an indication that the samples have very homogeneous optical properties.

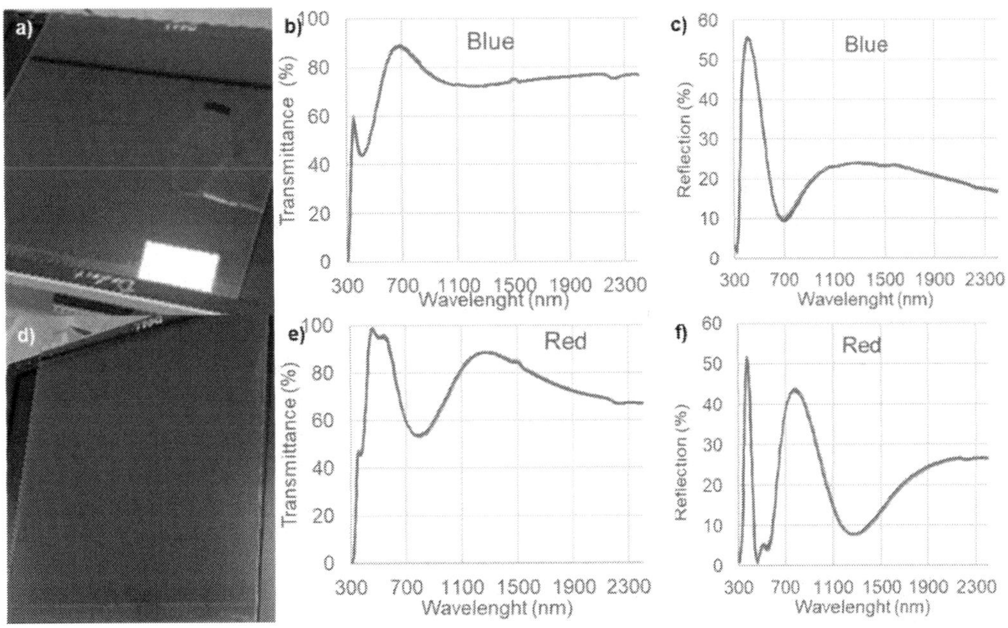

**Figure 3.** Appearance (a,d) and UV-vis-NIR transmission (b,e) and reflection (c,f) spectra of blue and red coated glass plates.

We compared the calculated optical properties with the measured optical properties of the manufactured sol-gel coatings. The reflection spectra of the blue and the red coating manufactured by sol-gel processing and by calculation were compared for the angles 0, 30 and 60°. Here a good comparison between both spectra could be shown. This means, that the optical calculation and manufacturing for the two prepared colors is quite in agreement. Furthermore we analyzed the color differences ($\Delta E$) between the simulated and the prepared blue sol-gel and red sol-gel coatings. For all angles the color differences between the prepared and calculated spectra was lower than 6, which is an exactable result. Additionally we calculated the efficiency of total light transmission for the colored glass plates in comparison to clear glass. Here both colors only showed a reduced light transmission of approximately 20% over the whole spectrum. Since the external quantum efficiency of silicon PV is in a wavelength region between 300-1100 nm where high transmission values are achieved, the total efficiency loss by use of the colored glass plates as cover for solar panels is expected to be even lower.

For demonstration purposes the cover glasses have been applied to custom-made PV panels (Figure 4) and have been tested on the roof of a test building for a period of more than two years. Both coatings and colored panels were found to be stable in these tests, and no decrease in electricity production over time was observed. Furthermore we analyzed the coatings durability and colorfastness in accelerated lifetime tests. Here we could validate the performance and durability of the coating for a period equivalent to 3 years. Additionally we analyzed the pencil hardness of the coating which resulted in a hardness of >9 H for the red and blue coating.

**Figure 4.** Colored coatings on cover glass of PV panels, integrated and demonstrated on the roof of a test building by TNO and Zuyd.

In conclusion, we showed that glass plates can be colored by use of an interference filter. The required coating stack can be designed using optical calculations and produced on glass plates via solution processed dip coating. This enables the low cost production of colored cover glass for solar panels. The structural color coatings can be used on BIPV panels for roof tiles, building façades and other structural elements, as well as on solar collector panels. BIPV panels with the structural color coating show high cell efficiencies and power output. The power output is up to 20% higher than for panels colored with traditional pigments. Furthermore we have analyzed the performance, durability and colorfastness of the colored PV panels in real environment and in accelerated lifetime tests. Here no degradation or performance loss could be observed in a period of over 3 years. The outside test is still ongoing.

**Acknowledgements**

The research work in this paper was carried out within Brightlands Materials Center, a joint research initiative between TNO and the province of Limburg. The work is funded by the Dutch government as well as the province of Limburg. Additionally, the work was partly performed within the EU Horizon 2020 project ENVISION under Grant Agreement No.767180.

## References

[1] Mardiana A and Riffat SB 2015 *J. Earth Sci. Climate Change* **S3** 001.
[2] Attoye DE and Aoul KAT 2020 *A Review of the Significance and Challenges of Building Integrated Photovoltaics*. In: Dabija AM (eds) Energy Efficient Building Design. Springer, Cham.
[3] ColorBlast® by Kameleon Solar. https://kameleonsolar.com/colorblast/ (accessed on 27 November 2020).
[4] Kinoshita S, Yoshioka S and Miyazaki J 2008 *Rep. Prog. Phys.* **71** 076401.
[5] Mertin S, Hody-Le Caer V, Joly M, Mack I, Oelhafen P, Scartezzini J-L and Schüler A 2014 *Energy and Buildings* **68** 764-770.
[6] Schüler A, Dutta D, de Chambrier E, Roecker Ch, De Temmerman G, Oelhafen P and Scartezzini J-L 2006 *Sol. Eng. Mater. Sol. Cells* **90** 2894-2907.
[7] Hody-Le Caer V, De Chambrier E, Mertin S, Joly M, Schaer M, Scartezzini J-L and Schlüter A 2013 *Renewable Energy* **53** 27-34.
[8] WO 2014/045144 Al
[9] WO 2014/045141 A2
[10] Buskens P, Arfsten N, Habets R, Langermans H, Overbeek A, Scheerder J, Thies J and Viets N 2009 *Glas. Perform. Days 2009* 505– 507.
[11] Lobmann P and Rohlen P 2003 *Glass Sci. Technol.* **76** 1.
[12] Sakka S 2003 *Handbook of sol-gel processing.*
[13] Sreemany M and Sen S 2004 *Mater Chem Phys* **83** 169-177.

# CO-CREATING SYSTEMIC CHANGES FOR A CIRCULAR ECONOMY IN/UNDER CONSTRUCTION, PRELIMINARY LESSONS FROM THE FLEMISH LIVING LAB ON CIRCULAR CONSTRUCTION

**Waldo Galle[1,2], Wim Debacker[2], Yves De Weerdt[2]**

[1] Department of Architectural Engineering, Vrije Universiteit Brussel (VUB),Pleinlaan 2, 1050 Brussels, Belgium

[2] VITO Transition Platform, Flemish Institute for Technical Research (VITO), Boeretang 200, 2400 Mol, Belgium

**Abstract.** Like any persistent challenge, the transition from a linear to a circular construction economy requires systemic changes. To divert from current practices that maintain the take-make-waste logic, we need new ways of doing and thinking within the value network of construction. In search for those enduring changes, a compass group advises the Flemish Living Lab on Circular Construction. This lab is a three-and-a-half-year project of study and experiment initiated by policy. During four participatory workshops, the compass group identified the most important system hurdles and necessary changes for achieving a circular construction economy, through subsequent steps of system analysis, translation and reflection. These outcomes have been synthesized and validated during the last workshop and are presented in this paper as a preliminary outcome of lab. Further, by bringing together forerunners of construction and co-creating together new knowledge, it is demonstrated how a well-guided compass group allows to bring into practice 'participatory system modelling'. By consequently translating the resulting insights into the requirements for a targeted call for experiments, it is finally shown that this co-created knowledge has been 'actionable' for the policy makers of Circular Flanders, and triggered dozens of partnerships to submit an experiment.

## 1. Context

### *1.1. The Flemish policy on Circular Building*

Since 2014 the policy program 'Material-aware construction through circular supply chains – a sustainable materials management prevention program for the construction sector 2014-2020' [1] has been aiming at an economy of closed material loops in the Flemish construction sector [2]. The program – an implementation of Flanders' Sustainable Material Management vision [3] – has been administrated by the Flemish governments' Agency for Public Waste, Materials and Soil, OVAM.

The program focuses on 5 innovation pathways; key domains through which it wants to make a difference with current practice including i. selective demolition and disassembly, ii. closing the loop of stony and iii. of other waste fractions, iv. environmental material performance of buildings, and v. design for change [1]. These domains not only consider how to manage materials, but take a critical perspective on what must be built, how it is conceived, and used [4]. The focus of the current Living Lab on both 'urban mining' and 'design for change', recently confirmed this perspective.

Yet today, new construction and renovation projects account for as much as half of the primary raw material consumption (including import) and one third of the annual construction and demolition waste generated in Flanders [5]. For this reason, circular choices – establishing service life extensions, effective reuse and efficient recycling – are still expected to reduce the current 37% of non-ETS greenhouse gas emissions [6]. According to the United Nations and the European Commission, greater material efficiency – including working in cycles – could even lead to savings up to 80% of material-related greenhouse gas emissions in residential buildings by 2050 [7], [8]. These figures make clear that the construction and real estate sector provides a relevant starting point for tackling climate change [9].

Therefore, the previous Flemish government approved in 2017 a starting memorandum for the 'circular economy' transition priority managed by Circular Flanders; i.e. the transition hub, inspirator and matchmaker for the circular economy in this Belgian region. The initiative is a partnership of public authorities, enterprises, civil society and knowledge institutions, supported by OVAM [10] and was recently confirmed by the responsible ministers of the new-elected Flemish government, stating that: "for the operation of Flanders Circular, from now on cross-policy domains and public-private will be deployed on the theme of circular economy" [11, p. 10].

In this paper, we review one policy initiative in this context: the Flemish Living Lab on Circular Construction. The aim of this review is twofold: while presenting as a preliminary outcome of lab the most important system hurdles and necessary changes that were co-created by compass group that is steering the lab, this paper also shows how this co-created knowledge proofed to be 'actionable' for the policy makers of Circular Flanders.

### *1.2. The Flemish Living Lab on Circular Construction*

The Living Lab on Circular Construction (in Dutch: Proeftuin Circulair Bouwen), is one of the youngest policy initiatives of the Flemish governments' Agency for Public Waste, Materials and Soil OVAM [4]. In its call for consortia, July 2018, OVAM wrote: "the ultimate goal of the Living Lab is – based on experiences in the field and on (systemic) research behind the scenes – to develop a policy and practice-oriented framework to accelerate the transition in the construction sector to a circular economy" [12]. Concretely, "the project team that coordinates the Living Lab and conducts related research must formulate recommendations to the relevant policy and practice actors."

The Flemish Institute for Technological Research (VITO), the Belgian Building Research Institute (BBRI), the Architectural Engineering department of Vrije Universiteit Brussel (VUB) and the department Government and Law of Hasselt University (UHasselt) jointly submitted their candidacy in September 2018 and have been selected for the research and coordination of the Living Lab in February 2019. Their design for and organization of the lab are characterized by a strong co-creation process supported by system analysis, and experimentation [4]. As requested in the call for consortia, the lab started in December 2018 with 'urban mining' as a first theme, and will continue from the spring of 2020 until the spring of 2022 with 'design for change' as a thematic focus.

The Living Lab sets out a framework to achieve a noticeable impact on the Flemish construction sector with a series of small and individual experiments, embedded in a larger transition arena. The lab has the ambition to become the focal point of knowledge and expertise on circular construction in Flanders, yet to make it as actionable as possible. It is the place where lessons from concrete projects are brought together and translated to others, and form the basis for policy recommendations. Therefore, the Living Lab also aligns its activities with two other initiatives (i.e. the Green Deal Circular Building and the Open Call projects) while offering space for more in-depth research and reflection [4].

## 2. Method

Like any systemic challenge, the transition from a linear to a circular economy, requires systemic changes; changes in what we do, how we organize ourselves and in what we value most [12]. These transitions are enduring and are expected to have a major impact on society. They are the result of economically, culturally, technologically, and socially mutually reinforcing developments and require public authorities, enterprises, civil society and knowledge institutions to play their part. More concrete, to divert from current practices that maintain the take-make-waste logic, new ways of doing and thinking by multiple actors within the value network of construction are necessary. All too often, actors point to each other – saying 'there is no demand' as well as 'there is no offer' – failing to see the bigger picture that connects them and grasp which enduring changes of their practices, structures and believes are necessary to transform together to a more sustainable equilibrium [13].

### *2.1. Compass group composition and workshops*

That is why, in the Living Lab, those actors have been put together. A key component of the Living Lab is its 'compass group' (Figure 1). In search for enduring necessary system changes, the compass group evaluates and advises the lab against the background of its changing context. The compass group is a driving force, as it helps determining which research issues should be addressed and which experiments must be carried out to further accelerate the transition to a circular economy in construction.

The group is composed of 16 forerunners, ambassadors and out-of-the-box thinkers of the sector with expertise or a strong interest in circular building, and vast knowledge of the sector's every-day practice. They include contractors, architectural designers, manufacturers, policy makers and researchers, and are supported in their reflection on the regional transition to a circular economy by the authors who guide, with expert knowledge about the topic, the group through a participatory co-creative process, also known as the 'network method' [13]. During four half-day workshops between September 2019 and May 2020, the compass group first finetuned its own tasks and drafted its guiding principles [14]. Subsequently, based on insights from their own everyday but diverse practices, they have been identifying together the most important system hurdles and necessary changes for achieving a circular construction economy through subsequent steps of system analysis, translation and reflection.

*Figure 1. The compass group, here at work during the third workshop (on 04/02/2020, Brussels) is a driving force of the Living Lab, as it helps determining which issues should be addressed and which experiments must be carried out.*

### *2.2. Participatory system modelling and Co-creation*

To understand what changes are necessary, a thorough system analysis can make things insightful. System analysis is a paradigm, interpreting what we observe not as individual events, but in the context of the network they are part of. It is a dynamic way of looking at the world, understanding phenomena in terms of functions and processes, and takes a holistic view, meaning that things are understood by putting them together rather than taking them apart [15].

Such analysis can be based on co-creative 'participatory system modelling'. First, bringing together matching skills and combined expertise are key to break through incumbent practices [16], [17] and easily explained as: "if you wouldn't need that input, you could have solved it on your own" [18]. Second, models of a system enable us to reason about, make sense of, and rethink our world in an abstract and comprehensive way. Both are brought together in the Living Lab.

### *2.3. Actionable knowledge and Experiments*

Yet ultimately, the objective is to translate the insights that are generated during a system analysis into concrete actions for those that can make a difference: the people on the field such as architects, contractors, manufacturers, project developers, investors, policy makers or citizens willing to play an active role in the transition towards a circular construction economy.

That translation happens when knowledge is made 'actionable'. The term actionable knowledge is often used within organizational management where it is defined as knowledge to initiate response to changes [19]. However, when the concept is applied to sustainability transitions, the impact of knowledge on each stakeholder its decision-making offers three more relevant qualifiers. They include "i. the logic of 'consequentiality' [for its] utilitarian value [...], ii. the logic of 'appropriateness' [for it] fits existing rules and routines, [and] a third decision-making logic, iii. namely the logic of 'meaningfulness' which considers at its core, the ideas of sensemaking and interpretation" [20].

However, when considering the complexity of the aspired transition it is almost impossible to develop 'solutions' to the identified problem. One way to overcome that uncertainty and initiate new sustainable practices is through learning-by-doing [21], [22]. Such an experimental approach creates an environment to test and fail, while learning further under real-life circumstances. The ambition to make possible such a learning cycle concludes the method of the Living Lab. The compass group has eight tasks that form together a continuous learning loop:

1) Collecting and sharing available lessons from practice.
2) Determining system bottlenecks/hurdles.
3) Defining underlaying causes; the major challenges for changing current practices.
4) Connecting the challenges with the available experimental space.
5) Follow-up of the experiments.
6) Translating new insights (also from research) into actionable knowledge.
7) Defining strategies to change practice(s).
8) Implementing strategies in (own) practice.

### *2.4. Implementation*

At the moment the Compass Group is more or less halfway the co-creation process and has been implementing various practical formats and frames, both simple and advanced ones. Some examples:

During the first and second workshop (06/09/2019, 7/11/2019), the facilitators and the authors of this paper asked all compass group members some straightforward questions; e.g. what bugs them each time they want to change current practice, which limits they bump into, or which opportunities they see. All their answers were listed and categorized according to the systemic iceberg model [23]. Besides events and trends, more structural issues and mental models became apparent; for example the lack of reliable information, and short-term profit being valued higher than qualitative craftsmanship.

During the third workshop (04/02/2020) we challenged the compass group by asking them which measures they would implement if they would take up the office of 'minister of circular construction'. We've asked them to relate these measures with output-parameters that will either rise or decline. By doing this exercise they acquired a basic knowledge of drafting causal loop diagrams. A causal loop diagram is one type of system models that shows and makes tangible the connection between effects and their 'origin'. At first answers were focusing on economic measures such as imposing circular targets through public procurement and taxing non-circular products. But when digging further into why such measures have not been working as expected so far, we discovered many unwanted feedbacks and the underlaying principles. By linking together several feedbacks into loops, systemic hurdles and lock-ins, as well as necessary changes and promising change measures could be made explicit [24].

## 3. Findings

### *3.1. Systemic hurdles*

Concretely, behind the easily cited demand-and-offer deadlock, five systemic hurdles and lock-ins have been identified, hindering or slowing down the transition of the Flemish construction sector to an economy of closed material loops:

The two first explain why the demand for circular building is limited:

1) Most decisions - and specifically investments - are based on a short-term cost-benefit considerations, fed by unbridled consumption and growing individualism; and
2) Few building clients and users are aware of, or find guarantees for, the positive effects of circular building on the economy, the environment (incl. climate) and society.

Three other hurdles greatly affect the supply-side of circular building:

1) Building practitioners rarely select unconventional, innovative solutions to avoid possible risks; a risk aversion which is fed by an unclear political vision and fierce competition;
2) There is also little trust and collaboration between the actors of the value network, resulting in a lack of insight in the whole life cycle of buildings and building products; and
3) The boundless diversity in construction and real estate make it difficult to replicate circular business models and make them valuable for entrepreneurs and society.

### *3.2. Necessary system changes*

With these systemic barriers in mind the compass group pinpointed to its best abilities and from its everyday knowledge of the Flemish construction practice the 'illnesses' of the conventional system and realized that traditional policy measures will have at the best only a marginal beneficial effect. Yet, they brought to focus what must be targeted in order to find an enduring 'cure'.

At the moment of our forth workshop (2020/05/29) however, another disease struck us all. In person co-creation was not feasible anymore as COVID-19 spread over our country too. But thanks to online calls and collaboration platforms such as Miro (Figure 2), we could look back at the causal loop diagrams and envisioned how a 'healthy' construction system must look like. From that vision and the identified hurdles it was possible to find consensus on three systemic changes:

1) FROM decisions based solely on financial benefits, the construction system must change TO long-term actions that are beneficial for the environment and society;
2) FROM a fragmented product chain, the construction system must change TO a highly intertwined value network; and
3) FROM limited awareness of its value, the system must change TO a broad understanding of why and how circular building supports a more resilient built environment.

IOP Conf. Series: Earth and Environmental Science **855** (2021) 012012 doi:10.1088/1755-1315/855/1/012012

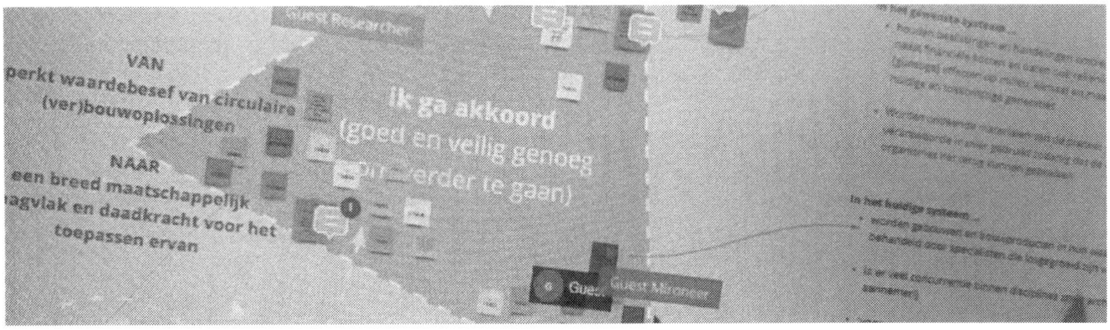


*Figure 2. Thanks to online calls and collaboration platforms such as Miro, the compass group could look back at the causal loop diagrams and envisioned how a healthy construction system should look like.*

### 3.3. Possible actions

Along with these systemic changes, 15 potential actions were co-identified as a means to break in into the current system of construction. During the fourth, as so far last workshop (2020/05/29), these leverages have been mapped and clustered by the members of the compass group according to their estimated impact and difficulty to implement. By doing so it was possible to set priorities; some action could be easily met, such as making life cycle costing mandatory in public procurement, or training students and professionals in the why, how, and what of circular construction. Other actions are more demanding, such as creating alternate financing for circular building solutions with added value for each stakeholder, or creating a quality framework for the management of building and building product data. After all, these last two leverage actions face quite some legal and economic barriers today.

### 3.4. Call for experiments

Yet, the co-identified actions are steppingstones to tangible, field-based experiments to demonstrate how barriers can be overcome and material loops will be closed. Monitoring these experiments will allow the compass group to acquire new insights that subsequently will be translated into an actionable transition agenda for policy and practice. No need to say that these experiments must go further than feasibility studies or the demonstration of technical solutions.

For this purpose, Circular Flanders and OVAM launched during the fall of 2020 a targeted call for experiments with a total budget of 2.8 million euros. These grants of up to 80% and 100,000 euro per experiment have been awarded by the end of that year to 29 of the 75 partnerships of local authorities, enterprises, research institutes and other organizations that submitted their ideas. Since experiments are defined bottom-up, sufficient motivation of the actors to successfully conduct the experiment and complete the associated reflection, are expected.

Following the identified hurdles, necessary changes and prior yet challenging actions, the call text set the expectations for successful experimental project proposals, including:

1) social and economic relevance and scalability of the project, i.e. the possibility of replicating the innovative concept on different projects or a larger scale;
2) the exemplary value of the project in terms of sustainability, urban mining and/or design for change in the broad sense (buildings, infrastructure,...);
3) quality of the partnership and roles, i.e. good specification of the added value and commitments of the various partners in the project;
4) the contribution to the learning, demonstration and dissemination of the principles of the circular construction economy in Flanders.

## 4. Conclusions and further work

By bringing together forerunners, ambassadors and out-of-the-box thinkers of the construction sector with expertise on circular building, and insight in the sector's everyday practice and co-creating together systemic hurdles, necessary changes and leverage actions, it is demonstrated how a well-guided compass group allows to bring into practice 'participatory system modelling'. By consequently translating these insights into the requirements for a targeted call for experiments, it is further demonstrated that the created knowledge has been 'actionable' for the policy makers of Circular Flanders and OVAM, and triggered dozens of partnerships who submitted for the call.

Further refinements and reflections are nevertheless indispensable. For example, whereas the first two hurdles are related to the building user, is group is unrepresented (or not represented directly) in the compass group, and a process and output monitoring system – i.e. to what extent the Living Lab contributes to systemic changes and realizes an effective reduction of the environmental burdens of the construction and real estate sector by closing material loops – is not in place yet.

Our journey is however far from finished. By defining, conducting and evaluating experiments which would not happen in the normal course of practice, the true possibilities and challenges of the circular construction economy are expected to become apparent. The concrete implementation of the bottom-up defined ideas in the field requires that technical and practical issues will be dealt with on the spot and will be answered almost instantly. In case of success stories, these experiments will be the first and best messenger of the outcomes for the broad field of stakeholders. In case of failure, these experiments will provided new insights on systemic hurdles, changes and actions.

## References

[1] OVAM, "Beleidsprogramma 'Materiaalbewust bouwen in kringlopen', het preventieprogramma duurzaam materialenbeheer in de bouwsector 2014-2020," Openbare Vlaamse Afvalstoffen Maatschappij, Mechelen, 2014.
[2] Ellen MacArthur Foundation, "Case studies: Flanders Materials Programme," 2019. https://www.ellenmacarthurfoundation.org/case-studies/belgium-flanders-materials-programme (accessed May 27, 2019).
[3] OVAM, "Vlaams Materialenprogramma," Openbare Vlaamse Afvalstoffen Maatschappij, Mechelen, 2012.
[4] W. Galle, W. Debacker, and Y. Weerdt, "The Flemish living lab on circular construction, from transition thinking to policy design," presented at the 4th International Conference on Public Policy, Montreal, Jun. 2019.
[5] M. Dubois and M. Christis, "Verkennende analyse van het economisch belang van afvalbeheer, recyclage en de circulaire economie voor Vlaanderen," Steunpunt Duurzaam Materialenbeheer, Leuven, 2014. Accessed: Sep. 18, 2014. [Online].
[6] BBL, "11 maatregelen voor een trendbreuk ik het Vlaamse klimaatbeleid," Bond Beter Leefmilieu, Natuurpunt, Greenpeace en WWF, Brussels, 2012. Accessed: Jan. 03, 2021. [Online].
[7] European Commission, "A new Circular Economy Action Plan for a cleaner and more competitive Europe," European Commission, 2020.
[8] IRP, "Resource Efficiency and Climate Change: Material Efficiency Strategies for a Low-Carbon Future," International Resource Panel. United Nations Environment Programme, Nairobi, Kenya, 2020. Accessed: Jan. 03, 2021. [Online]. Available: https://www.resourcepanel.org/reports/resource-efficiency-and-climate-change.
[9] Circular Flanders, "Gerichte Call Bouw 2020," 2019. https://vlaanderen-circulair.be/nl/aan-de-slag/gerichte-call-bouw-2020 (accessed Jan. 03, 2021).
[10] Circular Flanders, "The history of Circular Flanders," 2019. https://vlaanderen-circulair.be/nl/over-ons/historiek.

[11] Flemish government, "Mededeling aan de regering: een transversale werking voor de circulaire economie van Vlaanderen," Brussels, policy brief, 2020. [Online].
[12] OVAM, "Onderzoek en begeleiding proeftuin Circulair Bouwen." OVAM, 2018, Accessed: May 28, 2019. [Online].
[13] H. E. Wielinga *et al.*, "Netwerken met vrije actoren : stimuleren van duurzame innovaties met netwerken in de veehouderij," Wageningen UR, Wageningen, Jan. 2007.
[14] M. Q. Patton, *Principles-Focused Evaluation: The Guide*, 1st edition. New York: The Guilford Press, 2017.
[15] P. Vandenbroeck, *Working with wicked problems*. Brussels: King Baudouin Foundation, shiftN, 2012.
[16] F. Geels *et al.*, *Transitions to sustainable development : new directions in the study of long term transformative change*. New York (N.Y.) Routledge, 2011.
[17] E. Leising, J. Quist, and N. Bocken, "Circular Economy in the building sector: Three cases and a collaboration tool," *Journal of Cleaner Production*, vol. 176, pp. 976–989, Mar. 2018, doi: 10.1016/j.jclepro.2017.12.010.
[18] W. Galle, W. Debacker, De Weerdt, Yves, and N. De Temmerman, "Successful Collaborations in Transition Experiments, 14 guidelines from a value network perspective," presented at the 11th International Sustainability Transition conference, Vienna, Aug. 2020,
[19] J. C. Arnott, K. J. Mach, and G. Wong-Parodi, "Editorial overview: The science of actionable knowledge," *Current Opinion in Environmental Sustainability*, vol. 42, pp. A1–A5, Feb. 2020, doi: 10.1016/j.cosust.2020.03.007.
[20] A. Dewulf, N. Klenk, C. Wyborn, and M. C. Lemos, "Usable environmental knowledge from the perspective of decision-making: the logics of consequentiality, appropriateness, and meaningfulness," *Current Opinion in Environmental Sustainability*, vol. 42, pp. 1–6, Feb. 2020, doi: 10.1016/j.cosust.2019.10.003.
[21] R. Kemp, D. Loorbach, and J. Rotmans, "Transition management as a model for managing processes of co-evolution towards sustainable development," *International Journal of Sustainable Development & World Ecology*, vol. 14, no. 1, pp. 78–91, Feb. 2007, doi: 10.1080/13504500709469709.
[22] D. Loorbach and J. Rotmans, "The practice of transition management: Examples and lessons from four distinct cases," *Futures*, vol. 42, pp. 237–246, Apr. 2010, doi: 10.1016/j.futures.2009.11.009.
[23] Systems Innovation, "Iceberg Model," *Systems Innovation*, Aug. 02, 2016. https://www.systemsinnovation.io/post/iceberg-model (accessed Jan. 03, 2021).
[24] C. Goodier, S. Austin, R. Soetanto, and A. Dainty, "Causal mapping and scenario building with multiple organisations," *Futures*, vol. 42, no. 3, pp. 219–229, Apr. 2010, doi: 10.1016/j.futures.2009.11.007.

## Acknowledgments

The authors wish to express their gratitude for this learning opportunity to Circular Flanders and the responsible ministries. Special gratitude goes to the active members of the compass group: Almut Fuhr, Bernard Vanheusden, Brigitte Mouligneau, Christl Joris, Hilde Carens, Jeroen Vrijders, Joerie Alderweireldt, Jona Michiels, Jonas Voorter, Jonathan Verdonck, Karel Lootens, Petra Ronda, Philippe Van De Velde, Pieter Walraet, Roos Servaes and Sye Nam Heirbaut. Further we want to thank our colleagues and research partners for their support in this project: Romina Le Piane, Anne Paduart, Jef Bergmans, Aline Vergauwen and Niels De Temmerman.

IOP Conf. Series: Earth and Environmental Science **855** (2021) 012013 doi:10.1088/1755-1315/855/1/012013

# The Circular Retrofit Lab: a multi-disciplinary development of a building envelope according to circular design qualities

**Charlotte Cambier[1], Jeroen Poppe[1], Waldo Galle[1,2], Stijn Elsen[1,3], Niels De Temmerman[1]**

[1] Department of Architectural Engineering, Vrije Universiteit Brussel, Pleinlaan 2, 1050 Brussels, Belgium
[2] Transition Platform, Flemish Institute for Technical Research (VITO), Boeretang 200, 2400 Mol, Belgium
[3] Kaderstudio, Koolmijnenkaai 30-34 bus 6, 1080 Brussels, Belgium

**Abstract.** From 2016 until 2019, the Circular Retrofit Lab (CRL) was realised within the H2020-funded project Building As Material Banks (BAMB). The CRL was an experimental pilot project concerning the refurbishment of eight student housing modules in Brussels, Belgium. The lab aims to demonstrate how to implement circular building design in a retrofitting project. This paper discusses the design choices and the outcome of the CRL's building envelope. That envelope consists of a modular façade system of prefabricated panels. The innovative, adaptable and reusable building solutions applied in the CRL have been developed in collaboration with various construction stakeholders and included, exceptionally in the design and construction process, also industrial partners. In this study, we go deeper into the impact of the multi-disciplinary approach on the resulting implementation of a circular building envelope. Through a reflexive learning-history workshop, we can show how the involvement of various actors and unconventional circularity requirements increase the complexity of the project, but also how they positively impact the reactivation of existing buildings and proved to be a lasting learning opportunity to all partners involved.

## 1. Introduction

The construction sector, identified as one of the largest producers and consumers of respectively waste and natural resources, has a detrimental impact on the environment [1,2]. The increasing requirement of housing and other buildings due to the economy growth and the high growth rate in population have created a large demand for conventional building materials [3]. This enormous ecological footprint is therefore increasingly recognised as a serious concern and has made sustainable construction a priority in both political and industrial actions [4–6]. This has led to the introduction of the concept of circular economy, where the use of new resources and the production of waste is minimised by closing material flows [7,8]. However, the implementation of circularity in the construction sector is a challenging issue [9], as it requires radical changes at different levels: from organisational changes within the sector's supply chain to new design methods [9–11]. This complexity has as a result that there are, until today, few case studies built where circular principles have been applied. This situation urges for applicable solutions to fasten and upscale the transition towards a circular construction sector.

Within the H2020 funded project Building As Material Banks (BAMB), the Circular Retrofit Lab (CRL) was realised. This experimental pilot project concerns the refurbishment of eight student housing modules (Figure 1) out of about 350 modules built in the sixties at the university campus of Brussels (Vrije Universiteit Brussel) in Belgium. The lab aims to demonstrate how to implement circular building design in a retrofitting project and to bridge the gap between innovative circular building solutions and the daily construction sector.

**Figure 1.** The Circular Retrofit Lab after building completion. Photography: Simone Valerio

The CRL project illustrated how and to what extent reversible building design can coexist with or even lead to qualitative architecture, incorporating not only technical aspects but also user comfort, energy performance, and architectural value. The reversible building solutions applied in the CRL have been developed in collaboration with various construction stakeholders and included, exceptionally in the design and construction process, also industrial partners. To be able to upscale this experiment to the other 350 student houses on the university campus, and ideally to other renovation projects in Brussels and its surrounding area, it should be investigated what the positive and negative experiences were of the various partners.

It should be acknowledged that in previous years manifold of studies have observed, analysed and collected various challenges that construction stakeholders encountered while attempting to integrate circular principles during the building's design or construction phase [9,12–15]. Many challenges are related to finance; the most mentioned are the high(er) investment costs [12,13] and the high labour costs [14]. Furthermore, challenges linked to the nature of the construction sector are numerously mentioned, and these studies also highlight challenges related to reversible building design, reuse and recycling, and policy and regulations.

However, the CRL project was different from a building-as-usual project. It was an experimental living lab, where open innovation and collaboration from the beginning of the design process was encouraged. Even if many of these identified challenges can be related to the CRL's design and construction phase as well, an experimental environment allows testing and learning at low risk and helps to implement circular solutions [16,17] at larger scale. After all, implementation of CE in buildings is still in its infancy stage [12,18] and comprehensive studies on a complete renovation, where the project's aim is to maximise the implementation of CE principles and where an early and multi-disciplinary collaboration is mobilised, are limited. Most studies are theoretical and conceptual [19] or focus on construction waste minimization and recycling [11,20,21]. Consequently, there are little insights on how circularity thinking can be integrated into the decision-making during a building's design process [22,23]. In this study, it is researched how the early multi-disciplinary approach and the aim of the CRL project to integrate reversible design principles has influenced the design choices and the resulting building envelope. Following questions were set up:

- What design choices have been made differently than in a business-as-usual project?
- What are the underlying considerations of these design choices?

IOP Conf. Series: Earth and Environmental Science **855** (2021) 012013 doi:10.1088/1755-1315/855/1/012013

- What were the stakeholders' experiences and insights and what did they learn from this process?

## 2. Method

The different, or uncommon, design choices and their underlying considerations were derived from a design process analysis. The design drawings, construction pictures and BAMB reports were reviewed and the choices were discussed with the CRL partners. To be able to grasp which learned lessons, apart from the physical built lab, stakeholders will take with them and implement circularity in future projects, we organised a reflexive learning-history workshop with the researchers and industrial partners (on 5 September 2019). From this discussion during the workshop, a timeline map was developed, wherein the positive and negative experiences and insights that were observed are mapped.

## 3. Design of the Circular Retrofit Lab's envelope

Before identifying and analysing the uncommon design choices, the design of the CRL's facade is described. The modular concrete structure of the authentic student house was retained and reused as structure. On top of this structure, prefabricated, modular, demountable and reusable facade panels were mounted. The modules consist of a wooden frame filled with insulation, covered with a watertight layer and finished with a cement/wood fibre panel as inside finishing and a fibre-cement covering as facade cladding (Figure 2). The windows include aluminium window frames and triple-glazing glass. All elements are connected by reversible connections, and are thus easily demountable. [24]

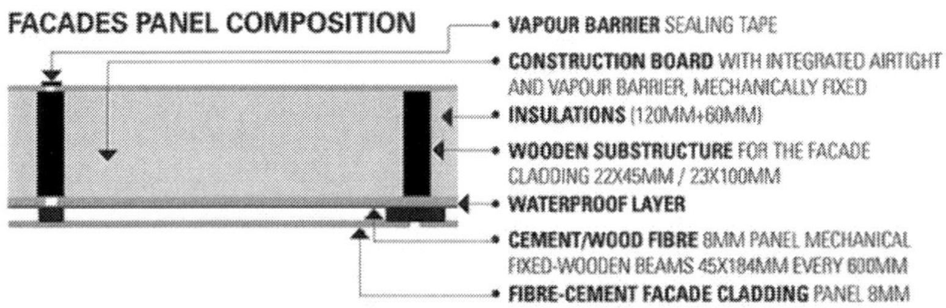


**Figure 2.** Composition of the façade panels [24].

To enhance the future reuse of the retrofitted student houses and of the modular facade panels, the designers used scenario planning to foresee different functions in the student houses [25]. Three types of facades panels were developed, each differing in presence and size of window openings. They have standardized sizes which allow them to interchange throughout the structure. As a result, the CRL is able to adapt to different uses by changing the configurations of the facade panels.

Because of the architectural heritage, the concrete structure must remain visible at the outside of the building. This posed challenges concerning the avoidance of thermal bridges and therefore the continuation of the insulation is provided at the inside of the structure.

## 4. Different design choices than in a business-as-usual building project

By analysing the facade design of the CRL more in detail, ten unusual design choices could be identified; design choices which are not common in business-as-usual building projects. These design choices were identified through discussions and by reviewing the construction pictures and the design drawings. These are available in the BAMB reports [24,25].

1. First, the reuse of the robust concrete structure and preserving its visibility was of main importance to keep in line with the initial design and architectural value of the student residences

(Figure 1). Moreover, the environmental impact (embodied energy) is lower compared to demolishing it and building a new structure [25].

2. Second, the facade panels follow the modularity of the concrete structure, and therefore they have the ability to be repeated over the 350 housing modules present on the campus. Having the same overall size but having dimensional margins allow them to interchange [24].
3. Furthermore, the facade panels, amongst other elements, were constructed in a demountable and reversible way. Besides reversibility being a circular design quality [26], LCA calculations supported the claim that both the cost and environmental impact was reduced [24].
4. In a usual project, the facade elements would have been glued, as in that case the visual value of the original student houses would have been considered as a priority. Also, most of the joints are easily accessible. This improves the potential to simplify maintenance and the installation and demounting process, without demolishing nearby elements.
5. Another uncommon design choice is to design the facade for different usage scenarios. Three panels types were developed, where each panel type has a different window opening [27]. The idea is that, when the student houses in the future transform into a dissemination space, a flexible workspace or an eco guest-house, the facade panels can be easily interchanged (and reused) to fit these new functions.
6. As materialisation for the prefabricated frame of the modular facade panel, joinery and wooden structure solutions were chosen. For the outer framework, wood has been selected based on its reversible properties, use of renewable materials and ease to prefabricate off-site. In these wooden frameworks, aluminium window frames were mounted. This structure solutions is rather uncommon but developed within the possibilities of the products of the partners.
7. Furthermore, the facade is covered with yellow-coloured cement fibre panels. The panel covering was selected in relation to different matters: reversibility, durability, compatibility with the visual identity of the student houses and consideration of the environmental impact. [24]
8. An additional uncommon choice is that the implementation and connection technique of the prefabricated facade panels is different for the east/west, the north and the south facade to illustrate their versatile implementation. The east/west facade panels are attached within the structural concrete frame using L-profiles and bolts. At the north facade, the prefabricated panels are bolted onto the concrete floor [24]. At the south facade, the prefabricated elements are installed in between aluminium profiles in a curtain wall system.
9. The architectural value of the concrete structure had as a consequence that the facade panels were installed inside the concrete structure instead of at the outside. This demanded special attention to fulfill the energy efficiency requirements and avoid the potential of indoor relative humidity impact. Not all thermal bridges could be avoided but the continuity of the insulation on the interior side [28] was maximised to optimise the building performance. [24]
10. Lastly, two openings were added in the concrete structure to provide vertical service shafts. This intervention was necessary to make the rest of the ground plan flexible and in order to increase the number of possible future plan layouts.

## 5. Underlying considerations

By analysing the BAMB feedback reports and by looking back at the design and building process, we could identify three major underlying considerations: the architectural value of the student houses, the aim of the project to implement circularity principles and the early multi-disciplinary collaboration.

### *5.1. Architectural heritage value*

The architectural value and identity of the existing student houses was a special point of attention during the redesign process. Although the student houses are not protected heritage, the design team and infrastructure department of the university decided to maintain the unique and revolutionary perspective of architect Willy Van Der Meeren with regards to modularity and adaptability and to keep the colourful appearance of the original design. This consideration had mainly an influence on the preservation of the concrete structure and the design of the panel covering, which are design choices concerning visual aspects.

### 5.2. *Circularity*
The CRL aimed to demonstrate how to implement circular building design in a retrofitting project and to bridge the gap between innovative circular building solutions and the daily construction sector. The project allowed to test different theoretical and methodological insights on circular building design in a practical and applied setting. As such, the project illustrated how and to what extent reversible building design can coexist with or even lead to qualitative architecture, incorporating not only technical aspects but also user comfort, energy performance, and architectural value. This aim has as a consequence that the circular building aspect permeated almost all design decisions (Table 1).

### 5.3. *Early multi-disciplinary approach*
One of the aims was to develop the experimental pilot project in a practical context, working together with different stakeholders from the construction industry. Additionally, this collaboration was set up already in early stages of the design process. This intense cooperation between research and practice, together with the low-risk context, have stimulated the sharing of information and allowed different learning cycles.

**Table 1.** Underlying considerations of the uncommon design choices in the CRL project.

| | Design decision | Underlying consideration |
|---|---|---|
| 1 | Reuse concrete structure | Circularity and Architectural heritage |
| 2 | Modular facade panels + dimensional margins | Circularity (and cost) |
| 3 | Demountable facade panels/ reversible connection | Circularity |
| 4 | Easily accessible joints | Circularity |
| 5 | Reusable facade panels | Circularity |
| 6 | Wooden and aluminium facade panels | Circularity and early multi-disciplinary collaboration |
| 7 | Panel covering | Circularity and Architectural heritage |
| 8 | Different connection techniques of the facade modules | Circularity and early multi-disciplinary collaboration |
| 9 | Insulation at the inside | Architectural heritage |
| 10 | Openings in concrete structure | Circularity |

## 6. Reflexive history-learning workshop: Insights and experiences of the partners
The experimental and low-risk context of the CRL research project was an enabler to engage circular design choices. However, since the project ended in 2019, we wondered what the different stakeholders had learned from this design, collaboration and learning process. To explore this, a reflexive history-learning workshop was organised where the partners could express their experiences, insights and learned lessons. We mapped which positive and negative experiences the partners encountered and which insights came along the process (Figure 3).

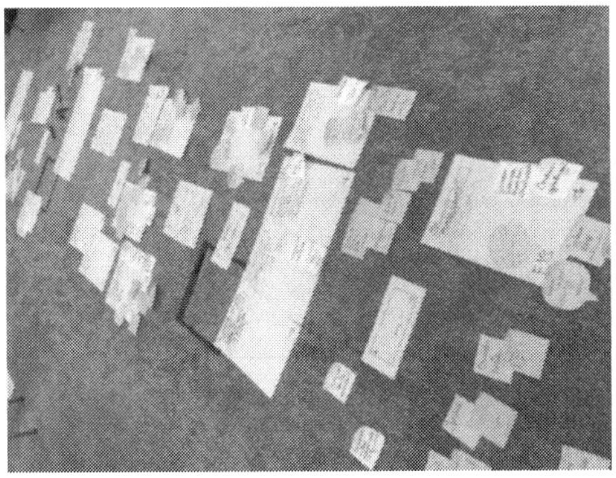

**Figure 3.** Mapping the positive and negative experiences and insights on implementing circular economy principles in the design and construction process, with a multi-disciplinary approach.

The polished version of the analogue timeline map is shown in Figure 4. The experiences and insights were categorised per year, from 2015 to 2020. The project itself was funded from 2016 until 2019, but the stakeholders also reflected on the preparatory process and the period after the completion of the renovation. This timeline map focuses on the design process of the CRL's building envelope. The green areas are the positive experiences, the red ones are the negative experiences and the blue ones are the gained insights. Remarkable is that the CRL project started with a lot of positive experiences and ended with mainly negative experiences. The positive experiences generally concern learning opportunities on circular building, establishing a network that shares the same interest and goals, and the possibility to take the learned lessons to future projects. The negative experiences were mostly due to time-pressure, mismatching or conflicting legislation and finance-related issues.

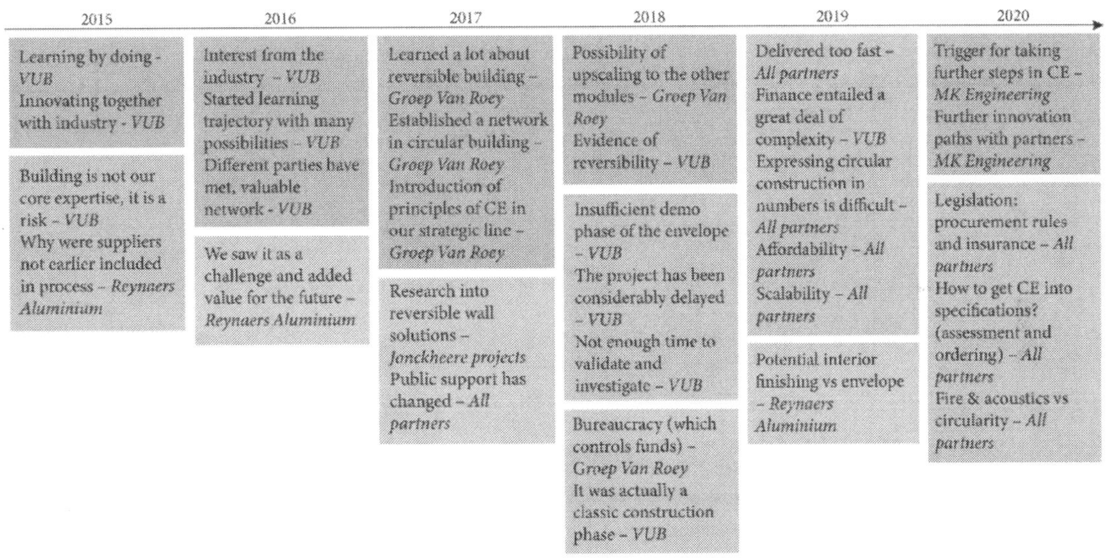


**Figure 4.** A timeline map containing the positive (green) and negative (red) experiences and the gained insights (blue) during the design of the CRL's building envelope, categorized per year and per partner.

## 7. Discussion

The encountered experiences and insights of the CRL partners (section 6) can be linked to the ten unusual design decisions and their underlying mechanisms (section 4 and 5). **The insights and (negative) experiences related to legislation and finance** confirm the main challenges that are described in the literature [9,13]. However, this study focuses on the influence of the early multi-disciplinary approach and the aim of the CRL project to integrate reversible design on final the design choices concerning the building envelope.

The experiences linked to the **early multi-disciplinary collaboration** show that it is considered as positive to establish a valuable network in circular building. Especially now that circular construction is still in its infancy, the various stakeholders saw it as an opportunity to learn from each other. It also had the advantage that much detailed information and technical knowledge, which usually is consulted during the last design phases, is available from the beginning of the design process (Figure 7) and which can be used to tailor the design. In this case, especially the materialisation and the connection techniques of the facade panels were a result of this intense collaboration.

The negative experiences linked to time-pressure ('Delivered too fast', 'Not enough time to validate and investigate', etc.) can be the result of several causes. They can be the result of the rising technical complexity by integrating circular design principles the project, such as reversibility. These principles are added to the already long list of design aspects such as acoustics, thermal comfort, fire prevention, etc. and other influencing factors such as uncertainties [29]. Another possibility is that because of the early collaboration, the first phases of the design process are extended [30] (Figure 5). Nevertheless, these time-related experiences should be further researched.

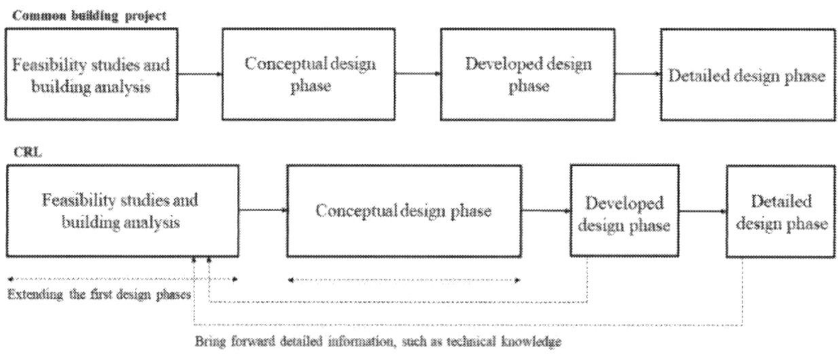


**Figure 5.** Early multi-disciplinary collaboration leads to better-informed design decisions in early design phases and results in extension of those phases (based on [30,31])

By completing this project, the participated stakeholders realise that circular construction is not insurmountable and that by aligning the goals of the various stakeholders at the start of the design process, a big difference can be made for the end result. After all, almost all connections are reversible and easily accessible and the facade panels are demountable and designed to be reused, while taking into account the architectural heritage of the building. This achievement led to that some partners of the CRL have taken concrete steps to review their business model by examining how they can incorporate maintenance, take-back or as-a-service contracts. Even if these business models were not applied in the CRL's design process, the project triggered the stakeholders to take further steps. Also sharing information, by e.g. focusing more on digitalisation and better preservation of information, has gained importance. Moreover, as also observed in other living labs [17], several partners plan to collaborate in the future to develop more innovative circular building solutions and projects.

## 8. Conclusion

The purpose of the study was to research how the aim of the Circular Retrofit Lab (CRL) project to integrate reversible design principles and the early multi-disciplinary approach has influenced the design choices and the design process. It was hypothesised that design choices have been made differently than in a business-as-usual project as the CRL was an experimental living lab, where open innovation and collaboration from the beginning of the design process was encouraged. Although only focussing on the building envelope, ten unusual design decisions could be identified. The largest part of these design decisions were made because of the aim to integrate circular principles. However, the impact of the multi-disciplinary approach on the resulting building envelope cannot be neglected. Through a reflexive learning-history workshop, we went deeper into this impact and reviewed the partners' experiences, both positively and negatively, and gained insights. This study has shown that the involvement of various actors and unconventional circularity requirements increase the complexity of the project, but also that it offers learning opportunities for all partners involved and that it stimulates to integrate circularity thinking in the decision-making process. These insights can be of value during the renovation of the other student houses at the campus. A limitation of this this study is that it has only examined the CRL as a case study. It would be interesting to examine other case studies to make more general conclusions on the impact of the early multi-disciplinary approach and of the circular aim on the design process.

## 9. References

[1] Romnée A and Vrijders J 2017 Circulair Bouwen: Naar een circulaire economie in de bouwsector
[2] European Commission 2011 Roadmap to a resource efficient Europe
[3] Behera M, Bhattacharyya S K, Minocha A K, Deoliya R and Maiti S 2014 Recycled aggregate from C&D waste & its use in concrete – A breakthrough towards sustainability in construction sector: A review *Constr. Build. Mater.* **68** 501–16
[4] European Commission 2020 EU Circular Economy Action Plan: A new Circular Economy Action Plan for a Cleaner and More Competitive Europe
[5] Demir Z 2019 Beleidsnota 2019-2024. Omgeving
[6] Circular Flanders 2019 Green Deal Circular Building
[7] Ellen MacArthur Foundation Circular Economy - UK, USA, Europe, Asia & South America - The Ellen MacArthur Foundation *Ellen MacArthur Found.*
[8] Schut E, Crielaard M and Mesman M 2015 Circular economy in the Dutch construction sector: A perspective for the market and government
[9] Adams K, Osmani M, Thorpe T and Thornback J 2017 Circular economy in construction: current awareness, challenges and enablers *Waste Resour. Manag.* **170** 15–24
[10] Hart J, Adams K, Giesekam J, Tingley D D and Pomponi F 2019 Barriers and drivers in a circular economy: the case of the built environment *Procedia CIRP* **80** 619–24
[11] Mahpour A 2018 Prioritizing barriers to adopt circular economy in construction and demolition waste management *Resour. Conserv. Recycl.* **134** 216–27
[12] Hossain Md U, Ng S T, Antwi-Afari P and Amor B 2020 Circular economy and the construction industry: Existing trends, challenges and prospective framework for sustainable construction *Renew. Sustain. Energy Rev.* **130** 109948
[13] Kirchherr J, Piscicelli L, Bour R, Kostense-Smit E, Muller J, Huibrechtse-Truijens A and Hekkert M 2018 Barriers to the Circular Economy: Evidence From the European Union (EU) *Ecol. Econ.* **150** 264–72
[14] Rios F C, Chong W K and Grau D 2015 Design for Disassembly and Deconstruction - Challenges and Opportunities *Procedia Eng.* **118** 1296–304
[15] Ghisellini P, Ripa M and Ulgiati S 2018 Exploring environmental and economic costs and benefits of a circular economy approach to the construction and demolition sector. A literature review *J. Clean. Prod.* **178** 618–43
[16] Bocken N M P, Schuit C S C and Kraaijenhagen C 2018 Experimenting with a circular business model: Lessons from eight cases *Environ. Innov. Soc. Transit.* **28** 79–95
[17] Andersson S and Rahe U 2017 Accelerate innovation towards sustainable living: exploring the potential of Living Labs in a recently completed case. *J. Des. Res.* **15** 234–257
[18] Lindgreen E, Salomone R and Reyes T 2020 A Critical Review of Academic Approaches, Methods and Tools to Assess Circular Economy at the Micro Level *Sustainability* **12** 4973

[19] Dokter G, Thuvander L and Rahe U 2021 How circular is current design practice? Investigating perspectives across industrial design and architecture in the transition towards a circular economy *Sustain. Prod. Consum.* **26** 692–708
[20] Ginga C P, Ongpeng J M C and Daly M K M 2020 Circular Economy on Construction and Demolition Waste: A Literature Review on Material Recovery and Production *Materials* **13** 2970
[21] Gálvez-Martos J-L, Styles D, Schoenberger H and Zeschmar-Lahl B 2018 Construction and demolition waste best management practice in Europe *Resour. Conserv. Recycl.* **136** 166–78
[22] Hildebrand L 2014 *Strategic investment of embodied energy during the architectural planning process* Ph.D. Thesis (Delft: TU Delft University)
[23] van den Berg M 2019 *Managing Circular Building Projects* Ph.D. Thesis (Twente, The Netherlands: University of Twente)
[24] Capelle T, Brancart S, Paduart A, Elsen S, Poppe J and Lanckriet W 2019 *Testing BAMB results through prototyping and pilot projects: Circular Retrofit Lab* (Brussels: Buildings as Material Banks)
[25] Henrotay C, Goens H, Durmisevic E, Beurskens P, Westerdijk R, Mul E-J, Paduart A, Elsen S, Brenner V, Mösle P, Schneider D, Özer M, Braun F, Keppler P, Heinrich M, Haberer M, Lang W, Debacker W, Vanassche S, Wang K and Hobbs G 2017 *D12 Feasibility + Feedback report: Testing BAMB results through prototyping and pilot projects* (Brussels: Buildings as Material Banks)
[26] VUB Architectural Engineering 2019 *Building a Circular Economy. Design Qualities to Guide and Inspire Building Designers and Clients.* (Brussels: Vrije Universiteit Brussel)
[27] Elsen S and Valerio S 2018 Final design detail facade panels BAMB CRL (DO_GEV COMP)
[28] Elsen S and Valerio S 2018 Final design detail components - column CRL (DE_H_01)
[29] Uotila U, Saari A and Junnonen J-M 2020 Uncertainty in the Early Phase of a Municipal Building Refurbishment Project—A Case Study in Finland *Buildings* **10** 137
[30] van Doorn A 2012 *het duurzame ontwerp project* ed M Lahr (Amsterdam: SUN)
[31] RIBA 2020 Plan of Work

**Acknowledgements**
This research was made possible thanks to the BAMB project, funded by the European Union's Horizon 2020 research program, and Fonds Wetenschappelijk Onderzoek (FWO) (grant No. 1S55518N).

IOP Conf. Series: Earth and Environmental Science **855** (2021) 012014 doi:10.1088/1755-1315/855/1/012014

# Legal instruments for the design of sustainable buildings

**A Kretschmer**, **M Zabek**, **K Tran** and **L Hildebrand**

Chair for Reuse in Architecture, Faculty of Architecture, RWTH Aachen, 52064 Aachen, Germany
akretschmer@rb.arch.rwth-aachen.de

**Abstract.** Currently the building sector is under high pressure: the lack of affordable housing in conurbations and the need to implement environmental performance strategies require concepts that incorporate user needs with planetary boundaries. Against the background of global population growth and a rising demand for resources, urban densification is regarded as a major global trend. The pressure on cities is already recognizable today and studies predict a growing tendency for the next three decades. This is a sensitive situation which provides not only the chance but also the urgent necessity to implement strategies to reduce climate change effects. Regulations are necessary to guide this process in the building sector. Timber construction as well as reused and recycled material show potential to be part of the solution. Against this background, the research deals with the status-quo of current legal instruments to design and construct sustainable buildings. This article provides an overview of the legal instruments in relation to sustainable building processes in the German construction industry with a focus on timber buildings. The relevant standards and guidelines for sustainable building design are investigated by literature review. This includes German Model Building Code (MBO), parts of the DIN standards, the German Building Code (BauGB) and the Building Usage Ordinance (BauNVO) as well as the Model Timber Construction Directive (M-HFHHolzR). The regulations are applied on a multi-residential building in Cologne and discussed whether regulations are enabler or obstacle for this specific project. The results of the study show the urgent need for more flexibility within current inner-city projects to pursue an environmentally friendly design and construction.

## 1. Introduction

### *1.1. Global Trends for Housing*

The United Nations Sustainable Development Goals (UN SDG) 11 - *sustainable cities and communities* name affordable housing for everyone as a standard while reducing the environmental impact of cities at the same time as one of the targets [1]. Against the background of global population growth and a rising demand for resources, such as land or raw material, urban densification is regarded as a major global trend. By 2050, 68 percent of the world population will live in cities [2]. The growing influx into cities and the trend for conurbations is a global phenomenon. Currently, 1 million dwellings are missing in Germany. The latest surveys predict a demand for 400,000 new apartments per year [3]. This leads to rising rents since capacities of affordable housing are already exceeded today. For example, in Berlin rents have risen by 80% in the last 14 years [4]. The inhabitants are subject to gentrification processes, which lead to homogeneous agglomerations in city centers. From a political point of view a sustainable, democratic society can only exist if no social classes are (spatially) excluded, making social housing a key topic. In 2016, the government adopted the Climate Action Plan 2050, making Germany one of the first countries to establish a long-term strategy as required by the Paris Agreement. Climate-friendly and resource-efficient construction is part of the strategy, but it also emphasizes the importance of social housing [5].

### *1.2. Environmental requirements for housing*

Within the EU in 2014, construction and demolition waste accounted for 33.5% of EU waste or 871 million tons [6]. The EU Construction Products Directive (CPD), drafted in 2011, sets out Circular Economy (CE) principles for the construction industry and calls for high-quality recycling of buildings. Annex 1, paragraph 7 includes requirements for sustainable resource efficiency and recyclability of buildings. None of the EU member states has yet transposed Annex 1 into national law and the status of the German regulation also remains that of a non-binding recommendation [7].
The UN SDG 12 *responsible consumption and production* aims to define goals to reduce mining of primary resources and waste generation through strategies for prevention, reduction, recycling and reuse [1]. Legal instruments have to provide definitions and procedures to be adapted but relevant German Codes do not focus on such objectives. Essential restrictions in the planning law of urban densification are primarily based on fire protection, but also include criteria such as stability, thermal, acoustic and visual protection, clearance areas to allow for lighting, traffic and accessibility. A major problem regarding ecological construction projects is the ever-increasing number of building regulations and standards, which may prevent the development of innovative alternatives.

## 2. Method and goal

The paper provides an overview of German building legislation in regard to environmental concerns for housing to support the integration of ecological parameters into the architectural design process. It discusses whether circular construction is possible under current legislation on the example of a case study. Knowledge of the standards helps navigate through the process and impacts design decisions. Relevant standards are named and are associated with the stages in the design process to which they refer to. First a literature review displays the different standards and categorizes them into mandatory and optional (figure 1). The standards are discussed regarding their feasibility on a case-study (figure 2). One construction strategy which was chosen in the case-study is the use of timber for the building structure and façade to reduce non-renewable resources. The case study is chosen as it represents a typical European situation of inner-city densification, an oblong plot between fire walls. The content of the standard is introduced, and possible design solutions are discussed.

## 3. Assessment

### *3.1. Standards, guidelines and certificates*

In the context of densification, three main sets of public building law apply: The Model Building Code (MBO), the Federal Land Utilization Ordinance (BauNVO), the Federal Building Code (BauGB) and parts of the DIN standards. The admissibility of projects under building planning law is regulated nationwide in the BauGB and the BauNVO. Nevertheless, the main part of decisions is the responsibility of the authorities of the federal states and their specific state building codes (BauO/LBO), which are based on the non-obligatory MBO. It divides buildings into five building classes (GK) which are specified by the height and area of the building and defines requirements for stability and fire protection. The higher the building class, the higher are the requirements for fire protection. In addition, there are further guidelines (VDI 4800) and certification systems for sustainable building (for example German Sustainable Building Council *DGNB* on building level or Environmental Product Declaration *EPD* on product level), which are not legally binding.

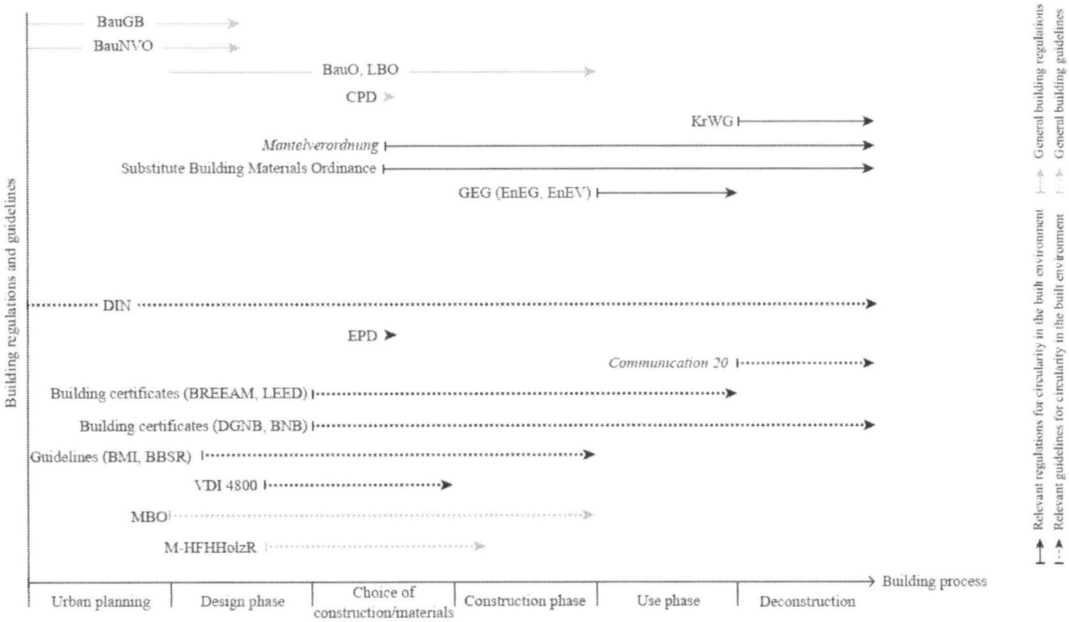


**Figure 1.** Guidelines in sustainable building processes.

**Figure 2.** Case Study, Cologne

### *3.2. The Case Study*

The case study is a six-storied building made from a timber construction in an urban context, in this case, Cologne. It is a fictional design with the idea to combine biotic material and computer-aided manufacturing technology. Further design principles include minimum consumption of land and environmental material selection as well as construction for disassembly. The legal instruments touched within the design and construction process of this case study are examined in the following. The building includes a total gross floor area of 2.400 m$^2$ on 6 floors with two interlocking atriums and a communal roof garden. The ground floor and the first floor are partly double storied and include communal, semi-public, and commercial used area, the remaining floors should serve as living space.

#### *3.2.1. Urban context*

According to the BauO of the federal state North-Rhine Westphalia, the top story of the case study is already at 17 m and thus no longer within the validity range of building class 4 (GK4). This could be improved, considering that a fuse to the highest floor in the event of fire could be up to 22 meters. Since building gaps in post-densification situations are already limited by two fire walls, building with lower demands on the lateral outer walls could be conceivable and is already permissible in several federal states.

#### *3.2.2. Construction and material choice*

For requirements in timber construction, the timber construction guideline (M-HFHHolzR) needs to be considered. The material cycle of timber is closed by nature itself, if harvested sustainably and not treated with pollutants. It can be either composted in smaller quantities or used in a cascaded utilization before undergoing an energetic use, making it a suitable material choice for sustainable buildings [9]. However, for buildings of GK 4 and 5, non-combustible insulating materials with a melting point of more than 1,000 degrees Celsius are currently prescribed in all building components, although cellulose fiber insulating materials in some cases meet fire protection requirements even better [8].

Unlike the design of the case study, the building regulations currently do not permit any visible timber components for GK 5, and load-bearing components in exterior areas must also be clad [8]. Fire protection currently takes place mainly by encapsulation of the building components, but could be achieved by sprinkler systems, which are hardly used in Germany despite their economic efficiency [8], or the ecological alternative of wood-immanent fire protection [9]. Facades have the building material requirement of "flame retardant", but combustible substructures are possible if there are no fire protection risks [8]. Non-load-bearing exterior walls, require less requirements (REI 30) up to the high-rise limit of 22 m. Since 2002, it has been possible for building class 4 (GK4) to use load-bearing components made of combustible building materials with fire protection cladding and non-combustible insulating materials. Timber load-bearing structures up to building class 4 have thus become possible without special approvals [8]. In this area, future adaptations of the regulations would be desirable, since insulating materials with A1 classification (mineral insulating materials such as expanded clay, foam glass or glass wool) have a lower potential in Circular Economy [9]. Current regulations set high requirements on the use of timber in building classes 4 and 5, making it a hurdle in the planning process and ultimately also in the transition into a CE.

#### *3.2.3. Use of recycled materials*

When reusing and recycling materials or products, EU regulations as well as German waste and construction law must be considered. The implementation of the EU Waste Framework Directive 2008/98/EC is reflected in the Waste Management Act (KrWG) 2012, which is binding as central federal waste law in Germany. However, it does not set any requirements for the quality and quantity of waste recycling or define specifications for the manufacture of new construction products regarding their recyclability. The KrWG is supplemented by legal ordinances, such as the Commercial Waste Ordinance, the Waste Recovery and Disposal Ordinance and the Landfill Ordinance. Further regulatory powers in the field of waste law are the responsibility of the state legislature [10]. Legal ordinances generally contain exceptions, such as the Industrial Waste Ordinance, which came into force in 2017. It requires specific waste to be kept separately. However, if separate collection of the respective waste fraction is *economically unreasonable* or *technically impossible*, these obligations do not apply. *Communication 20* of the federal states working group on waste (LAGA) is a set of rules designed to simplify the assessment of the environmental compatibility of recycled building materials in all federal states but has so far only been implemented in 11 of the 16 federal states in Germany. Since 2006, the federal government has been trying to create a comprehensive national legal framework for groundwater, building material substitution, landfill and soil protection, the so called *Mantelverordnung*. It is intended to uniformly regulate and tighten the recycling of building materials and the requirements for the production and use of mineral substitute building materials. Like *communication 20*, it is oriented towards civil engineering and does not cover the use of recycled building materials in building construction [11]. To date, there is no current legally binding version of it, but in 2017 a new Substitute Building Materials Ordinance (*Ersatzbaustoffverordnung*) was adopted by the Federal Cabinet, which includes the above-mentioned topics in a revised form. One innovation includes the definition of scenarios of the end of waste of materials [12]. Currently, recycled materials are still only accepted based on individual case approvals in the respective federal state. The standards of the DIN or EN norms regarding serviceability must be fulfilled without exception, even for reused products or materials [7]. As a rule, manufacturers do not recertify their products in order avoid warranty. Consequently, time- and labour-intensive tests offer little economic incentive.

## 4. Conclusion and Discussion

The German regulation allows buildings to be built with wood in a recyclable construction method. Circularity is provided if encapsulations can be removed non-destructively and no composite materials are used. Since in almost all of the construction projects carried out to date within the scope of application deviations have been applied regarding the reduction of the encapsulation class or the use of combustible insulating materials and solid wood components as supporting structure [9], it can be figured that updating the planning instruments, especially the regulation for wood M-HFHHolzR, is urgent. Certain requirements of the urban context like the BauO North Rhine-Westphalia are not fulfilled in the developed case study in a regular way, although special permissions are likely to be granted due to the increasing demand and the difficult development in case of densification. An adjustment of the MBO is therefore desirable. It can be stated that multi-story timber construction is possible in building class 4 (GK4), building with timber in building class 5 (GK5) has so far been difficult to carry out in Germany and requires special permission, although international examples have already proven its feasibility. Reusing building materials according to European Conformity (CE) marking principles currently involves legislative and administrative obstacles. Here, further relaxation of standards and acceleration of test procedures as well as more information on used products would be desirable. The paper derives the urgency to build more with renewable raw or recycled materials and to update planning instruments in such a way, that ecological buildings do not require special permits and cause too much planning effort.

## 5. References

[1] DGNV. *UN-Basis-Informationen 52. Die Ziele für nachhaltige Entwicklung*. Berlin, Germany: Deutsche Gesellschaft für die Vereinten Nationen e.V.; 2015. p 8-9.
[2] Department of Economic and Social Affairs. Population Division. *World Urbanization Prospects 2018. Highlights*. New York, United States: United Nations; 2019. (ST/ESA/SER.A/421).
[3] Oebbeke A. *Bericht von der Jahrespressekonferenz des Hauptverbandes der Deutschen Bauindustrie anlässlich des Tages der deutschen Bauindustrie*. Berlin, Germany; 2015.
[4] Hofmeister S. *Bezahlbare Wohnungen und ihre Qualitäten*. In: Detail. Cost-Effective Housing, 4.2018. München, Germany: DETAIL; 2018. p 1.

[5] BMUB. *Climate Action Plan 2050. Principles and goals of the German government's climate policy.* Berlin, Germany: Bundesministerium für Umwelt, Naturschutz, Bauen und nukleare Sicherheit; 2016. p 44.
[6] European Commission. *Measuring progress towards circular economy in the European Union – Key indicators for a monitoring framework.* Strasbourg, France: European Commission; 2018.
[7] Heisel F, Schlesier K and Hebel D E. *Prototypology for a circular building industry: the potential of re-used and recycled building materials.* In: IOP Conf. Ser.: Earth Environ. Sci. Graz, Austria: IOP Publishing; 2019. p 012023. (323).
[8] Cheret P, Schwaner K and Seidel A. *Handbuch und Planungshilfe. Urbaner Holzbau. Chancen und Potenziale für die Stadt.* Berlin, Germany: DOM publishers; 2014. p 154-160, p 164-170, p 114-130.
[9] Hillebrandt A, Riegler-Floors P, Rosen A and Seggewies J K. *Atlas Recycling. Gebäude als Materialressource.* München, Germany: Edition DETAIL; 2018. p 91, 67, 60.
[10] Deloitte. *Construction and Demolition Waste management in Germany.* 2015. (Resource Efficient Use of Mixed Wastes). V2
[11] BMUB. *German Resource Efficiency Programme II.* Berlin, Germany: Bundesministerium für Umwelt, Naturschutz, Bauen und nukleare Sicherheit; 2016.
[12] Bundeskabinett. *Verordnung zur Einführung einer Ersatzbaustoffverordnung, zur Neufassung der Bundes-Bodenschutz- und Altlastenverordnung und zur Änderung der Deponie-verordnung und der Gewerbeabfallverordnung.* Berlin, Germany: Bundesministerium für Umwelt, Naturschutz und nukleare Sicherheit (BMU); 2017.

IOP Conf. Series: Earth and Environmental Science 855 (2021) 012015 doi:10.1088/1755-1315/855/1/012015

# Upscaling non-residential bio-based circular construction in the United Kingdom

**B Dams[1], D Maskell[1], A Shea[1], S Allen[1], V Cascione[1], P Walker[1] and S Emmitt[1]**

[1] Department of Architecture and Civil Engineering, University of Bath, Claverton Down, Bath, BA2 7AY, United Kingdom.

**Abstract.** Non-residential circular construction projects using bio-based materials have been realised in the United Kingdom. Case studies include the Adnams Distribution Centre, the University of East Anglia's Enterprise Centre and the British Science Museum's hempcrete storage facility. The bio-based buildings utilise the natural properties of bio-based materials to insulate and regulate internal environments, particularly with reducing fluctuations in temperature and relative humidity, which can be harmful to sensitive stored products and artefacts. Projects have been successful on both on environmental and physical performance levels; however, they have not led to a subsequent proliferation of non-residential large-scale circular projects within the UK using emerging bio-based materials. This study examines why and uses analysis based upon exclusive interviews with key figures associated with bio-based case studies. Challenges faced include the ability to upscale production by manufacturers of bio-based materials, problems surrounding initial costs, gaining accreditation for materials, the vested interests present in the construction industry and levels of knowledge among clients and construction professionals. Potential upscaling solutions identified include long-term financial savings on running costs and high staff productivity, policies regarding grants, incentives and planning applications and local economic regeneration.

## 1. Introduction

A circular economy strives to eliminate waste product and reuse or recycle source materials upon completion of first use. This is in contrast to the linear economy model which ultimately involves waste product requiring disposal and new, increased demand for source material. The construction industry needs to play a significant role in reducing the use of virgin material, reducing the energy required in processing resources and furthermore a reduction in waste resulting from non-recyclable product [1]. Circular economy, in the context of the construction industry, represents the aim to construct in a sustainable manner, with renewable natural resources being re-used following the design life of a project; however, the application of circular principles to the built environment is currently not considered mainstream [2]. Bio-based materials are suitable for circular construction; they are renewable when sustainably and responsibly managed, and can be recycled, resulting in minimal waste and reduced demand for resource material [3]. Bio-based materials are used for thermal performance, sound insulation, thermal mass, indoor air quality and the prevention or reduction of condensation [4].

Large non-residential buildings using emerging, non-traditional bio-based materials have been realised in the United Kingdom (UK). However, despite on-budget delivery and success from environmental, operational and building physics perspectives, the realised projects have not resulted in a significant increase in large-scale projects, with circular construction procurement still being considered an emerging field [5]. The aim of this study is to evaluate the numerous challenges regarding the upscaling of circular bio-based construction and identify potential opportunities and approaches to promote upscaling. The contribution of the paper is to input into the wider Circular Bio-based Construction

Industry project [6] which seeks to further bio-based construction adhering to circular principles by means of creating guides, tools and platforms for construction industry professionals and clients, in order to disseminate knowledge of bio-based materials and advise how logistical and financial issues associated with the implementation of bio-based solutions can be addressed. Interviews were conducted with senior professionals who had first-hand experiences in the conceptualisation and implementation of large-scale non-residential circular buildings using emerging bio-based technologies, with a focus in this paper upon the use of hemp-lime. Issues are formulated into three core categories; finance, knowledge and policy.

### *1.1 Finance*

There is a perception that green building inevitably comes with an increase in capital cost [7] in an industry driven by finance [5]. Environmentally certified natural products can attract premiums in price [8]. Stakeholders and decision-makers involved with project funding typically focus upon initial costs and managing budgets [9]. Bio-based materials need to be a profitable option for construction industry professionals to have confidence in their use [10].

### *1.2 Knowledge*

The advantages of using bio-based materials in a construction project include hygroscopic properties (absorption of moisture from the air), a renewable supply chain and biogenic carbon sequestration within plant-based material [11]. Hygroscopic materials adsorb and desorb moisture, a property known as 'moisture buffering' [12], which attenuates peaks and troughs in humidity, leading to improved indoor environmental quality and potential for improved energy performance [11], [13]. Bio-based materials can sequester more carbon than is embodied in construction, therefore having a negative carbon footprint [14]. However, knowledge of beneficial hygroscopic properties of emerging bio-based materials may not be universally, or extensively, known by construction industry professionals [10].

### *1.3 Policy*

Policy-makers have a key role to play to promote sustainability in construction. Sustainable construction has been addressed at national and international levels; government policies have required planning applications to meet carbon targets [9], [15]. A tax on carbon in the construction industry is one option for punitive legislation [16] to promote circularity in construction and has been adopted in numerous countries, though the aim of reducing carbon emissions inevitably impacts upon the operational costs of the construction industry [7]. Bio-based subsidy measures are also in operation [17], but they can be focused upon energy and bio-fuels, with subsidy for bio-based materials limited to research and development [18]. There can also be adverse effects upon construction industry supply chains due to political events [19].

## 2. Methodology

Buildings can typically be considered a prototype [20] and this can be considered even more so with large-scale projects using emerging bio-based materials; projects are unique and cannot be considered numerous or mainstream. It was therefore decided to choose a select group of senior professionals for their extensive, first-hand experience in the realisation or development of pioneering bio-based construction projects (as summarised in table 1). Interviewees were selected so that insights could be gained from the perspectives of a private sector client, a public sector client, an academic and a policy-maker involved with case study projects in the United Kingdom. The core categories of finance, knowledge and policy formed the basis of adaptable questioning tailored to the interviewees' perspective. Interviews were recorded and transcribed by the authors. Analysis arising from responses to questions extending from the three core categories of finance, knowledge and policy is subsequently presented. Personal involvement on the realisation of a particular bio-based project commenced the interview process. Questions then progressed in a flexible manner on to wider views concerning the present climate of emerging bio-based materials within the UK and wider construction industry and thoughts on upscaling circular bio-based construction.

**Table 1**. UK Case study experience of interviewees.

| Interviewee | Case study experience |
|---|---|
| A | Adnams Distribution Centre, Suffolk |
| B | Enterprise Centre, UEA, Norfolk |
| C | Enterprise Centre, UEA, Norfolk |
| D | British Science Museum, Wiltshire |

*2.1 Case Studies and materials*

The case study buildings, shown in Figure 1, are the Adnams Brewery Distribution Centre near Southwold, Suffolk (private sector); the British Science Museum Hempcrete storage facility in Wroughton airfield, Wiltshire and the University of East Anglia's (UEA) Enterprise Centre in Norwich, Norfolk (both public sector). The buildings were selected as prominent examples of buildings designed to be circular (disassembly, reassembly, expansion and modification capabilities) and use of innovative non-traditional bio-based materials, locally sourced wherever possible.

**Figure 1**. Bio-based case studies in the United Kingdom. a) The Adnams distribution centre, Southwold, Suffolk; aerial view (top) [21] and view from the road. b) British Science Museum hempcrete storage building, Wroughton airfield, Wiltshire; world war two concrete hangar (top) and the hempcrete storage construction within a hangar [22]. c) University of East Anglia Enterprise Centre, Norwich, Norfolk.

Built in 2006, the Adnams Distribution Centre has hemp-lime block walls, the largest green roof in the UK and structural glulam beams (the longest in the UK at up to 41.6 m long and 2 m deep). Sequestered within the bio-based elements is an estimated 150 tonnes of $CO_2$ [23]. Hemp-lime was used to regulate the internal environment, reducing building operational costs and maintaining suitable temperatures for storing produce prior to national distribution. The British Science Museum's storage building was constructed inside an existing concrete hangar dating from the 1930s and features prefabricated dried hemp-lime walls within timber panels connected to a structural steel frame along with cross-laminated timber floors and wood fibreboard ceilings [22]. The purpose of hemp-lime was to reduce fluctuations in relative humidity (RH) and protect valuable historic artefacts. Hemp is a hygroscopic fast-growing plant and can be considered readily available in Europe and highly renewable [24]. Hemp-lime (also known as hempcrete) consists of shives from the core of hemp plant stems mixed with a binder made from air lime with pozzolanic, cementitious or hydraulic lime additions [25]. Hemp-lime can be sprayed or rendered in-situ while wet, or dried and pre-cast as compressed blocks, as used in the Adnams project [23], or in panels as used in the Science Museum store. Constructed on the UEA campus and opened in 2015, the Enterprise Centre has a glulam structural frame with studs and joists sourced from local timber, cellulose fibre insulation made from recycled waste paper, sprayed recycled paper ceilings and cladding solutions including locally sourced thatch and straw. Straw is known for excellent thermal insulation properties [26] and bales possess a negative carbon effect with each kg of straw able to sequester 1.35 kg of $CO_2$ [27].

## 3. Results

Interviewee responses are arranged into the three core categories of finance, knowledge and policy.

### *3.1 Finance*

An established barrier for bio-based construction is the initial higher cost; however, responses showed there are financial savings to be made over the course of the design life of a building due to lower operational running costs. Extending economic viability throughout the supply chain is also an important factor. Non-food crops for bio-based construction needs to be grown locally and be part of a coordinated program to boost regional economies, using local businesses, trades and crafts. Contractual and risk issues may result in circular or bio-based elements being value-engineered out of projects. A major economic issue for emerging bio-based materials and technologies is upscaling production and whether materials can be supplied in the quantities required for large-scale construction. An important step towards upscaling is gaining certification and accreditation, but there are potentially large costs involved. Data and certification resulting from research is important for the confidence of clients, contractors and insurers to work with bio-based materials. The construction industry can be described as conservative and risk-averse. Large, powerful companies working in conventional materials may be financially motivated to maintain the status quo.

> *There was considerable discussion around initial cost, and we estimated it was 20-25% more expensive than doing a conventional build. But it's a long term perspective; a typical warehouse of equivalent size would be running heating in winter and cooling in summer, but we have none of those expenses - we're saving six figures a year in energy costs. Construction companies at the time, and I suspect it's true today, will look to value-engineer environmental and sustainability aspects out of the project.* (Interviewee A)

> *In the civil service, we were looking at ways to make non-food crops profitable. An important point was that farmers and people in rural areas could make a profit. We*

*are very good in the UK at doing research, but we need to jump from the research to making the product; scaling up is a big problem. You can't promote a product if there isn't enough of it to be sold. We wanted to build data on bio-based materials. People could not get insurance for their materials, that's why the government gave finance for testing. The construction industry is conservative; the plastics, concrete and steel industries all have big lobbying groups. Biobased industries do not; they are disparate and small. It is hard for new innovators to push through with new products, even though we've put lots of finance into the research.* (Interviewee B)

*The process is driven by cost and not by environmental, aesthetic or human factors. The whole cost consultancy community is telling me we can't do timber-framed and modular pre-fabricated business unit designs on a comparable budget to concrete and steel. We need to get there on price, otherwise we're never going to prove this can be done commercially. We have issues around material supply chains in the UK; we buy globally from the open market based on price. We need to drive local economies, drive jobs in local businesses, use local suppliers wherever possible and address the carbon impact of the transportation of materials. We did encounter problems with contractors not having confidence in the materials, and the inevitable happens which is they try and cost risk into every aspect, which makes many sustainability outputs almost unachievable.* (Interviewee C)

*A hempcrete building costs about 10% more to build than the cost of a normal museum storage building of that size. But costs were mostly in the design rather than the materials. The solution was considered to be more important; a slightly elevated cost was accepted. The hempcrete building costs two-thirds less to run than an equivalent, traditionally built store on-site.*
(Interviewee D)

### *3.2 Knowledge*

In addition to sustainability and environmental considerations, bio-based materials chosen for realised projects were there to perform a function; the regulation of the internal environment, reducing fluctuations in temperature and Relative Humidity (RH). Interviewees were happy with the performance of materials and from a building physics perspective, the bio-based materials can be judged to be a success. An opportunity for the uptake and promotion of bio-based materials is the concept of wellness and creation of a natural and appealing environment in which to work and live, resulting in low levels of turnover and staff absences, and high levels of health, well-being, morale and productivity. The bio-based projects also considered circular construction and engineered-in the capability to expand, modify or even relocate. There have been problems with Life Cycle Assessments (LCA) arising from where materials originate globally and the carbon involved in transportation. Case studies of bio-based material use and university research involvement are deemed to be effective methods of dissemination to construction industry professionals. Databases of information on bio-based materials have been established, but these need to be promoted, maintained and merged to form a central body of knowledge. The question of who should be knowledgeable; client or construction industry professionals, and thus influencing the bio-based circular agenda in design, is a subject of debate. The case studies have not led to an increase in industrial scale bio-based construction. Interviewees broadly agree that rather than progress being made, the industrial scale bio-based agenda in the UK has, if anything, slipped backwards over the course of the past five years. Challenges can be faced within client companies, with people promoting the use of bio-based materials potentially conflicting with reluctant colleagues who are perhaps lacking knowledge of the benefits of using bio-based materials or are primarily concerned with initial costs. It is clear that a bio-based project requires an advocate at a senior level within client organisations.

> *We have been successful in maintaining an ambient temperature optimal for our product of between 13°C – 16°C. The material is performing as we hoped it would perform. I think it was very beneficial that we were seen to be bringing sustainable values to life and the project gave members of staff a sense of pride. We made sufficient space at one end of the building to extend in the future with the same technologies; in essence, it's the first part of a modular building. We worked with the University of Bath on the properties of the hemp-lime blocks, glulam beams and sedum roof. The (bio-based) drive has to come from architects, structural engineers, the supply chain into companies - rather than from the companies. At the time, professionals were relishing the opportunity, because it was such a landmark project. Whether you would get the same response today, I don't know; a later sustainable university project had challenges we did not have. Every project needs a champion in the boardroom, who has credibility and can argue the long term case. Without someone who can paint the picture, finance directors won't do it.* (Interviewee A)

> *Things such as growing crops locally, not using pesticides, knowing where the material comes from are all very important. When Brussels gave finance for biofuels, the unintended consequences were people buying biofuels from South America, and the cutting-down of forests. The circular economy has been hijacked by waste in some ways; it's easier, because you can say that you're using waste material from the food industry; but knowing where it's come from is much more important. Rather than keep setting up new bio-based data directories, we should put existing directories together and keeping directories current.* (Interviewee B)

> *You're not just delivering energy and carbon savings, you're also creating a better workplace and people are more productive according to their wellness. We would monitor the carbon miles associated with materials that went into the building. We had been working with the UK Building Research Establishment (BRE) and felt there was a fairly good knowledge base, but there is still not enough case study material around. We need informed clients, and then that needs to run right through a project. I think the attitude at the time of the project was that this is the future. But at the moment, there's no impetus in this country, the sector is not moving as fast as it should be. The project hasn't triggered a flow of projects like it afterwards, which it should have done. The UK has a problem with technology translation; it does world class research, but then doesn't implement it. We are not as well connected into various movements in Europe as we were. You need to be an ambitious leader and never take 'no' for an answer.* (Interviewee C)

> *Museums are interested in RH because that damages objects far more than temperatures. We attended a university lecture and heard about hempcrete; it seemed reasonable to have a very hygroscopic material as part of a low energy solution controlling RH. There was opposition; museums are very conservative. We must have been very persuasive! The hempcrete store is modular, so it could be disassembled; if we needed to move it out of the hangar to another, it would be feasible. Currently, we are in the process of building a massive store on site. We wanted to build it out of hempcrete, but couldn't find a contractor in the UK who*

IOP Conf. Series: Earth and Environmental Science **855** (2021) 012015 doi:10.1088/1755-1315/855/1/012015

> *would take it on. They're afraid of it, they don't know how it works, there isn't enough information and we just couldn't find anybody comfortable with doing a building of that size - so we've gone backwards in a way.* (Interviewee D)

*3.3 Policy*

The construction industry, and innovative bio-based material use within it, has not necessarily been at the top of governmental agendas in looking at societal needs and the issues of energy and climate change. Major topical issues of the day can form an obstacle or promote the restructuring of governmental policy - and client - priorities. Legislation at the national level to guide and direct change can be punitive; banning, taxing or penalising, or positive; incentives and subsidies. All interviewees were of the opinion that government legislation was needed to drive a bio-based, circular agenda in the construction industry. There are pros and cons attached to both approaches. Banning products has been effective at changing behaviour and practice in other industries. There may be a role for punitive, and educative, legislative measures in the construction industry. Positive legislation was broadly welcomed; however, there are risks associated with positive legislation. One opportunity for sustainably-sourced bio-based materials is the potential for building design to be sympathetic with the surrounding natural environment, especially in green-belt areas of natural beauty with strict planning restrictions.

> *We have had the aftershocks of the financial crisis of ten years ago, and we see another crisis now with Covid-19. Finance is tighter; are people prepared to invest in innovative technologies? I'm always in favour of positive legislation. Trying to level the playing field between bio-based materials and conventional materials would be the right route; rather than put industries out of business by pricing them out of the market, I think price-in the new technologies. Bio-materials also eased planning permission issues because it was something different, good and sustainable.* (Interviewee A)

> *There was a big push to go for industries like nuclear, solar and wind, which were worked on a lot, while bio-mass was demoted. Policies were tried that were green around the edges, with a lot done on renewable and clean energy. However, climate change was always the poor cousin of energy. With electricity, we were trying to encourage renewable energy, but a lot of finance was put on nuclear power. We are going to have to grow more crops in the UK to be bio-based because it doesn't make sense to bring it from somewhere else; the further away the biomass is, the more expensive. We need legislation; banning worked with pesticides, incinerators with noxious emissions and lead in petrol. People didn't want to change, so we had to ban. You have to be very careful about policy. Giving finance for a particular thing, all of a sudden there is a focus on that; everybody wants to do it and people with no skills jump on the bandwagon.* (Interviewee B)

> *You need institutions who are procuring projects, whether private or public sector, to understand the impact of using non bio-based materials and high carbon emission materials. I think we need quite strict legislation to drive this through the supply chain; otherwise, it won't happen.* (Interviewee C)

> *It has to be legislated; the only way we will stop driving petrol cars is if we're no longer allowed to. Big commercial builders are never going to move to bio-based materials until they are made to move.* (Interviewee D)

## 4. Discussion

It is clear from the interviews that emerging bio-based materials and circular construction face barriers to breaking through into mainstream construction practice; however, opportunities can be identified.

### *4.1 Finance*

There are major issues within client organisations concerning the initial cost of materials and whether they would be available in the required quantity. A major consideration is whether client organisations are financially and structurally able to think long term with regards to recovering initial outlays by reducing energy costs and having high levels of staff attendance, retention and productivity. Measured financial incentives such as interest free loans at governmental level may be a route to enable client organisations to plan for projects over a much longer time-scale, allowing greater initial expenditure on emerging bio-based technologies and circular design, and subsequently long term energy and personnel-related savings facilitating repayment. Loans would also assist innovators and material developers to test, certify and market their product. There are clear long term financial benefits to circular construction, especially if adaptability and modification is considered in the design process to accommodate future company needs and expansion. However, these may not be enough to satisfy the balance sheets of client companies structurally having to think and work on shorter time frames.

### *4.2 Knowledge*

Gathering data from experimentation carried out on newly developed materials, and knowledge of where the constituents within the material have globally originated will aid the growth of data and assist in achieving product accreditation. Dissemination of data and certifications to professionals will aid the promotion of the products' use and reassure clients of properties and performance. Even with incentives to plan long-term, client organisations still require leadership from senior personnel who are informed and enthused about using bio-based materials and designing innovative structures which are designed for adaptability and can be modified to reflect the future changing requirements of the business. More research and living laboratory case studies are required to demonstrate the benefits and feasibility of bio-based construction and cases need to be accessible, not solely flagship and premium high-end projects. The authors are engaged in further research into wall systems using insulating and load-bearing bio-based materials which can be disassembled and re-used. High levels of wellness among occupants, with high productivity, low turnover, low absenteeism and high-density use and occupation are measures of a successfully realised building compatible with circular construction principals. Personnel data from case studies should form part of further research and be gathered and assessed; if it is found that bio-based buildings are conducive to high levels of staff performance and productivity, this information would require dissemination along with the long-term financial savings.

### *4.3 Policy*

Policy can be identified as being key to upscaling bio-based construction. Conventional materials such as concrete and steel would realistically continue to be required to meet the construction needs of society. Therefore, punitive measures for an industry already in a difficult position as a result of years of austerity and the Covid-19 virus, would surely be inappropriate and would not gain wider support. This study submits that a long-term, measured program of positive legislation should promote bio-based circular construction. A clear way forward is to grow more non-food crops such as hemp within the country where a construction project is taking place. Subsidies to increase growth would be required to stimulate supply and create regeneration, boosting regional economies and local employment, businesses and supply chains. Countries must become more self-sufficient with growing crops for non-food purposes and landowners and farmers can be financially incentivised towards non-food crop growth. A further consideration is the extent to which free, market-led economies should be regulated or encouraged to promote a sustainable, bio-based agenda. Central policy should be accompanied by appropriate regulation and protection of virgin forest and areas of natural beauty. Local planning authorities could be incentivised to look favourably upon planning applications which are designed for adaptable, modular construction, future expansion and for blending into the environment with the use of bio-based cladding and green roofs; this may be a solution for greenbelt development. If one punitive measure may be considered, it is that of placing conditions upon the granting of planning applications that a degree of circularity in design must be demonstrated.

## 5. Conclusions

The upscaling of circular bio-based construction with sustainably managed resources would impact significantly upon reducing the carbon footprint of the construction industry. Semi-structured interviews conducted with senior professionals with experience in bio-based construction identified barriers to upscaling in the form of initial costs, opposition to bio-based projects within client companies, absence of knowledge, limited case studies, absence of accreditation and material availability.

Policy both at national and regional levels can be viewed as crucial to drive the growth of bio-based circular construction. Though caution should be exercised, positive legislation can encourage bio-based material use. Potential solutions and recommended courses of action for policy-makers and circular construction practitioners can be found in promoting long-term economic cases to client organisations with low running costs, promoting the wellness agenda with high staff productivity, regenerating local economies with farms locally growing non-food crops, boosting local businesses, skills and labour with transforming crops into products and emphasising sustainability as a requirement in planning applications. Every successfully realised project would increase case studies and increased university research experimental data would facilitate accreditation and certification of emerging materials.

## 6. References

[1] Verspeek S and Van Der Burgh F 2019 Performance of bio-based facades *3rd International Conference on Bio-Based Building Materials* (Belfast, UK) pp 1–6

[2] Adams K T, Osmani M, Thorpe T and Thornback J 2017 Circular economy in construction: Current awareness, challenges and enablers *Proc. Inst. Civ. Eng. Waste Resour. Manag.* **170** 15–24

[3] Leipold S and Petit-Boix A 2018 The circular economy and the bio-based sector - Perspectives of European and German stakeholders *J. Clean. Prod.* **201** 1125–37

[4] Maskell D, Thomson A and Walker P 2018 Multi-criteria selection of building materials *Proc. Inst. Civ. Eng. Constr. Mater.* **171** 49–58

[5] Sönnichsen S D and Clement J 2020 Review of green and sustainable public procurement: Towards circular public procurement *J. Clean. Prod.* **245**

[6] Interreg 2019 Circular Bio-based Construction Industry *Interreg 2 Seas Mers Zeeen*

[7] Tsai W H, Yang C H, Huang C T and Wu Y Y 2017 The impact of the carbon tax policy on green building strategy *J. Environ. Plan. Manag.* **60** 1412–38

[8] Lombard R 2017 It's not easy being green: Difficulties in sourcing sustainable building materials *QScience Connect* **2017** 10

[9] Sourani A and Sohail M 2011 Barriers to addressing sustainable construction in public procurement strategies *Proc. Inst. Civ. Eng. Eng. Sustain.* **164** 229–37

[10] Markström E, Bystedt A, Fredriksson M and Sandberg D 2016 Perceptions of Swedish Architects and Contractors for the Use of Bio-Based Building Materials *COST Action FP1407 2nd Conf. – Innov. Prod. Technol. increased wood Prod. Recycl. reuse* 29–30

[11] Shea A, Lawrence M and Walker P 2012 Hygrothermal performance of an experimental hemp-lime building *Constr. Build. Mater.* **36** 270–5

[12] Cascione V, Maskell D, Shea A and Walker P 2019 A review of moisture buffering capacity: From laboratory testing to full-scale measurement *Constr. Build. Mater.* **200** 333–43

[13] Kraniotis D, Nore K, Brückner C and Nyrud A Q 2016 Thermography measurements and latent heat documentation of Norwegian spruce (Picea abies) exposed to dynamic indoor climate *J. Wood Sci.* **62** 203–9

[14] Lawrence M, Shea A, Walker P and de Wilde P 2013 Hygrothermal performance of bio-based insulation materials *Proc. Inst. Civ. Eng. Constr. Mater.* **166** 257–63

[15] Maskell D, Silva C F da, Mower K, Cheta R, Dengel A, Ball R, Ansell M, Walker P and Shea A 2015 Properties of bio-based insulation materials and their potential impact on indoor air quality

[16] Shi Q, Ren H, Cai W and Gao J 2019 How to set the proper level of carbon tax in the context of Chinese construction sector? A CGE analysis *J. Clean. Prod.* **240** 117955

[17] Philippidis G, Bartelings H and Smeets E 2018 Sailing into Unchartered Waters: Plotting a Course for EU Bio-Based Sectors *Ecol. Econ.* **147** 410–21

[18] Philp J 2015 Balancing the bioeconomy: Supporting biofuels and bio-based materials in public policy *Energy Environ. Sci.* **8** 3063–8

[19] Malik A, Ajonbadi H A and Jimoh I 2019 Investigating the Potential Economic Impact of Brexit Decisions on Business Performance in the United Kingdom : A Case Study of the UK Construction Industry *Int. J. Manag. Account. Econ.* **6** 347–67

[20] Bryden Wood 2018 *Platforms. Bridging the gap between construction + manufacturing*

[21] EBuilding 2018 Adnams: Sustainability Leader *Eco Constr. Blog*

[22] Leskard M 2018 Case Study: The Hempcrete Museum Store *Stud. Conserv.* **63** 381–3

[23] Kennett S 2006 Innovative Adnams distribution centre minimises environmental impact *Building.co.uk*

[24] Korjenic A, Zach J and Hroudová J 2016 The use of insulating materials based on natural fibers in combination with plant facades in building constructions *Energy Build.* **116** 45–58

[25] Kinnane O, McGranaghan G, Walker R, Pavia S, Byrne G and Robinson A 2015 Experimental investigation of thermal inertia properties in hemp- lime concrete walls 942–9

[26] Lawrence M, Heath A and Walker P 2009 Determining moisture levels in straw bale construction *Constr. Build. Mater.* **23** 2763–8

[27] D'Alessandro F, Bianchi F, Baldinelli G, Rotili A and Schiavoni S 2017 Straw bale constructions: Laboratory, in field and numerical assessment of energy and environmental performance *J. Build. Eng.* **11** 56–68

**Acknowledgements**

The authors gratefully acknowledge the interviewees for their time and acceptance of the invitations to be interviewed and for sharing their insights and experiences. Interviews took place as part of the Circular Bio-based Construction Industry (CBCI) project, which is funded by the European Union Regional Development Fund Interreg 2 Seas Mers Zeeen (2S05-036).

# Design potentials between empirical data and architectural imagination: an exploration

**O Paans**[1]

[1] Technische Universität Berlin, Strasse des 17. Juni, 13405 Berlin, Germany

**Abstract.** This article discusses ways in which we can connect empirical data and design skills to create more circular and sustainable urban areas. The case of Hellersdorf-Süd (Berlin, DE) is used to describe a design approach that is context-driven and combines data and design skills to formulate coherent and relevant design proposals. Based on this exposition, this article presents and exploration of the future urbanist practices in which sustainability and circularity on a conceptualized larger scale. As such, it provides a speculative account of a future course of action to achieve urban sustainability and systemic circularity.

## 1. Introduction

The recent focus on creating a "circular economy" by implementing strategies for re-use and recovery in various industries is not just a technical notion. Instead, it is shift in thinking about production processes and inputs. Formerly linear production processes become circular, or at least as circular as possible. If we focus on circular building and its associated cycles of construction, renovation and demolition, we can observe that currently, there is a tangible focus on re-using components. To a degree, this emphasis is the logical consequence of the fact that the building industry leans heavily on prefabricated components that can easily be re-used – for example glass sheets or aluminium window frames. On the technological side, new logistic tools allow for a fine-grained mapping and tracing of components, along with appraisal and the connection of possible projects in which they can be applied.

Against this background, I would like to raise two points that merit attention:

1) *Information does not generate objectives*
First, no dataset, report or piece of information can inform us about the spaces we should design, even if those tools are extremely useful supports once basic scenarios have been defined. A technologically optimized solution might not be a liveable space at all; and conversely, a perfectly liveable space might be designed very unsustainably.

2) *Scaling up in circularity and sustainability*
Second, we have to think about circularity and sustainability on an urban scale. In practice, this means overcoming the focus on components and individual buildings and zooming out to a conception of sustainability that encompasses components or the built environment in its entirety.

IOP Conf. Series: Earth and Environmental Science **855** (2021) 012016 doi:10.1088/1755-1315/855/1/012016

Why these two points? The problem in its most basic form is that we have an overload of information and/or data when designing [1]. Yet, each physical space is the result of a process in which some information has been selected for usage, while other sources have been omitted. Moreover, modern design practices require the conception of an idea with a great amount of detail, an effort that is strenuous on the mind and the imagination alike [2]. Yet, the seeming certainty and efficiency that data-driven or instruments like Blockchain or BIM appear to offer obfuscate the fact that architectural spaces are not just one-to-one implementations of technologically optimized scenarios or the result of one-directionally applying information to a given context. While it gives an illusion of control, data cannot by itself inform us about the goals to be set [3].

Thus, the challenge is to connect the *availability of information* with the *architectural imagination*, as physical spaces result from of multiple creative, technological and social influences. Phrased differently: while data can inform us, it cannot set the objectives. This point was readily recognized in the 1960s and 1970s with the emergence of information technology and the emergence of decision-theoretic design models but seems to have disappeared somewhat into the theoretical background [4].

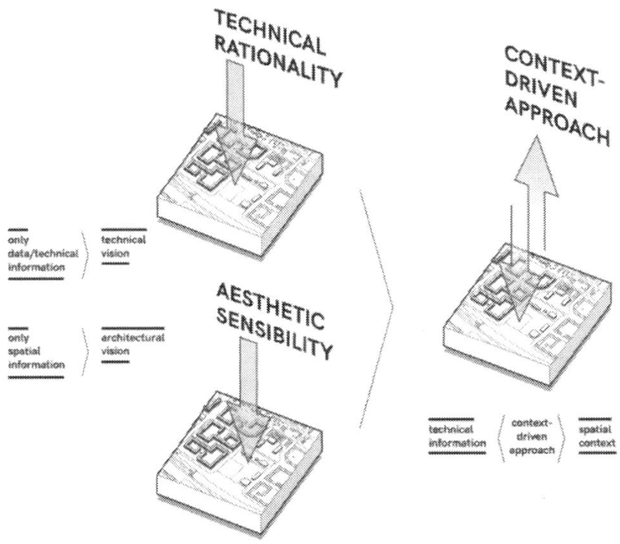


**Figure 1: Two opposite scenarios on the left, and the proposed context-driven approach on the right**

If we would conceive urban spaces as the outcome of only technically optimized scenarios, supported by empirical data, we would end up with a "technical vision" – with spaces that are optimized from a single point of view – namely a technical rationality. This is not just a hypothetical possibility. With the emergence of technologies like BIM or Blockchain, there is the potential and the temptation to define spaces solely as the results of technological optimization. Moreover, there is the clear and present temptation to let technological capabilities decide what would be the best space. In those cases, "best" equals "optimized according to technological means." If we imagine the other extreme, we end up with (urban) spaces that are optimized solely according to their architectural or aesthetic features. It would amount to a purely "architectural vision", with little or insufficient regard for issues like sustainability, circularity or re-use. In those cases, aesthetic sensibility claims pride of place over functional or sustainability-related issues. In both cases, the approach is top-down: either it is a form of technology-driven rationality or it is an aesthetic sensibility is unilaterally projected on an (existing) architectural context.

So, when we consider the relation between data-driven sustainability and spatial design, the task is to make design capabilities and skills communicate with information-driven technologies to avoid the top-down approach and possible mismatches. When considered separately, neither technical rationality nor aesthetic sensibility can accomplish the transition towards more sustainable urban areas in which careful design and re-use is the norm. Therefore, we require approaches that combine the availability of information with a robust architectural approach, without falling into either one of the extremes sketched above. We present one possible approach here, and by way of preliminary, call it a *context-driven approach*.

## 2. Context-driven approach: connecting information and spatial design

So, we require a way to connect *information* and *spatial design*. Again, we present here but one way of establishing this connection. While re-use or circular construction is just one strategy in the transition towards urban sustainability, the broader method presented here can be followed in different contexts, and its rules of application can be adapted to numerous different cases [5].

The city district (*Bezirk*) Marzahn-Hellersdorf was conceived as a prestigious "housing for the masses" project by the GDR government. Its construction started in 1977 and continued until the fall of the Berlin Wall in 1989. It was intended to provide high-quality, standard housing. The characteristic *Plattenbau* method allowed for a swift construction, making Marzahn-Hellersdorf an impressive architectural achievement in its own right [6]. It is one of the larger boroughs of Berlin, forming a significant part of its Eastern edge. Its large scale of buildings and the spacious urban plan caused outdoor spaces to be rather undefined and oversized [7]. In addition, increasing levels of private car ownership turned sizeable parts of the public space into parking areas. Individual car ownership was deemed a privilege, and consequently, the whole neighbourhood is designed with frequent care use in mind. In an attempt to mitigate the uniform and industrial outlook of the *Plattenbau* buildings, the fronts have been renovated and replaced with a variety of materials and colours after the fall of the Iron Curtain. These measures are largely cosmetic; they do not alter the urban structure. This is not to say that they are unimportant. One must keep in mind, however, that such changes cannot be expected to solve structural sustainability problems entirely.

**Figure 2: Location of Hellersdorf-Süd (black) and street view**

From a sustainability point of view, the existing urban structure presents some recurring problems. The large surface area of pavement contributes to the formation of Urban Heat Islands (UHI's). These UHI's negatively impact the local microclimate, especially during prolonged periods of heat. Canalized storm water drainage causes the larger area to get rid of rainwater quickly, resulting in drought during the long continental summer. The concrete components of many buildings trap the heat, necessitating the use of air-conditioning. Renovating all buildings at once would be a costly affair – especially when investments cannot be spread. The renovation efforts combined with the mature greenery and large open spaces have caused a positive turnaround for Marzahn-Hellersdorf as a whole. Young families are actively seeking housing in this area, as they provide generous spaces for a new generation [8].

When defining design proposals for this urban area, it turned out that there was so much information on urban sustainability (including re-use and renovation) that it was hard to define scenarios that were not just arbitrary collections of measures. After all, in order to define a realistic scenario, one must be able to compare them according to their merits. But if one collection of measures is just as good as the other, evaluating the advantages of a design proposal becomes increasingly difficult.

The core of the problem was that the information suggested no definite directions for future actions. This problem is not entirely unknown. Already in the 18th century, the Scottish philosopher David Hume remarked that you cannot derived an "ought" (i.e. a description of how things *should* be) from an "is" (i.e. a description of how things *are*). But the "ought-is" problem raises questions: what is the role of all that information when it cannot tell us where to go? And what is the criterion for using a certain piece of information and excluding another? In the absence of a clear answer to this problem, the design process was set up as a sequence of steps intended to clearly demarcate the utility and relevance of the information at our disposal:

A) In the first step, high-level information is combined to create an overall picture that depicts trends or recurring phenomena. For instance, we know that the annual global energy consumption increases, and we know that much of this energy is produced by burning fossil fuels.
B) Given known sustainability goals (for instance, reducing fossil dependency), this situation necessitates action.
C) In turn, this leads to a design objective – every design proposal has to contribute to reaching the goals indicated in point (B). For instance, reducing fossil fuel dependency is the design goal that emerges when we combine (A) and (B).

Process steps A–C are generic in the sense that they could be applied to every urban area. The architectural context is not included, only the *goals* of future scenarios are being defined. In a subsequent series of steps, these design objectives were linked to an existing architectural context. Here, however, the ought-is problem emerges once again: there is so much information that one must select strategies and scenarios. Because of the overload on info, a balance must be struck between technologically optimization and spatial qualities.

One of the ways to do this is to analyse the material one works with – the problems and promises of the existing spatial context. Therefore, this approach became the next step in the process:

D) Analyse a given urban area and identify its spatial potentials and shortcomings with regards to the goals set in (C)
E) Identify how the spatial layout functions with regard to the potentials and shortcomings identified in (D)
F) Select a limited number of sources for each problem, so that it is analysed as a graspable subproblem of a larger compound problem
G) Define a catalogue of design options on a high conceptual level to see where solutions overlap and can be combined.

The advantage of the approach described in (D)-(G) is that design processes can be made context-driven. The physical and organizational features of the existing urban context determine which design strategies are most suitable for the conception of spaces. This method is neither completely architectural, nor completely technical. Instead, it takes the characteristics and potentials of physical spaces as point of departure. However, it does so in such a way that information has a clearly demarcated role to play in view of the goals to be realized.

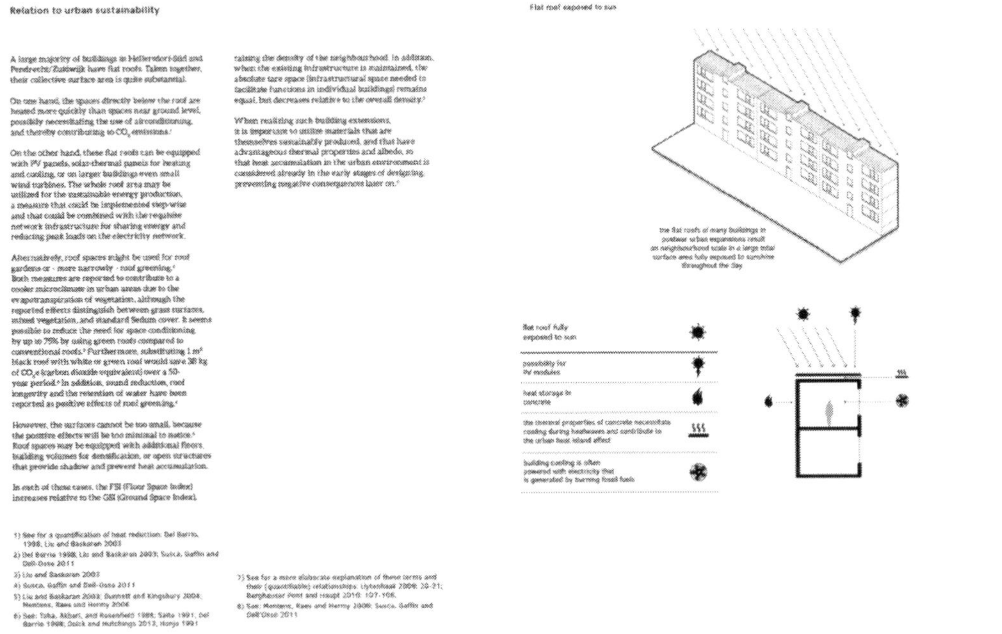


**Figure 3: Catalogue page with selected sources on the left and the spatial translation on the right**

The art is to define what spatial characteristics could be utilized to fulfil the sustainability objectives defined in (C). Once these characteristics are mapped and understood, they overview provide an insight in the possibilities and limitations of the typologies and infrastructure of the neighbourhood, both on the level of urban structure and the characteristics of individual buildings.

The proposals in the catalogue (step G) are underpinned by carefully collected empirical data, yet the fact that this information is translated into spatial parameters links to the core competences of urban designers and architects alike – namely *to conceive and shape functional, lively and inhabitable spaces*. One of the main advantages of this approach is that the translation between information and the possibilities it offers for spatial design practice is at the core of the entire approach. As such, realizing sustainability objectives becomes amenable to a design approach. This approach sets also a limit on the sources to be selected: anything that is not translatable into a spatial strategy is considered as being "off the board" or at least beyond the scope of a design assignment. As such, it allows designers to work with "wholes", as Kevin Lynch once called them [9]. The spatial ensemble is a creative combination of different solutions and possibilities that forms a coherent unity in the spatial sense.

## 3. Example of a context-driven design strategy

Phrased in process terms, this description is somewhat abstract. An example shows how steps D-G were implemented with regard to a specific problem, namely the ubiquitous presence of large parking lots with impervious concrete surfaces. The emphasis on motorized traffic and mobility resulted in an urban fabric that is infrastructure-oriented. Consequently, traffic lanes and parking places are oversized and present everywhere.

To define a range of feasible design solutions for large parking lots, different types of knowledge need to be connected. A working knowledge of the phenomenon of urban heat islands is necessary; insight in its causes and consequences; knowledge about thermal properties, albedo and pavement materials; knowledge about the scale and urban effects of thermal emissions; an estimation of the feasibility of possible solutions; knowledge about ecology and its role in combating urban heat islands. These facts and insights must be aligned with each other in coherent design proposals [10]. The proposal as a whole must respond adequately to the actual issues that have been identified through analysis or experience.

By representing the problem visually as a diagram (step D and E), the way that the problem "functions" can be analysed easily. By adding sources and additional information (F) the functioning of the problem is explained, quantified and described in more detail. This is especially useful in designing on an urban scale, as one situation may occur in multiple locations. In the concluding step (G), all knowledge is combined in simple, yet effective proposals. In this example, one could opt for an open pavement (and respond to water run-off and heat emission); alternatively, one could choose for a cover to shield the pavement form sun – the cover could catch the sunlight and be made of PV panels; one could reduce the total area of pavement by "greening" the parking places etc. These options are very straightforward and can be combined in compound solutions.

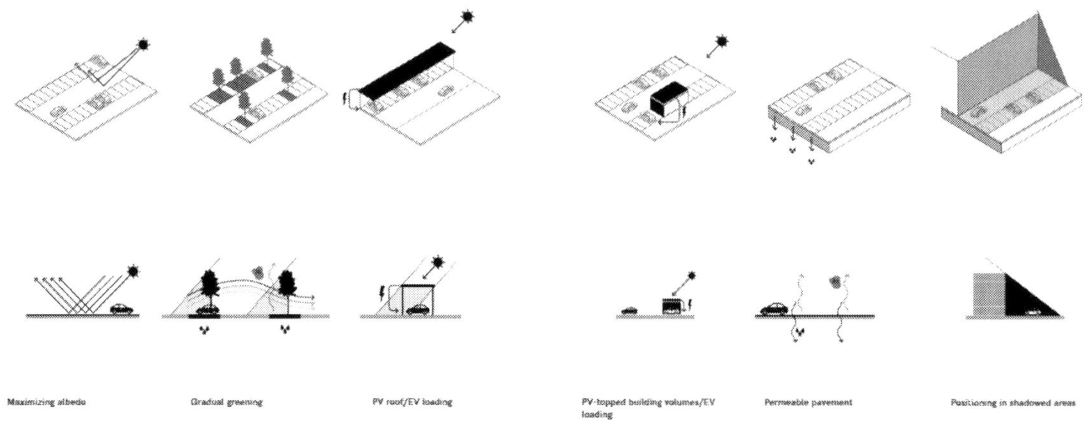


**Figure 4: Design options for a given problem**

More importantly, however, their spatial consequences can directly grasped. The fact that these ideas can be applied in similar contexts, but in a different location makes them highly context-adaptable. Not only provides this approach building blocks that are supported by empirical data, it does so in a way that does not describe how the end result has to look according to a technical or aesthetic norm. The final appearance of any solution cannot be dictated on technical grounds but must fit the context.

All factors that are relevant in arriving at a solution are listed as discrete components of a design proposal, thereby locating them firmly within the competence of design professionals. In every case listed in the catalogue, the body information about a given problem is incomplete. However, the point is not to be comprehensive, but to collect and comprehend enough information to develop a sufficient understanding of the issue to be solved in relation to the spatial context. Thus, the quality of the design response determines whether more information is required, or whether it was sufficient to address the issue. The design perspective is generalist. It is this generalist tendency that constitutes in part the potential of designing. It selects from an unimaginably large set of fact and factors just some elements and lines them up in a coherent and context-sensitive design proposal [11]. During designing, incompleteness plays no crucial role in evaluating the merits and shortcomings of a proposal. This allows designers to "see the proposal on its own terms". It is treated as a potential that can be changed, worked out and improved – but that nevertheless looks promising – every design stage simultaneously brings its own context for evaluation along [12].

## 4. Scaling circularity up for the 21st century

As of yet, the focus on re-using either waste materials or components is a limit on the thinking of circularity as such. From natural processes, we observe that waste and wealth are intimately connected, and that circularity as such is the core phenomenon that makes natural processes function [13]. So, when thinking about upscaling circularity, the first question to ask is "What the largest circular system that we can imagine?" Doubtlessly, the answer must be the conglomerate of abiotic and biotic processes that shape landscapes and climates. These processes still exert their influence on urbanism and landscape architecture, as the impact of climate change on urbanism clearly shows. For instance, urban heat islands and urban dust domes are consequences of a mismatch between the built environment and natural phenomena. Smaller local phenomena like thermal emission, heating of air, evaporation and lack of water retention converge in larger phenomena.

To integrate circularity from the very start into (urban) design processes requires a return to mapping the most fundamental problems that become uncontrollable phenomena once scaled up to an urban scale. The catalogue introduced mapped a number of discrete phenomena and analysed how they came about as a consequence of spatial design choices. But if a local phenomenon (let's say, impervious pavement with all its disadvantages) is widely utilized throughout an urban structure, it becomes a larger problem by virtue of its scale. Thus, a scenario where landscape processes (both biotic and abiotic) and urban structure are actively aligned will enable the unconstrained unfolding of abiotic and biotic processes. On the level of materials and waste production, this approach has already been considered [14] but not yet on the large-scale integration of landscape-forming processes and urban tissue beyond so-called "landscape urbanism" [15].

By integrating the large-scale cycles of climate and landscape-forming processes into the built environment, currently unsustainable typologies, traffic infrastructures and production zones can be transformed in accordance with the largest cyclical systems that we know. Instead of having to cool air-conditioning buildings, typologies can be rethought to cool and ventilate naturally; instead of travelling to work, jobs can becomes locally oriented; instead of importing food, it is locally produced; instead of depending on enlarging and merging companies, decentralized production becomes the norm; instead of having fixed ecological zones, ecological principles become the foundation for urban organization.

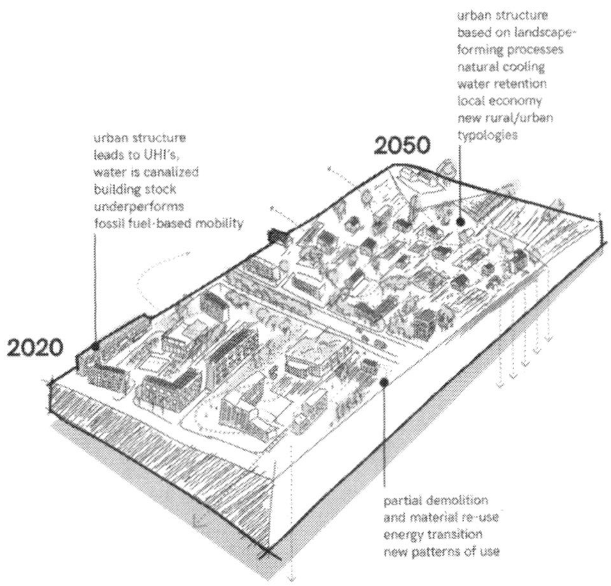


**Figure 5: Concept for transforming urban edges and integrating circularity**

One possible design scenario in a shrinking area like Southern Limburg could be the systematic restructuring of urban edges. In this process, large-scale urban developments from the 1970s and 1980s on city edges could be gradually demolished, and the resulting materials could be repurposed for the design of new public spaces. The transport sector could slowly be shifted from fossil fuel-based system to an EV system, based around urban edges and capitalizing on the fact that already within short distances, numerous companies, services and products services can be found.

In a subsequent phase, new typologies based on the demographic needs of the region could be developed, together with a range of facilities that integrates a 21$^{st}$-century production landscape with a transformation program of increasing biodiversity, greening to reduce $CO_2$ impact and stimulate evapotranspiration to reduce heat islands. The resulting urban-rural typology could be implemented along urban edges and adapted to the adjacent landscape types [16]. This would support the shift towards a regional economy, local production and the creation of job opportunities close to home. In particular, this new regional economy would need to realize a "broad prosperity" that encompasses economic, ecologic and socio-cultural aspects.

This design approach aligns the design objectives of both landscape architecture and urbanism. By stacking various transitions (climate adaptation, urban sustainability, letting natural processes unfold, renewable energy, local economy) the very notions of circularity could be implemented in the way the landscape-urban relation is conceived. The resulting urban tissue would benefit from the existing fine-grained infrastructure of urban conglomerates and villages, creating an ecologically balanced hybrid of landscape and urban areas – a thoroughly circular and sustainable in-between typology that functions in accordance with climatic and landscape processes. It would amount to a vital durability in accordance with the largest circular systems we know.

The transition process could take place over a period of around 30 years, slowly dismantling and transforming urban areas, and slowly building up new urban-rural typologies and an associated regional economy. This timeframe could be linked to the population decline that is expected after 2030 in order to prevent the construction of superfluous buildings.

The most important strategic step in this transition process would be to secure a broad commitment to from all stakeholders, along with clear objectives at the very beginning of the process. This would ensure that the goals of the roadmap are clear, but there is significant leeway to select different scenarios to realize the objectives [17]. In such an open process, different routes can be taken, according to strategic and financial opportunities, without being bound to a fixed order. In particular, it is important to clarify the roles of stakeholders at the very start of the process. This clarification serves to secure commitment of stakeholders, as it is clear what their role is, and what they can expect.

The previous section provided a very broad outlook and a "designer's view" of the various transitions that have to be made. As design strategy, rethinking the relation between landscape and urban structure was proposed in order to integrate circularity from the very start into spatial design processes and moving it beyond the level of re-using components towards a truly urban scale. But above all, it showcased an opportunity to upscale and radicalize the concept of circularity in thinking about the urban-rural connection, the ubiquitousness of landscape-forming processes and the necessity to embrace and harness these processes as drivers of design thinking. In this manner can circularity become context-adaptable and can its potentials be utilized to the fullest.

**Acknowledgments**
The author would like to express his sincere gratitude for having had the opportunity to have stimulating and inspiring discussions around the topics of urban transformation, circularity and sustainability with Marc Heusschen (HeusschenCopier) and Erol Oztan (Erol Architecten), who both generously agreed to engage in thinking about the future. Thanks are also due for the anonymous reviewer who pointed out a number of things that were not mentioned in the text and that were outside the scope of this paper.

IOP Conf. Series: Earth and Environmental Science **855** (2021) 012016 doi:10.1088/1755-1315/855/1/012016

[1] Paans O and Pasel R 2019 Describing Liminal Knowledge in Architectural Design: Knowing What We Do Without Knowing Everything *Int. J. of Design Education* **13** 3 pp 13–25; Brey A, Innerarity D and Mayos G 2009 *The Ignorance Society and Other Essays* (Barcelona: Zero Factory)

[2] Tuan Y-F 1977 *Space and Place. The Perspective of Epxerience* (Minneapolis: University of Minnesota Press) p 101

[3] Paans O Pasel R Ehlen B *NAF/NAAR Proc. Series 2019-1: Reflecting Histories and Directing Futures* ed A Tofte M Rönn E Wergeland (Oslo: The Nordic Academic Press of Architectural Research) pp 203–228

[4] Asimow M 1962 *Introduction to Design* (Englewood Cliffs, NJ: Prentice Hall); Eastman C 1968 *Explorations of the Cognitive Processes in Design* (Pittsburgh, PA: Carnegie Mellon University); Newell A 1967 *Heuristic Programming: Ill-Structured Programming* (Pittsburgh, PE: Carnegie Mellon Univ., 1967); Rittel H and Webber M 1973 *Policy Sciences*, 4 pp 155–169; Simon H and Newell A 1972 *Human Problem Solving: The State of the Theory in 1970 in Human Problem Solving* (Englewood Cliffs, NJ: Prentice-Hall); Simon H 1973 *Artificial Intelligence* 4 pp 181–201; Simon H 1969 *The Sciences of the Artificial* (Cambridge, MA: MIT Press).

[5] Paans O 2020 *Creating Knowledge Through Architectural* Design (Berlin: Technische Universität Berlin)

[6] Rubin E 2016 *Amnesiopolis. Modernity, Space, and Memory in East Germany* (Oxford: Oxford University Press)

[7] Tilroe A 2004 *Idealen in beton. Verkenningen in Midden- en Oost Europa* ed Wagenaar C (Rotterdam: nai010 uitgevers) pp 17–22

[8] Senatsverwaltung für Stadtentwicklung 2010 *Stadtumbau Ost in Berlin. Ein Zwischenbilanz.* Retrieved from https://www.stadtentwicklung.berlin.de.

[9] Banerjee T and Southworth M 1996 *City Sense and City Design. Writings and Projects of Kevin Lynch* (Cambridge, MA: The MIT Press) p 658

[10] Ullmark P 1996 *Nordisk Arkitekturforskning* **1** pp 53–60. 55

[11] Paans O and Pasel R 2019 Describing Liminal Knowledge in Architectural Design: Knowing What We Do Without Knowing Everything *Int. J. of Design Education* **13** 3 pp 13–25

[12] Pirolli P 1992 *Knowledge and Processes in Design. DPS Final Report* (Berkeley: University of California at Berkeley) p 29

[13] McDonough W and Braungart M 2002 *Cradle to Cradle. Remaking the Way We Make Things* (New York, NY: North Point Press)

[14] Hajer M and Dassen T 2014 *Slimme Steden – De opgave voor de 21e-eeuwse stedenbouw in beeld* (Rotterdam: nai010/PBL Uitgevers); Tillie N ed 2014 *Urban Metabolism. Sustainable Development of Rotterdam* (Rotterdam: Municipality of Rotterdam/IABR/FABRIC/JCFO/TNO)

[15] Waldheim C 2016 *Landscape as Urbanism: A General Theory* (Princeton, NJ: Princeton University Press)

[16] Sieverts T 2000 *Zwischenstadt. Zwischen Ort und Welt, Raum und Zeit, Stadt und Land.* (Basel: Birkhäuser).

[17] Bergevoet T and van Tuijl M 2013 *De flexibele stad – Oplossingen voor leegstand en krimp* (Rotterdam: nai010 Uitgevers) p 108-9

IOP Conf. Series: Earth and Environmental Science **855** (2021) 012017 doi:10.1088/1755-1315/855/1/012017

# *Circularity in Selected EU Countries: The Case of Construction and Demolition Industry*

*N Dytianquin[1], J Gregersen-Hermans[2], N Kalogeras[34], J van Oorschot[5], M Ritzen[6]*

**Abstract.** The construction and demolition industry (CDI) became a priority area for the EU circular economy as the industry has the largest waste stream by volume. Most of the CDI waste, however, ends up in incinerators for energy production or as downcycled content for road surfaces despite its enormous potential for recycling and reuse. This study focuses on the application of circularity in the CDI. We recognize that the construction industry provides a prominent decision context for studying circularity since many scrap materials resulting from demolished buildings and houses could be recycled, reused and refurbished in newly constructed edifices from foundation to rooftops. Adopting the fundamental circularity concepts, the study intends to compare how selected countries as early adopters of circularity in the EU apply distinct concepts of the circular economy in the CDI. The study investigates the manner circularity in CDI is achieved in the selected countries using determinants for project success in integrating the three dimensions of sustainability to identify and determine best practices. These determinants are: 1) stage of circularity as described in the waste hierarchy and circularity ladder, 2) building design principles, 3) life cycle stages specific to construction, and 4) availability and extent of policy or strategy on circularity in the EU countries where the selected projects are located. The EU countries covered are: the Netherlands, Belgium, Germany, France, and Denmark who are among the top 10 waste generators in Western Europe. The projects chosen are: (i) Superlocal project in the Netherlands; (ii) the Circular Retrofit Lab (CRL) project in Belgium; (iii) Open Air Library in Germany; (iv) Rehafutur Engineer's House project in France; and (v) Upcycle Studios in Denmark.

## 1. Introduction

The construction and demolition industry (CDI) became a priority area for the EU circular economy as the industry has the largest waste stream by volume contributing 374 million tons based on 2016 EU-wide data. This study focuses on the application of circularity in the CDI. We recognize that the construction industry provides a prominent decision context for studying circularity since many scrap materials resulting from demolished buildings and houses could be recycled, reused and refurbished in newly constructed edifices from foundation to rooftops. The objective of the study is to investigate how CDI-specific determinants affect project success in attaining sustainability by comparing selected EU countries (projects) in attaining and balancing environmental, economic and social sustainability performance in CDI. These determinants are: (1) circularity policy at the national, municipality or regional levels; (2) stages of circularity as described in the waste hierarchy or circularity ladder framework; (3) building design principles; and (4) the life cycle phases of the construction industry. In the end, the study intends to identify best practice and draw lessons for balancing the three sustainability dimensions and the different business models evolving from the different projects using these determinants.

## 2. The Context

The EU Action Plan on the Circular Economy (EU Commission, 2015) formalized the EU mandate for member states to transit towards a circular economy in line with the EU's commitment to reach the UN Sustainable Development Goals (SDGs) by 2030. A circular economy (CE) at the EU-level is envisioned to: (i) protect businesses against scarcity of resources and price volatility that accompanies it, hence generating innovative and sustainable competitive advantages; (ii) save energy and avoid

[1] *Center of Sustainable International Business, International Business School Maastricht, e-mail:* norman.dytianquin@zuyd.nl
[2] *Center of Sustainable International Business, International Business School Maastricht, e-mail:* jeanine.hermans@zuyd.nl)
[3] *Center of Sustainable International Business, International Business School Maastricht, e-mail:* nikos.kalogeras@zuyd.nl)
[4] *Commodity Risk Management Expertise Center Dept. of Marketing & Consumer Behaviour Wageningen University & Research, e-mail:* nikos.kalogeras@zuyd.nl)
*Center of Sustainable International Business, International Business School Maastricht, e-mail:* nikos.kalogeras@zuyd.nl)
[5] Academy Built Environment, SURD Research Group, Zuyd University Heerlen, *email:* john.vanoorschot@zuyd.nl
[6] Academy Built Environment, SURD Research Group, Zuyd University Heerlen, *email:* michiel.ritzen@zuyd.nl

irreversible damages caused by resource usage that exceeds the planet's renewal capacity; and (iii) create business opportunities in more efficient ways of producing and consuming, thereby creating local jobs, fostering skills and promoting social cohesion and integration. The circular economy, in summary, cuts across many EU common policies and priority areas from energy, climate change and the environment, food safety and quality, farming and sustainable agriculture to social cohesion, research and innovation, and entrepreneurship and promotion of small and medium enterprises (SMEs).

Several sectors were targeted in the EU circular economy action plan. The priority sectors that were identified were plastics, food waste, critical raw materials, construction and demolition, and biomass and bio-based products. This paper will focus on the application of circularity in the CDI, which became a priority area as the industry according to Wahlström et al. (2020) has the largest waste stream by volume contributing 374 million tons based on 2016 EU-wide data (Eurostat, 2019). In 2018, this reached 36% of all waste in the EU as shown in Figure 1. Most of the CDI waste, however, ends up in incinerators for energy production or as downcycled content for road surfaces despite its enormous potential for recycling and reuse, hence of 'closing the loop'. Earlier, CDI was a target for the EU energy policy regarding its operational efficiency in energy usage.

The EU action plan on circularity requires translation into national action, requiring stakeholder involvement from member states, cities and municipalities, companies and citizens. The action plan for a circular CDI is linked to the revised Waste Framework Directive (WFD 2008/98/EC, amended 2018/851) where a mandatory target of 70 percent recovery rate was set by 2020. Recovery rates vary by member state based on Migliore, Talamo & Paganin (2020) as shown in Appendix Table 1, reaching 90 percent for all 28 members states in 2016. The countries chosen for this study were the Netherlands, Belgium, Germany, France and Denmark. Figure 2 shows that on a per capita basis, these countries were among the top 10 waste producers in Western Europe with the Netherlands ranking fourth (after Finland, Luxembourg and Sweden); Belgium, sixth; France, seventh; Germany, eighth; and Denmark, ninth.

## 3. Literature Review

The evaluation of selected projects in terms of their environmental, social and economic impacts will be conducted in terms of determinants and antecedents in addressing these three sustainability dimensions. The topics to be reviewed in this section will cover the stages of circularity, the building design principles, the life cycle of the construction industry and policy-making on circularity in the selected countries which form the dimensions for choice of projects in CDI used in the study. Literature review also covers the criteria or indicators used in environmental, economic and social impact assessments to be used in evaluating project success in achieving sustainability.

### *3.1 The Stages of Circularity*

There is no universal definition of CE based on literature surveys of De Angelis (2018) and Sillanpää & Ncibi (2019) although a global definition of circularity awaits some consensus. The difficulty of finding a clear definition is attributed to the concept of the circular economy being intertwined with sustainable development, corporate social responsibility (CSR) and business models. While the former survey looks for common principles of CE found from originators in the various disciplines of economics[7]; industrial ecology[8], biomimicry[9], architecture[10], and management[11], the latter work decries the limited attention to the social dimension to complete the triptych of sustainable development (economy-environment-social) and CSR's 3 P's (people, planet, profit) found in existing definitions.

[7] See Boulding (1966), Mäler (1974), Tietenberg (1984), Pearce & Turner (1990), Daly (1992)
[8] See Frosch & Gallopoulos (1989), Allenby (1992), Bringezu (2003), Wells & Seitz (2005), Chertow (2007), Linton et al. (2007), Lifset & Boons (2012), Bloomsma & Brennan (2017)
[9] The study of innovative solutions in nature and natural processes. See Benyus (2002), Kennedy (2007), Marshall & Lozeva (2009), Habib (2011), Green et al. (2015), Das et al. (2015)
[10] See Stahel (1986), Lyle (1994), McDonough & Braungart (2002)
[11] See Lovins et al. (1999), Hawken et al. (2000), Pauli (2010)

Both above-mentioned surveys use the metaphors of spaceships and sailboats, dating back the conceptualization of CE to Kenneth Boulding's (1966) publication which allegorized a closed economy as a spaceship with limited finite resources subsisting on waste conversion. The sailboat analogy, in turn, refers to Ellen MacArthur for whom a foundation for circularity was created. She circumnavigated the world in 2004 in record time, alluding to her boat as her world, requiring minimum resources to make the boat lighter and faster, hence just relying on available supplies and avoiding unnecessary stops to restock. It was in 1990 though when the word CE was formally introduced by Peace & Turner (1990) in the second chapter of their book describing the path for economic growth within ecological limits. The early academic writings on CE resembled those of closed-loop supply chains described in industrial ecology although focusing more on technical and engineering rather than business perspectives.

De Angelis (2018) further explored the link between CE and business models where the latter were a crucial constituent. A circular business model is one that combines notions of *value proposition, value creation and delivery* and *value capture* within the Ellen MacArthur Foundation (EMF)'s ReSOLVE framework, which represent measures needed to innovate business models according to CE principles. The acronym stands for Regenerate, Share, Optimize, Loop, Virtualize and Exchange. Rizos et al. (2017) summarize multiple CE definitions and interpretations as found in Appendix Table 2. Evident from the surveys of the origins of CE is that the circular economy is conveniently juxtaposed against the alternative linear economy that is often characterized as a 'take-make-use-dispose' paradigm as shown in Figure 3. What is common among the different constructs of CE is the idea of '*closing the loop*' which means the disposal stage at the end feeds back into the production system to create circularity. Thus, the final state of waste disposal is linked to the concept of circularity as it is this waste that is looped back.

Grant, Trautrims & Wong (2017) developed a hierarchy of waste management that comprises the stages of circularity, namely: reduce, reuse, recycle, recover and dispose as shown in Figure 4. Reduce means using fewer resources. Reuse involves use of a product or some parts that can be shared, refurbished (restoring a product to good condition but not comparable to brand new), repaired (fixing a fault that made the product inoperable) or remanufactured (replacing worn non-functional product or component to 'like new' or 'better than new' condition, hence upcycling). Recycle refers to product recovery where waste is separated into materials that may be reprocessed or fitted into new products. Recover is creating energy from waste that cannot be reused or recycled and hence, incinerated. Disposal, in turn, is the final destination for waste as landfill, which may be problematic when materials are toxic or non-biodegradable. In this framework, waste prevention is preferable to waste disposal. Ghisellini et al. (2018) & Kirchherr et al. (2017) extend the hierarchy to comprise the 9Rs (Refuse, Reduce, Reuse, Repair, Refurbish, Remanufacture, Repurpose, Recycle, Recover) which is a more comprehensive construct.

De Groene Zaak & Ethica (2015) amended this hierarchy and introduced a *circularity ladder* concept which probes into the environmental sustainability of different CE business models. A distinction is made between product life extension and sharing models and resource recovery business models where the former potentially reduces the amount of waste generated by slowing resource loops and funneling resource flows whereas the latter simply diverts existing waste towards material and energy recovery. Compared to the waste hierarchy model which ranks waste management options according to environmental impacts associated with the end of life phase, the circularity ladder takes into account the environmental impact of the entire product life-cycle. The circularity ladder is depicted in Figure 5. The EU defines CE using the circular stages approach as a "*production and consumption model which involves reusing, repairing, refurbishing, and recycling existing materials and products to keep materials within the economy wherever possible...waste will itself become a resource, consequently minimizing the actual amount of waste. It is generally opposed to a traditional, linear economy, which is based on a 'take-make-consume-throw away' pattern*" (EU parliamentary research, as quoted in Sillanpää & Ncibi, 2019, pp.8-9). Using this definition, the EU member states designed their circular strategies around the stages model as discussed later below.

*3.2 Building Design Principles*

Tebbatt Adams et al. (2017) note that the application of CE principles to CDI are heavily fixated on construction waste minimization and recycling, despite the existence of other circular strategies. Cheshire (2016), however, suggests that other stages of circularity can be applied in CDI by adopting approaches such as: building in layers, designing out waste, designing for adaptability, designing for disassembly or deconstruction, and selecting materials to allow a building to be more adaptable as its components can be reconfigured for repair or replacement as shown in Figure 6. These stages were identified as retain, refit, refurbish, reclaim/reuse, remanufacture and recycle/compost. The inner three circles depict that retaining existing buildings is the most resource-efficient option followed by refitting and refurbishing while the outermost circles pertain to building materials that could be reclaimed or remanufactured and only recycled or returned to nature as final end. The design principles associated with CE probes into the lifespans of the building components and possibilities of dismantling them for later reuse.

As described in Cheshire (2016), Mossberg (2018), and van Vliet (2018) the philosophy behind *building in layers* is partitioning building elements according to their different lifespans such that their independence allows different layers to be peeled off without damaging the adjoining layers. This is depicted in Figure 7 where a building structure is fragmented into 6S representing site, structure, skin (façade and roof), services, space plan and stuff (Brand, 1994), where the site and structure which is independent of the skin has the longest life span while the services are accessible layers that are replaceable when required, and the space plan and stuff are the short-lived components. *Designing out waste* principle employs lean design to explore waste reduction by reusing or recovering building components, producing off-site (i.e., pre-fabrication), materials optimization, waste efficient procurement and deconstruction, taking account of factors such as embodied carbon content and water use. *Designing for adaptability* involves reconfiguring buildings during their lifetimes by adapting their lay-out to changing markets from residence to retail to leisure and office spaces using multi-space concepts and flexed floor-to-ceiling height requirements. In *designing for disassembly,* parts of the building can be moved or transferred to another site by dismounting in case a site is no longer available as intended. This means that buildings and materials can be valuable assets independent from their site, thus prolonging their value and for extractable materials to be turned into 'materials banks'. Finally, *selecting materials* distinguishes between technical and biological materials where the latter can easily be returned back to nature whereas the former is created using an industrial process and is harder to recycle at end of life.

Figure 8 depicts the benefits of circular built buildings versus linear economy buildings with reduced construction time and less wastage on site, increased rental yields from healthy interiors, higher residual values and longer life of material components, leading to more valuable assets. These benefits outweigh the costs of demolition and disposal of materials and depreciation and loss of rental yield in non-circular buildings.

*3.3 Life Cycle Phases of Construction*

Apart from the stages of circularity, one has to consider as well the stage of construction where circularity principles can be applied. Wahlström et al. (2020) identified the lifecycle in the built environment as: (i) *material production phase* – which refers to selection of building materials as having high recyclable or renewable or hazardous content, durability or life span, and the production processes of these materials having low environmental impacts (see Windapo & Ogunsanmi, 2013); (ii) *design phase* – which looks at resource use versus functionality of a building with deconstruction in view (i.e., easy to disassemble, durability, flexibility, upgradability, reparability and adaptability of structures; (iii) *construction phase* – where stocks of construction materials are recorded, avoiding material surpluses, creating a materials passport and selective sorting of construction waste; (iv) *use phase* – pertains to building maintenance and use intensity of buildings including flexible functionality for different users such as shared work or living spaces, and lifetime extension through retrofitting and rehabilitation of structures; and (v) *end of life phase* – which is the demolition or deconstruction phase involving removal and safe handling of hazardous materials, source sorting for material streams, and auditing of pre-demolition material.

*3.4 Selected EU Member State Country (City/Regional) Strategies on Circularity*

The available literature on country strategies on circularity mostly focus on city-wide and regional strategies in a specific country. Gravagnuolo, De Angelis & Iodice (2019) attribute this to the predicted rise in population that would reach 8.6 billion in 2030 according to the United Nations, accompanied by greater urbanization. As the EMF (2017) attributes cities to account for an average of 70% of greenhouse gas emissions, 75% of natural resource consumption and 50% of global waste production, city involvement in the transition to CE is pivotal for a country to attain its waste recovery targets. The EMF (2017, p. 7) defines a circular city as '*aim(ing) to eliminate the concept of waste, keep(ing) assets at their highest value at all times, and enabled by digital technology. A circular city seeks to generate prosperity, increase live-ability, and improve resilience, for the city and its citizens, while aiming to decouple the creation of value from the consumption of finite resources.*"

As such, the concept of *urban metabolism* is introduced where cities are likened to living organisms where energy and resources flow that allow it to function. Kennedy et al. (2010) define the concept of urban metabolism as the sum total of technological and socio-economic processes occurring in cities that result to growth, energy production and waste elimination. This is in line with the EMF & McKinsey (2012) conceptualization of CE as "an industrial system that is restorative or regenerative by intention or design". A more useful typology akin to urban metabolism was introduced by ESPON (2020) called the *urban circular collaborative economy* (UCCE) involving exchange of unused or underused resources as portrayed in Figure 9. The UCCE combines the triple bottom-line elements of sustainable development— people, planet and profits— into determining environmental, economic and social impacts in the CDI that will be addressed in a later section. The city-wide or region-wide circular strategies in the built environment moreover, are replicated in other cities and regions of the country, making them distinctively country-specific until another participant in CDI experiments with constructing techniques that pertain to another stage or another design principle or another construction life cycle phase in the circularity ladder.

In this paper, the country, regional and/or municipality-led strategies of the selected EU countries of the Netherlands, Belgium, France, Germany and Denmark are summarized in Appendix Table 3. The countries were chosen on the basis of the different stages and design principles of the circularity loop that were targeted to reduce waste and considering that these are among the top 10 waste generators in Western Europe. Based on the comparative table, the Netherlands and Denmark developed a well-defined country strategy to achieve a circular economy in the CDI sector. France developed a country circular economy strategy for all sectors in general where CDI is a subset. Belgium formulated a circular economy strategy for the CDI for the capital region Brussels. Finally, Germany developed a circular economy strategy in general but not specific for the CDI although a regional plan on CE in the construction industry exists.

*3.5 Environmental, Economic and Social Impact Assessments*

Building construction has a pivotal role to play in sustainable development as noted by Zabihi et al. (2012), Yilmaz et al. (2015), Zavadskas et al. (2018), and Zhang et al. (2014) requiring integration of technological, economic, social and environmental benefits. Following the triple bottom-line paradigm of sustainable development in the circular economy, the following section will compare the selected projects on the three dimensions of environmental, economic and social impacts using qualitative methodology of simple ranking based on the case project descriptions in the absence of more detailed environmental impact assessment, life cycle assessment, social impact assessment and economic feasibility studies. A case study approach was one methodology suggested by Best et al. (2018) in studying transitions to circular economies along with scenario-based modelling and statistics and indicators, which are not possible to do under this study given time constraints.

*3.5.1 Environmental Impact*

Singh et al. (2016) describe the common tools used to assess the environmental impact of projects, which are: environmental impact assessment (EIA), ecological footprint (EF) and life cycle assessment (LCA). Performed usually during the planning stage of a project, an EIA identifies and predicts the impact on the environment and outlines mitigation measures to eradicate negative effects of the project on the environment such as carbon or greenhouse gas (GHG) emissions, water use, energy demand, solid waste, exposure to hazardous and toxic materials and other forms of pollution. In Yijun et al. (2011) circular economy theory is applied to EIA. On the other hand, EF, based on the Global Footprint Network (see https://www.footprintnetwork.org/) was introduced in the 1990s to measure how much resources, especially from nature, are consumed and how much waste is generated, hence accounting for the demand for and supply of natural resources. On the demand side, EF aggregates all productive areas for which consumers, society or a given product compete, hence measuring the ecological assets required by users to produce the natural resources they consume and to absorb its waste, specifically carbon emissions. On the supply side, EF measures the bio-capacity or productivity of ecological assets to absorb the waste that an economy generates. The third tool LCA is an international standardized methodology for accounting for the environmental footprint of a product or service within the requirements of ISO 14040 and 14044, as cited by OECD (2019). LCA involves data collection on inputs (resources) and outputs (emissions and waste) that constitute inventory analysis, which are translated into indicators of environmental, health and resource availability impacts that lead to some quantification of the environmental load of products or services throughout their lifetime. In this regard, Ortiz et al. (2009) review research conducted on LCA in the construction or building sector. Meanwhile, Haupt & Hellweg (2019) developed a framework for measuring environmental sustainability in a circular economy that measures the environmental value retained by circular products through the different circularity stages.

*3.5.2 Economic Impact*

Rizos et al. (2017) describe the economic effects of CE as having general employment effects in the recycling business although caution should be placed on net employment as there are also jobs that will be displaced by CE especially in sectors involved in production of primary extractive materials. Other economic effects cited as in Hysa et al. (2020) are trade effects in terms of exports of new recycled materials coupled with reduction of imports of primary materials; value chain multiplier effects in terms of input purchases and delivery channels as well as cross-value chain cooperation; changes in consumption usage and demand patterns; higher value added and greater residual values at end of life of buildings; increase in investments in innovations and new technologies related to recycling and remanufacturing; and savings on maintenance costs of buildings. While most circular projects in the built environment are subsidized, the potential for self-financing is important for replication of similar CE projects, otherwise these become reliant on availability of subsidies.

*3.5.3 Social Impact*

Padilla-Rivera et al. (2020) surveyed how social aspects have been considered and incorporated in the circular economy and suggested thematic areas in terms of labor practices and decency of work, human rights, society and product responsibility. The social impact of the circular economy concerns mostly stakeholder participation as shown by Persson et al. (2004) and Bal et al. (2013) but social impact assessments suggest other indicators relating to income equality, cultural diversity, involuntary settlement, gender and race disparities; social cohesion and inclusion, and preservation of cultural heritage as outlined in IADB (2018). The social life cycle assessment (S- LCA) by UNEP (2009) also covers social and socio-economic aspects along the life cycle of products and services along extraction, processing, manufacturing, distribution, , use, reuse, maintenance, recycling and disposal stages that directly affects stakeholders positively or negatively but many of these indicators overlap with economic and environmental indicators such as local employment, supplier relationships, health and safety of consumers and workers. In this paper, these indicators were used in the economic impact instead of social.

## 4. *Selected EU Projects for Comparison*

The selection of EU projects were driven by the theoretical construct of determinants to project success in CDI presented in the literature review involving circularity stages, building design principles, life cycle stages of construction and availability of national strategies on circularity in the building environment. There was no selection criteria made for the type of building (whether residential or public or commercial) for as long as the project involves construction and demolition activities. Further, the more important consideration for project selection was availability of data in terms of project description, interim or final assessments as cited by official websites of the project sponsors, social media including architectural digests and magazines, or industry associations. Other selection bases were that the project must be located inside the five selected EU member states and represent as much as possible diverse elements of the determinants, e.g., different circularity stages, different building designs, different phases along the construction life cycle. The projects must also allow for expedient comparative analysis in line with the requirements of sustainable development.

The projects chosen are: (1) Superlocal project in the Netherlands (Figure 10); (2) the Circular Retrofit Lab (CRL) project in Belgium (Figure 11) ; (3) Open Air Library in Germany (Figure 12); (4) Rehafutur Engineer's House project in France (Figure 13); and (5) Upcycle Studios in Denmark (Figure 14). Appendix Table 4 compares the project descriptions. Appendix Table 5 compares the determinants for balancing the triptych of sustainability of selected projects on the basis of the determinants which are circularity stages, construction life cycle and designing principles and country strategies and policies on circularity. These determinants are rated from a scale from 1 to 3 with 3 being strong, 2 is moderate and 1 is weak as found in the project case descriptions.

The key points of this table are: (1) the Superlocal project of the Netherlands is distinct for dismounting a whole apartment studio hull and moving it to a new site which is a feature of designing for disassembly, as is the development of bricks that are stacked instead of cemented to facilitate disassembly at end of life for reuse; (2) the Retrofit Lab is prominent for designing for adaptability with reversible solutions being brainstormed by stakeholders including architecture students, contractors and product manufacturers; (3) the Open-Air Library project in Germany will be remembered as a community project that was involved in the design using donated beer crates and reused façade of a historic warehouse as construction materials as well as reclaiming unused abandoned industrial site for the project; (4) the Rehafutur Engineer's House in France exemplifies the selection of materials as it experiments on use of natural or biological resources for insulation while preserving cultural heritage; and (5) the Upcycle Studios in Denmark showcases the use of upcycled materials from construction debris, careful segregation of toxic from non-toxic elements and designing for adaptability where functionality of space is flexed between housing and business configurations.

## 5. Methodology for Comparing Environmental, Economic and Social Impacts

To compare the selected EU projects on the three dimensions of sustainability, a qualitative approach of multi-criteria decision analysis was used where criteria for each sustainability dimension on environmental, social and economic impacts were ranked by importance and then projects were scored based on each criteria used for assessing impacts of the three dimensions. The criteria used for economic, environment and social dimensions were chosen based on indicators cited in impact assessments discussed under Literature Review Section 3.5 of the paper. The literature review which established the determinants of project success was the initial step of the methodology which led to the choice of projects and also described elements of environmental, economic and social impact assessments and identified relevant indicators that were used for evaluating sustainability performance of the selected projects.

In multi-criteria decision making (MDCM) analysis, Odu (2019) outlines several weighting methods broken down into subjective, objective and integrated methods. In this paper, the subjective method of direct rating or ranking was chosen for convenience and expediency as the other methods require more sophisticated statistical techniques, hence, more data requirements. The importance of the criteria in each sustainability dimension (environment, social and economic) were ranked from 1 to 3 where 3 is the highest, depending on whether the criteria is a necessary and sufficient condition to produce the required sustainability effect. A necessary condition must be present for an event to occur while sufficiency establishes conditions that will produce the event. A necessary condition must be present but it alone does not provide sufficient cause for the event to occur. According to Dul (2015), a necessary condition allows an outcome to exist whereas a sufficient condition ensures and produces the outcome. The author uses the analogy of a high score on the Graduate Record Examination (GRE) test for a student to enter graduate school (the outcome) where an adequate score is necessary for the outcome but not sufficient as other admission requirements like a good motivation letter, TOEFL score, recommendation letter, evidence of student performance and grades, etc. are required although a low GRE score would certainly not lead to admission independent of how the student performs on the other requirements. A sustainability criteria is a 3 if it is both necessary and sufficient to produce the sustainability effect, a 2 if it is sufficient but not necessary, and a 1 if it is necessary but not sufficient.

All indicators of each sustainability dimension are likewise scored from 1 to 5 with 5 indicating the highest if it relates specifically to the main construct of circularity as applied in the CDI. A Likert scale from 1 to 5 will be used where the higher the rating, the more the project addresses the sustainability/circularity yardstick or indicator of the sustainability dimension. The scale allows some range of scoring that is not too close than if only 3 scores were used. A score of 5 means that the project carries this indicator as a dominant project objective or intention, a 4 means that the indicator is not dominant but the project intends to address the issue in project implementation; a 3 means the indicator exists in the project but it is not clear as to whether it was intentionally designed or is just a side effect or whether this was indeed achieved even if intended in the absence of final impact assessments; a 2 means that the indicator is just a side effect of the project but was not originally designed nor intended; and a 1 means that the indicator was not addressed at all by the project nor is it a project objective. The scores are then multiplied by the rank of the indicator for each dimension and then summed up to get the total score for the project for that sustainability dimension. Then the ranks of each project on all dimensions are averaged implying equal weights for each dimension of environment, economic and social. The lower the calculated average for rank, the more balanced the project is in attaining the three goals of sustainability and circularity, hence representing best practice. The environmental, economic and social impacts of the projects were compiled by the authors from the sources listed in Appendix Table 6.

### *5.1 Environmental Impact*

In this paper, a combination of indicators from all three tools of EIA, EF and LCA relevant to circularity in CDI will be employed. The indicators for environmental impact are: 1) *reduction of GHG emissions*; 2) *savings on water use*; 3) *savings on energy consumption*; 4) *embodied pollution*; 5) *solid waste reduction*; 6) *product life cycle extension*; 7) *use of biotic materials*; 8) *reduction of exposure to human toxicity*; and 9) *land use*. Appendix Table 7 provides a description of these indicators. The indicators ranked by importance to environmental impact are: 3 for reduction of GHG emissions, water use, energy use, embodied pollution, solid waste reduction; and 2 for use of bio-based materials, land use, product life cycle extension and human toxicity as these are sufficient to achieve the outcome but not necessary conditions. The ratings of the selected projects based on these indicators for the environmental dimension are shown in Appendix Table 8. The Rehafutur project in France scored the highest on environmental impact followed by the Dutch Superlocal Project, the Danish Upcycle Studios, and the Belgian Circular Retrofit Lab, with very close scores. The German Open Air Library ended up with the lowest score. All projects scored equally on solid waste reduction and product life cycle extension which are the main indicators of the waste disposal directive applied to CDI. The projects varied in their scores on reduction of greenhouse gas emissions and energy savings with the highest scores for the Danish and French projects where these two environmental indicators where specifically targeted. Also both projects included human toxicity of building materials into account. The French and Dutch projects scored higher than the Danish project because of the use of bio-based materials as dominant feature along with land use considering that the project was built on UNESCO heritage site for Rehafutur and in the case of Superlocal, land threatened by demographic transition. Land use also boosted the score of the German project which converted abandoned industrial site with fallow land to a green public or community space. So based on environmental impact, the projects are ranked from 1 to 5 with 1 being the best as follows: 1 Rehafutur House of France; 2 Superlocal project of the Netherlands; 3 Upcycle Studios of Denmark; 4 Circular Retrofit Lab of Belgium; and 5 Open Air Library of Germany.

*5.2 Economic Impact*

The economic impact of the projects will be similarly scored and evaluated as in the environmental dimension. As many of the projects do not have ex post evaluation studies, the following indicators based on ex ante case descriptions will be used for assessing their economic impact along with available mid-term impact reviews and press releases: 1) *employment generation*; 2) *trade effects*; 3) *business creation and multiplier effects on the value chain*; 4) *financial self-sufficiency*; 5) *consumption patterns*; 6) *savings on maintenance costs*; 7) *higher residual values*; 8) *increase in innovation and new technologies and associated investments*; and 9) *fiscal effects*. These economic indicators are described or defined in Appendix Table 9. Appendix Table 10 compares the economic impact of the selected EU projects using these indicators. By importance, the indicators for economic dimension were ranked from 1 to 3 with 3 indicating the highest as the indicator is necessary and sufficient to produce the sustainability outcome. Employment generation, business creation and value chain multiplier effects, financial self-sufficiency, consumption patterns, and increase in innovation and new technologies were ranked the most important at 3; followed by trade and fiscal effects, savings on maintenance costs, and higher residual values ranked at 2 by importance. For economic impact, the projects ranked as follows: 1 Upcycle Studios of Denmark; 2 Rehafutur House of France; 3 the Circular Retrofit Lab of Belgium; 4 Superlocal project of the Netherlands; and 5 Open Air Library of Germany. The projects scored equally or fractionally the same in the indicators for higher residual values and business creation and value chain multiplier effects, and in innovation effort and associated investments in new technologies except for the German library project. The Danish project scored high because of the fiscal effect with no amount of subsidy used for financing, having been designed as a commercial project whereas the Dutch, Belgian and French projects were funded by the EU while the German project was partly financed by the federal state although most of the funding were raised from public donations. The employment effects varied with the Danish project intentionally using upcycled materials to be sourced locally to provide local employment whereas the Dutch, Belgian and French projects created jobs during project implementation and potentially in the future with the application of the business models. Trade effects also varied with conscious decisions to reduce reliance on imports for the Danish, Belgian and French projects compared to the Dutch project which entailed importation of machinery from China for the development of the detachable bricks.

*5.3 Social Impact*

In this study, absence of data precluded assessment of other indicators commonly found in social impact assessments such as gender and race equality, cultural diversity, or human rights violations. For the comparison of the selected EU projects, the following indicators for the social dimension are used: 1) *Stakeholder participation*; 2) *Community engagement*; 3) *Preservation of cultural heritage*; 4) *Development of social capital*; 5) *Regional inequality*; 6) *Formation of human capital*; 7) *Demographic transition and change*; and 8) *Social cohesion*. Appendix Table 11 provides a description of these social indicators. The social indicators ranked by importance are as follows: 3 for stakeholder participation, community engagement, development of social capital, regional inequality, formation of human capital, and social cohesion. The remaining two indicators on preservation of cultural heritage and demographic transition and change are ranked a 2 by importance as they can produce social outcomes but are not requisite conditions. The results of the assessment of social impacts of the projects are tabulated in Appendix Table 12. The German Open Air Library ranks the highest with a 1 as it excelled in all the social impact indicators, being a social project from the start. This is followed by 2 Rehafutur House in France;3 the Superlocal project in the Netherlands; 4 Circular Retrofit Lab and 5 the Upcycle Studios. While both projects have elements of cultural preservation and demographic transition and change, the French project edged out the Dutch project in terms of issues of regional inequality considering the geographical significance of the project site. Meanwhile, the Danish project was a brainchild of the architectural company and did not involve much stakeholder participation except with the product manufacturers.

Based on the rankings on the three dimensions of environmental, economic and social impacts, Table 1 summarizes the projects to obtain an average rank.

Table 1. Summary of Ranks on Environmental, Economic and Social Impacts of Selected EU Projects

| Project | Environmental Impact | Economic Impact | Social Impact | Average Rank |
|---|---|---|---|---|
| Superlocal Project, NL | 2 | 4 | 3 | 3 |
| Circular Retrofit Lab, BE | 4 | 3 | 4 | 3.7 |
| Open Air Library, DE | 5 | 5 | 1 | 3.7 |
| Rehafutur House, FR | 1 | 2 | 2 | 1.7 |
| Upcycle Studios, DK | 3 | 1 | 5 | 3 |

The results show that the Rehafutur House in France achieved a neat balance, hence, best practice of environmental, economic and social dimensions of sustainable and circular construction. This was followed by both the Danish Upcycle Studios and Dutch Superlocal projects which only varied in their rankings for social and economic impacts where the former was stronger on economic but weak on social and conversely, the latter was stronger on social and weaker on economic effects. The Circular Retrofit Lab of Belgium and the German Open Air Library tied in last place although the Belgian project was more balanced in addressing the three dimensions compared to the German project which only scored strongly on social impact.

### *6 Findings and Conclusion*

The CDI sector generates the most waste among all economic sectors and has been the target of sustainability and circularity efforts of the EU. While the overall aim is to reduce waste, the paper has shown that the success of a project on balancing sustainability goals of environment, economic and social objectives are determined by the stages of circularity the project operates in, the building design principles it applies, the phase of the construction life cycle where it is focused and the overall policy framework that can be national, municipal/city-wide or regional that guides the project conception.

Depending on the stage of circularity, the higher the rung a project targets in the waste hierarchy or circularity ladder from merely reuse or recycle to refurbish and remanufacture, the more economic benefits accrue, specifically business creation and value chain multiplier effects, greater residual values, increase in innovation effort and new technologies along with associated investments, and possible trade effects by reduced imports of primary materials and potential rise in exports of circular building materials. These augments the employment effects that already exists at lower rungs of the ladder. Meanwhile the environmental effects are enhanced the higher a project goes through the waste hierarchy or even the circularity ladder. More than simply reducing solid waste and greenhouse gas emissions or saving water and energy, more environmental benefits accrue with product life cycle extension and use of natural or bio-based materials.

The building design principles discussed in the paper also influence the triple bottom line. Building in layers, building for adaptability and building for disassembly have the most potentials as shown by the projects to attain more economic, environmental and social objectives. In the Danish and Belgian projects, designing for adaptability/disassembly or what in the Belgian case is called reversible building solutions create possibilities for the sharing economy business model where different reconfigurations of a building create flexibility for office and living spaces, fostering not only income potentials for entrepreneurs but also formation of social capital. A key first step though is selecting materials for the built environment and evaluating their embodied pollution in addition to toxicity levels on human health, which was evident in majority of the projects, thus enabling a system of materials passport.

In the construction life cycle, the design phases and use phases are critical in framing the building design principles discussed above while the use phase looks at business model applications like the sharing economy and reversible functionalities as shown in the Danish and Belgian projects, respectively, where more economic and social benefits accrue. In the material production, construction and end of life stages of construction, more environmental impacts are achieved as waste reduction and embodied pollution of materials are performed more in these stages.

The cross country comparison of projects show that projects that involve use of biotic resources in the environmental indicators (i.e., the Rehafutur House in France, Superlocal project in the Netherlands and CRL project in Belgium) edged out the other projects that do not. Land use is another feather in the environmental cap as shown in the French, Dutch and Belgian projects but also in the German project where abandoned fallow land was reclaimed. Checking for the toxicity of building materials is also integral to environmental impact which boosted the score of the Danish project as many demolished materials were constructed in the 1950s to 1970s when lead and asbestos were commonplace building fixtures. Embedded or embodied pollution in materials should also be considered and not just the visible reductions in energy and greenhouse gas emissions that circular projects offer as evident in the French, Dutch and Belgian pilots.

In the economic realm, financial self-sufficiency of the projects is critical for circular projects to thrive as availability of subsidies limit replication of similar initiatives. The Upcycle Studios in Denmark showcase the importance of market solutions in attracting consumer acceptance when offering circular building materials that are equivalently aesthetic and price-competitive as conventional alternatives.. The German project in a way underscores the momentum that the private sector (social or community-led) cooperation can exert on circular initiatives which later on in the project cycle drew supplemental federal support to fill in the financial gap needed for the project. While subsidies help in triggering pilot circular prototypes as displayed by the Dutch, French and Belgian projects, more self-sustaining funding will be required for these circular projects to thrive. Even in the absence of the self-financing solution that the market offers, the Danish landfill tax system to finance recycling technologies would achieve the same effect as government revenues collected from landfilling will be diverted to finance circular projects in CDI and hence be self-continuing. Multiplier effects from commercialization and scaling of innovation and newly discovered technologies on the value chain such as those from eco-materials used for insulation in Rehafutur, the stackable bricks of Superlocal, the reversible wall panels of the Retrofit Lab, and the upcycled wood, glass and concrete of Upcycle Studios amplify the job creation effects on the economies of the sampled countries.

The social aspect which is often the ignored pillar of the sustainability triptych should transcend just mere stakeholder participation but a deeper community engagement and ownership of the project as validated by the German Open Air Library where residents not only donated the beer crates and books but took over running the management and maintenance of the project. The downside to this private communal endeavor is the 'tragedy of the commons' that public property are prone to, with incidences of vandalism inflicted on the library. However, as economic theory suggests, Coasean bargaining solutions as opposed to Pigouvian state solutions may prevail as attested to by residents assuming the maintenance responsibility. Many of the projects exploited the opportunities for developing social capital through co-ownership and formation of clusters and networks of users, suppliers and producers of circular building materials as shown in the Dutch, Belgian and French projects, as well as fostering social cohesion among residents of the demolished buildings of Superlocal or the artisans and farmers of Rehafutur for processing the eco-materials. The element of a sharing economy or business model also bolsters the social impact as shown by the flexible residential/workspace townhouses of Upcycle Studios or the reversible functionalities of building space that CRL in Belgium endorses.

The projects selected for this comparative study comprised countries that reflect country-wide, city-wide or region-wide strategies of circularity with only the Netherlands and Denmark articulating a national circular strategy in CDI. France composed a national strategy on CE but more encompassing several sectors including CDI. Belgium adopted a circular strategy for the construction industry but was not national but focused on the capital region. Finally, Germany has a regional circular strategy but not specific for construction. This lack of political impetus shows why the German project lagged behind the other projects and was more a grassroots attempt, hence more dominantly displaying its social enterprise nature. The point here being made is that for the circular mentality to take root in society, political ascendancy is needed and is an antecedent to steer stakeholders to this mindset.

The basic limitations of the study is the *ex-ante* nature of the analysis, relying only on project or case descriptions and in some cases mid-term reviews. There are also partial overlaps in the determinants such as between circularity stages and life cycle stages that could possibly lead to double counting or over- and under-evaluation of some projects. The scoring and ranking system may also be subjective although the findings would not have been significantly altered unless the multi-criteria decision making analysis were more developed mathematically or scientifically. The choice of projects also do not speak volumes about the national strategy as a better project could have boosted the ranks or the scores of a particular country. To achieve a better comparison among EU experiences on circularity in CDI, several projects like 3 to 4 projects and not just one prototype project of that country would be a better gauge. There is also need for further study on multi-criteria decision analysis in the context of circularity like the use of objective or integrated methodologies to properly weigh the different criteria on environmental, social and economic impacts. Another area for future study is to include more EU/European countries (i.e., Finland, Spain, Italy, Sweden, Austria, UK, Switzerland, Norway) and the Eastern European countries (Slovenia, Hungary, Czech Republic, Poland) which are the largest waste generators and which have themselves defined circular strategies in CDI at the state, regional or metropolitan level. Finally, a more global comparison of experiences in circularity in CDI with Asian countries (e.g., Japan and Korea), countries on the American continent (e.g., Canada, Mexico, USA), Africa (South Africa) and Oceania (Australia) may unlock significant insights.

Figure 1. Waste Generation by Economic Activity and Households, EU-27, 2018

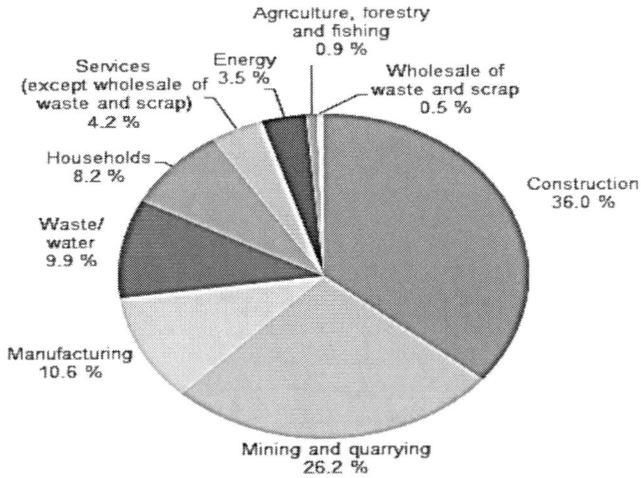


Source: Eurostat (online data code: env_wasgen)

Figure 2. Per Capita Waste Generation in the EU-27 (in kg)

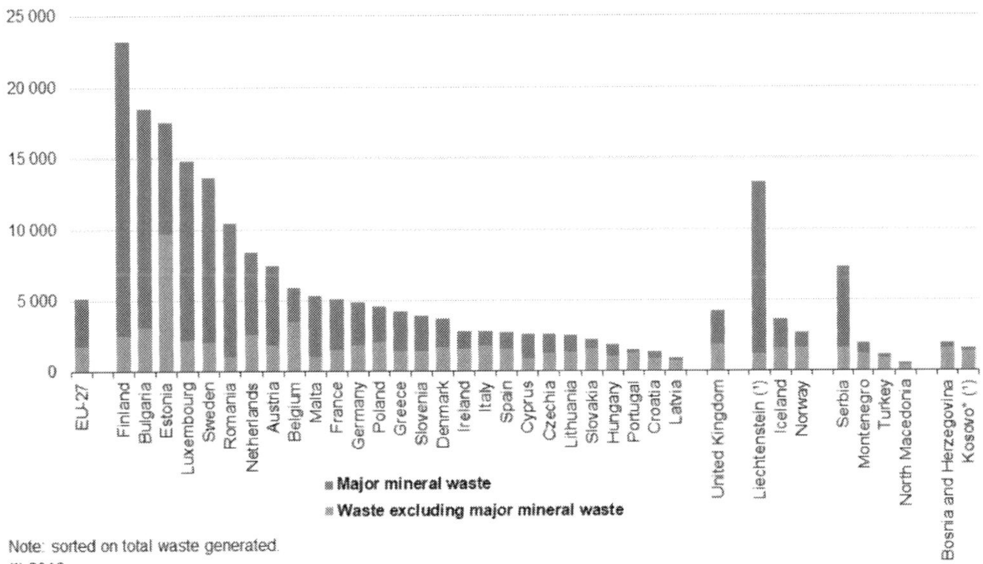


Note: sorted on total waste generated.
(¹) 2016.
* *This designation is without prejudice to positions on status, and is in line with UNSCR 1244/1999 and the ICJ Opinion on the Kosovo declaration of independence.*

Source: Eurostat (online data code: env_wasgen)

Figure 3. Circular vs. Linear Economy

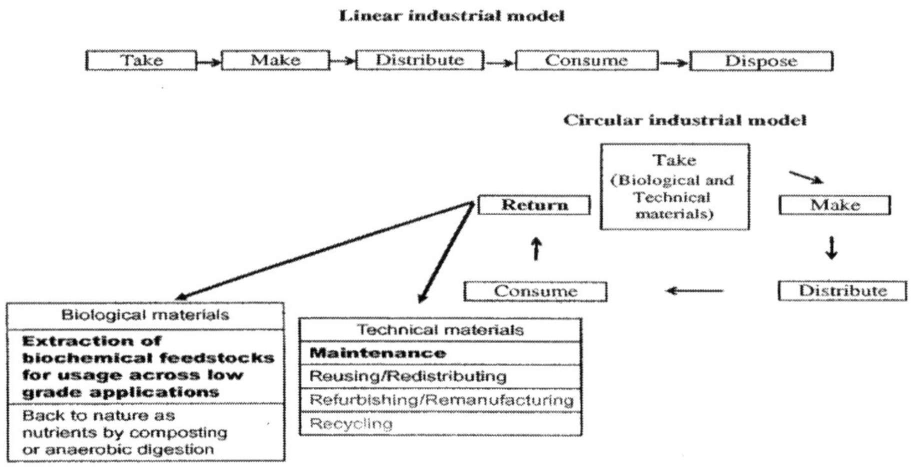


Source: De Angelis (2018), p.21.

Figure 4. Hierarchy of Waste Management

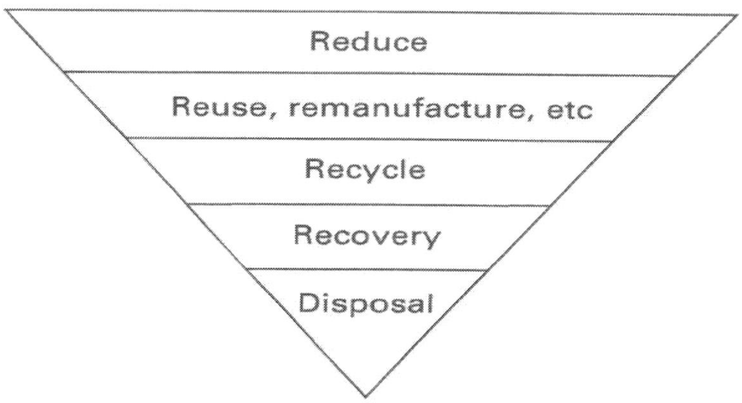


Source: Grant, Trautrims & Wong (2017), p. 183.

Figure 5. Circularity Ladder Concept

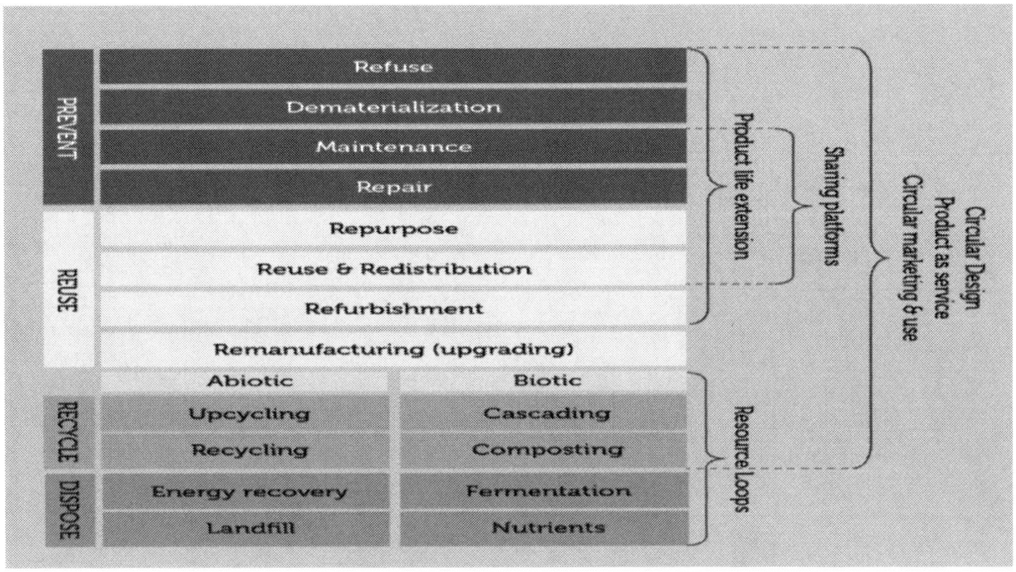


Source: De Groene Zaak & Ethica (2015)

Figure 6. The circular economy stages in the building environment

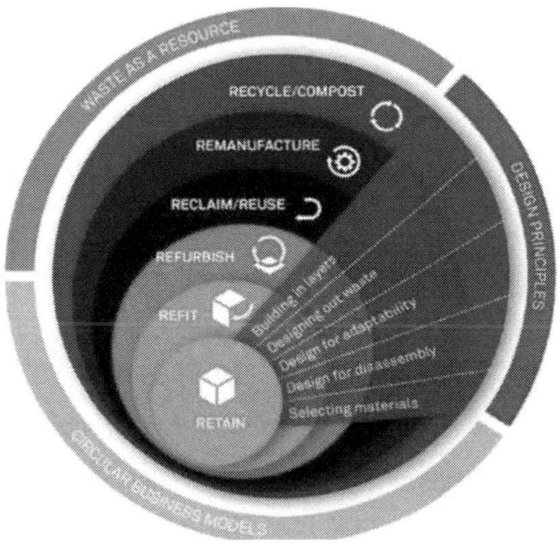


Source: Cheshire (2016)

IOP Conf. Series: Earth and Environmental Science **855** (2021) 012017 doi:10.1088/1755-1315/855/1/012017

Figure 7. Building in Layers

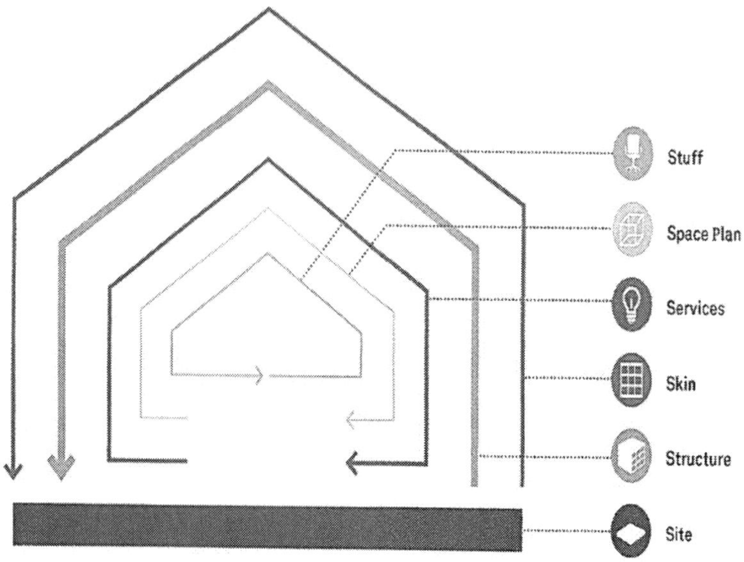


Source: Brand (1994)

Figure 8. Benefits of Circular Economy Buildings

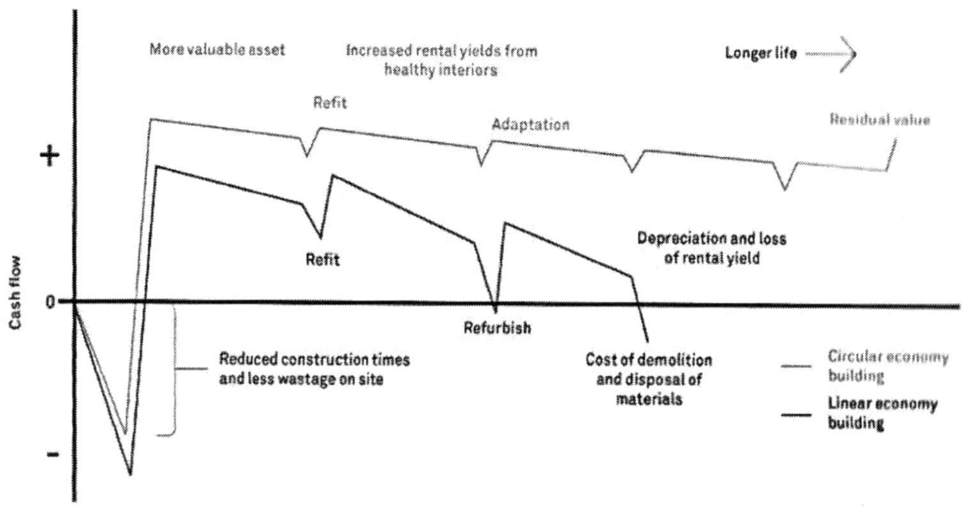


Source: Cheshire (2016)

Figure 9. Concept of Urban Circular Collaborative Economy

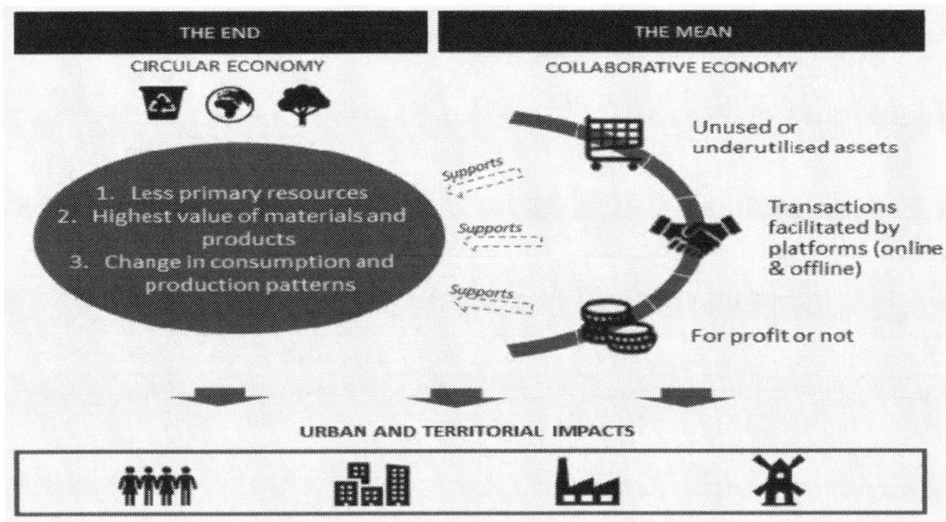


Source: Valdani, Vicari & Associati (VVA) Economics and Policy as cited in ESPON (2020), p. 8.

IOP Conf. Series: Earth and Environmental Science **855** (2021) 012017 doi:10.1088/1755-1315/855/1/012017

Appendix Table 1. Recovery Rates of Construction and Demolition Waste in the EU (%)

| Geo\time | 2010 | 2012 | 2014 | 2016 |
|---|---|---|---|---|
| EU (28 countries) | 79 | 86 | 89 | 90 |
| Euro area (19 countries) | : | : | : | : |
| Belgium | 17 | 18 | 32 | 95 |
| Bulgaria | 62 | : | : | 90 |
| Czechia | 91 | 91 | 90 | 92 |
| Denmark | : | 91 | 92 | 90 |
| Germany | 95 | 94 | : | : |
| Estonia | 96 | 96 | 98 | 97 |
| Ireland | 97 | 100 | 100 | 96 |
| Greece | 0 | 0 | 0 | : |
| Spain | 65 | 84 | 70 | 79 |
| France | 66 | 66 | 71 | 72 |
| Croatia | 2 | 51 | 69 | 76 |
| Italy | 97 | 97 | 97 | 98 |
| Cyprus | 0 | 60 | 38 | 57 |
| Latvia | : | : | 92 | 98 |
| Lithuania | 73 | 88 | 92 | 97 |
| Luxembourg | 98 | 99 | 98 | 100 |
| Hungary | 61 | 75 | 86 | 99 |
| Malta | 16 | 100 | 100 | 100 |
| Netherlands | 100 | 100 | 100 | 100 |
| Austria | 92 | 92 | 94 | 88 |
| Poland | 93 | 92 | 96 | 91 |
| Portugal | : | : | : | 97 |
| Romania | 47 | 67 | 65 | 85 |
| Slovenia | 94 | 92 | 98 | 98 |
| Slovakia | : | : | 54 | 54 |
| Finland | 5 | 12 | 83 | 87 |
| Sweden | 78 | 81 | 55 | 61 |
| United Kingdom | 96 | 96 | 96 | 96 |
| Iceland | 75 | 100 | 86 | 99 |
| Liechtenstein | : | : | : | : |

Data from Eurostat, dataset (cei_wm040)

Appendix Table 2. Definitions and Interpretations of the Circular Economy

| Source | Definition/interpretation |
|---|---|
| Sauvé et al. (2016) | Circular economy refers to the "production and consumption of goods through closed loop material flows that internalize environmental externalities linked to virgin resource extraction and the generation of waste (including pollution)". |
| Preston (2012) | "Circular economy is an approach that would transform the function of resources in the economy. Waste from factories would become a valuable input to another process – and products could be repaired, reused or upgraded instead of thrown away". |
| EEA (2014) | Circular economy "refers mainly to physical and material resource aspects of the economy – it focuses on recycling, limiting and re-using the physical inputs to the economy, and using waste as a resource leading to reduced primary resource consumption". |
| Mitchell (2015) | A circular economy is an alternative to a traditional linear economy (make, use, dispose) in which we keep resources in use for as long as possible, extracting the maximum value from them whilst in use, then recovering and reusing products and materials. |
| Heck (2006) | The utilisation of sustainable energy is crucial in a circular economy. The transition to a circular economy would require addressing the challenge of establishing a sustainable energy supply as well as decisive action in several other areas such as agriculture, water, soil and biodiversity. |
| Su et al. (2013) | The focus of the circular economy gradually extends beyond issues related to material management and covers other aspects, such as energy efficiency and conservation, land management, soil protection and water. |
| Bastein et al. (2013) | The circular economy transition "is an essential condition for a resilient industrial system that facilitates new kinds of economic activity, strengthens competitiveness and generates employment". |
| EEA (2016) | "A circular economy provides opportunities to create well-being, growth and jobs, while reducing environmental pressures. The concept can, in principle, be applied to all kinds of natural resources, including biotic and abiotic materials, water and land". |
| Ghisellini et al. (2016) | The radical reshaping of all processes across the life cycle of products conducted by innovative actors has the potential to not only achieve material or energy recovery but also to improve the entire living and economic model. |
| ADEME (2014) | The objective of the circular economy is to reduce the environmental impact of resource consumption and improve social well-being. |
| Ellen MacArthur Foundation (2013a; 2013b; 2015a) | Circular economy is "an industrial system that is restorative or regenerative by intention and design. It replaces the 'end-of-life' concept with restoration, shifts towards the use of renewable energy, eliminates the use of toxic chemicals, which impair reuse, and aims for the elimination of waste through the superior design of materials, products, systems, and, within this, business models". The overall objective is to "enable effective flows of materials, energy, labour and information so that natural and social capital can be rebuilt". |
| European Commission (2015a) | The circular economy is an economy "where the value of products, materials and resources is maintained in the economy for as long as possible, and the generation of waste minimised". The transition to a more circular economy would make "an essential contribution to the EU's efforts to develop a sustainable, low-carbon, resource-efficient and competitive economy". |

Source: Rizos, Tuokko & Behrens (2017), p.6.

Appendix Table 3. Country (City-Municipality and/or Regional) Strategies on Circularity in the CDI Sector

| Country | Strategy Document | Goal | Strategy/ Approach |
|---|---|---|---|
| Netherlands | Holland Circular Hotspot (2018); Breen et al. (2018); Pötting et al (2018); Schut et al. (2015) | • Make Dutch economy fully circular by 2050,<br>• Reduce use of primary raw materials by 50%<br>• Halving CO2 emission in construction industry by 2030<br>• Realizing high grade (or higher grade) recycling in all construction submarkets (civil engineering & utility construction, (C&U), ground water & road construction (GWR)<br>• Construction of 1 million energy neutral houses in circular way in period of 10 years per De Bouwagenda<br>• Develop knowledge, experience & tools for people in whole construction chain<br>• Research into stimulating socio-innovation<br>• Develop uniform circularity measuring method | • Encourage use of granulated/recycled concrete mixed with new<br>• Apply circular criteria on public land used for housing development projects<br>• Training for large contractors on circular procurement and construction<br>• Optimal use of materials for all phases in the construction cycle<br>• Use many infinite sources as possible with more and higher grade reuse at end of life phase<br>• Make use of finite sources as efficiently as possible<br>• Pioneer research and execute practical projects from experiment to demonstration to market introduction of new products and services for circular construction<br>• All public tenders will be circular by 2030<br>• Mandatory materials passport by looking at value added of all materials and resources used in buildings<br>• Government subsidy and financial support at individual business level for circular business and revenue models<br>• Circular construction as comprehensive part of education in 2021 involving architectural training and circular commissioning training programs |
| Belgium | Fontaine & Dewulf (2018); Fontaine & Dewulf (2019); Breen et al. (2018) | • Circular design, building maintenance, reuse and remanufacturing of construction waste in the territory<br>• Design buildings that allow them to adapt to new uses while limiting environmental impact, promote maintenance of material resources, build from recovered elements<br>• Integrate circularity in territory programming<br>• Stimulate/support requests of clients<br>• Stimulate/accompany the offer with professionals<br>• Develop a sector of re-employment materials<br>• Prepare/create a favorable logistics framework<br>• Prepare/create a favorable regulatory framework | • Transitioning businesses to more circular use of materials and subsidized collaborative work spaces (i.e., Greenbizz program)<br>• Pilot projects for reversible construction<br>• Business park created for CE companies<br>• Co-creation workshops with stakeholders<br>• Adopted "Be circular" to become resource efficient, maximize local resources to boost entrepreneurship and create employment opportunities<br>• Support new retail businesses and start-ups/ existing businesses to adopt CE principles |

| | | | |
|---|---|---|---|
| France | Ministry of Ecological and Solidarity Transition (2018); Breen et al. (2018) | • Make waste collection in built environment more effective by combatting use of landfill and enable good sorting of construction materials so these can be recycled<br>• Move to assessment/ inventory process for reuse and recovery of waste from renovation and demolition sites | • Make a building stock bank for future construction materials<br>• Material recovery from exchange, repair, reuse, and recycling<br>• Waste reduction<br>• Networking of and synergies among actors<br>• Raise awareness and train contractors on CE |
| Germany | The Circular Economy in Southern Germany (2019); Breen et al. (2018); Lah (2016) | • Raising resource efficiency in production and consumption<br>• Reduction of using non-renewable resources in construction and replacement by re-growable raw materials<br>• Extending the life span of buildings and reconstruction using passivhaus (energy efficient) standards | • Increased application of recyclate materials in construction instead of natural resources at no additional costs<br>• Quality insurance of recyclate materials<br>• Information campaign promoting usage of recyclate materials |
| Denmark | Deloitte (2015); Ministry of Environment and Food and Ministry of Industry, Business and Financial Affairs (2018); Breen et al. (2018) | • The recycling of construction and demolition waste (CDW) does not cause contamination of environment and maintains high recycling rate<br>• Identifying, separating and safe handling of problematic substances in buildings before demolition<br>• Improve quality of CDW using new technologies for specific waste streams aiming for high grade recycling (upcycling)<br>• Focus on quality recycling of CDW | • Develop new technological solutions for waste management by recycling and embracing CE principles where new technologies aim for high-grade recycling for specific waste streams (concrete, bricks, wind turbine blades, impregnated wood, district heating pipes, tar paper, etc.)<br>• Final recycling and upcycling of resources<br>• Screening by developers to assess whether buildings or structures contain polychlorinated biphenyls (PCBs)<br>• Sorting of CDW into 10 waste fractions (natural stone, non-glazed tiles, concrete, mixtures of the first 3, iron and metal, gypsum, rock-wool, soil, asphalt and mixture of concrete and asphalt_<br>• Stakeholder involvement in construction and renovation of buildings and structures (producers of construction products, contractors and architects)<br>• Use of landfill tax with lower rates for recyclable materials to reduce amounts to be landfilled and finance development of recycling technologies |

Source: Compilation of authors from sources cited in column 2 of the table.

Appendix Table 4
Selected EU Projects Case Descriptions

| EU Projects | Project Case Descriptions |
|---|---|
| Superlocal, Netherlands<br>Figure 10. Superlocal Project, Netherlands<br><br>Source: www.superlocal.eu | The Superlocal project in the Netherlands entails the demolition of 10-storey apartment blocks into components ranging from complete apartments to concrete aggregates for use in building new houses (superlocal.eu) and construction of 100 new homes in a district facing population decline and threat of vacant occupancy. The project used three construction techniques: 1) removal of an entire studio unit and reusing the hull as shell for new houses; 2) remanufacturing concrete bricks that can be stacked on top of each other without cementing them but using steel tension cables or rods to form a concrete wall, thus facilitating disassembly at end-of-life for reuse; and 3) retrieving concrete rubble from demolition to be upcycled into new circular concrete that reduces use of new cement, water and plasticizer usage. The project was funded by a €4.8 million EU subsidy under the Urban Innovative Actions program and is a collaborative effort with 12 associations that range from surrounding municipalities, universities, construction companies, utilities provider, and homeowner's association to experiment on new building techniques in the circular construction of homes. This project combines retain, reuse, and refit stages in circularity with elements of designing for disassembly and designing for adaptability using Cheshire's (2016) framework. The Superlocal project in the construction lifecycle is oriented towards the material production, design, construction and end of life phases. |
| Circular Retrofit Lab, Belgium]<br>Figure 11. Circular Retrofit Lab, Belgium<br><br>Source: www.bamb2020.eu | The Circular Retrofit Lab (CRL) project in Belgium is a pilot project of the Buildings as Material Banks (BAMB) initiative, involving renovation and transformation of eight student houses that were built using prefabricated concrete support modules which could be arranged in spatial ways leading to urban cluster configurations through reversible solutions (bamb2020.eu/topics/pilot-cases-in-bamb/retrofit-lab/). Renovation involves internal, external and spatial transformation of concrete structural modules. The Retrofit Lab is a pilot project owned by the Vrije Universiteit Brussel (VUB) for circular reconstruction that engages co-creation process with stakeholders (architecture students, contractors and construction material suppliers) along the redesign, rebuild, reuse, and dismantling stages of on design and test solutions with hands-on workshops for students and to test integration of materials passport. Reversible solutions for circularity. The team in charge of the project organized roundtables with industry stakeholders to debate internal partition walls and facades using modular, pre-fabricated and kit-of-part design approach to foster flexibility in assembly, manufacturing efficiency and scalability in implementation. New and reversible functionality of partition walls to living room separation was explored together with product manufacturers. The project employs refit, refurbish, remanufacture principles of Cheshire (2016) with strong design for adaptability and building in layers applications. The Retrofit Lab could be positioned in the material production, design, construction and use phases of the life cycle. |
| Open Air Library, Germany<br>Figure 12. Open-Air Library, Germany<br><br>Source: www.archdaily.com/39417/open-air-library-karo-architekten | The Open Air Library in Germany is a community endeavor that reclaimed land on abandoned industrial site using beer crates as building material and donations of books and other materials from residents (archdaily.com/39417/open-air-library-karo-architekten). The library was constructed in 2005 in an abandoned fallow industrial site in Magdeburg, East Germany. Started as a public initiative with supplemental funding from the federal government, the library used beer crates lent by local residents as building material. The residents filled the shelves of the library with books they donated which now houses a collection of 30,000 as of the latest count. The shelves are never closed and anyone can just take a book without registration provided they are returned or replaced with another book. The outdoor space was designed with local residents with a stage for communal functions such as theater plays for the elementary school, concert gigs for youth bands and other cultural events. Besides the social aspect, the architecture holds historic significance as it re-used the facade of the old Horten warehouse of the city of Hamm which was demolished in 2007. The project exemplifies the reclaim and reuse stage of circularity and applies designing as waste principle. In terms of the construction lifecycle, the Open Air Library involves the design, construction and use phases. |
| Rehafutur Engineer's House, France | The French project Rehafutur Engineer's House is unique in that renewable eco-materials such as flax fiber, wood fiber, hemp bricks, sheep wool, recycled textile and the like were used as insulation materials [Derbal, Lucas, Brachelet, Defer & Antczak (2016); CE100 (2016)]. The |

IOP Conf. Series: Earth and Environmental Science **855** (2021) 012017 doi:10.1088/1755-1315/855/1/012017

| | |
|---|---|
| Figure 13. Rehafutur Engineer's House, France<br><br>Source: www.ed2e.com | project also involves preservation of cultural heritage as it renovated a typical mining house into offices in the Loos-en-Gohelle region in Northern France which is a UNESCO-listed historical site. This is one of the projects under the EU-sponsored CAPEM (Cycle Assessment Procedure for Eco-Impacts of Materials) consisting of improving the building thermal insulation. Because of the historical value, the external façade could not be touched so that retro-fitting could only be done in the interiors. Reuse of existing materials were applied by moving two marble fireplaces as ornamental pieces in public rooms, by relaying 62 square meters of hundred-year-old spruce floorboards after careful dismantling to install the new floor insulation, by reusing 18 square meters of multi-colored cement tiles, and by repaving the parking spaces and access paths with 350 square meters of rubble. The project applies the retain, refit and refurbish phases of Cheshire (2016) with selection of materials as the major design principle. In the construction lifecycle, the project involves material production, design, and construction stages. |
| Upcycle Stuidos,<br>Denmark<br>Figure 14. Upcycle Studios, Denmark<br><br>Source: www.lendager.com | Finally, the Upcycle Studios in Denmark consists of twenty terraced houses that were constructed using upcycled waste materials consisting of glass, wood and concrete refuse (urbannext.net/upcycled-studios/). Based on the concept of raw urban context where houses blend with surroundings while economy business model as principles of new housing, where ownership is cast aside in favor of access to resources. This is facilitated by the use of the houses at all hours of the day not just by the owners but by creative freelancers and self-employed entrepreneurs while serving as one dwelling promoting sustainable living, the Upcycle Studios uses symbiosis, resource efficiency and the shared for large families as houses can be split into two separate apartments. The owners can earn extra income by renting out one apartment either permanently a la Airbnb, or temporarily while waiting for their household memberships to enlarge at a time when only few rooms are used. The Upcycle Studios, built by the Lendager Group, is located in the Ørestad Syd district of Denmark, requiring balancing of housing and business elements. Similar to the Rehafutur house in France, the Upcycle Studios blends new building with history by reusing walls, residual wood, crushed concrete and windows from abandoned buildings and construction waste that are non-toxic certified. About 75 percent of the windows originate from abandoned buildings in North Jutland, Denmark; 1,400 tons of upcycled concrete are forged from durable concrete waste from the construction of the Copenhagen Metro; and the wood and floors are manufactured from off-cuts of Dinesen which is Denmark's leading manufacturer of plank floors. The Upcycle Studios combine refit-refurbish, reuse, remanufacture stages of circularity and applies the designing principles of selecting materials, building in layers, and designing for adaptability. In the construction life cycle, the project manifests aspects of material production, design, construction, and use phases. |

IOP Conf. Series: Earth and Environmental Science **855** (2021) 012017 doi:10.1088/1755-1315/855/1/012017

Appendix Table 5
Comparison of Selected Projects on Basis of Circularity Stages, Construction Life Cycle and Design Principles

| | *Superlocal* Netherlands | *Retrofit Lab* Belgium | *Rehafutur House* France | *Open Air Library* Germany | *Upcycle Studios* Denmark |
|---|---|---|---|---|---|
| *Circularity Stages* | | | | | |
| • Retain | 3 | 3 | 3 | 2 | 2 |
| • Refit | 3 | 3 | 3 | 3 | 3 |
| • Refurbish (Upcycle) | 2 | 3 | 2 | 2 | 3 |
| • Reclaim/Reuse | 3 | 3 | 3 | 3 | 3 |
| • Remanufacture | 3 | 3 | 3 | 1 | 3 |
| • Recycle/Dispose (Downcycle) | 3 | 2 | 3 | 2 | 3 |
| *Construction Life Cycle* | | | | | |
| • Material production | 3 | 3 | 3 | 1 | 3 |
| • Design | 3 | 3 | 3 | 3 | 3 |
| • Construction | 3 | 3 | 3 | 3 | 3 |
| • Use | 2 | 3 | 3 | 3 | 3 |
| • End of life | 3 | 3 | 2 | 1 | 3 |
| *Design Principles* | | | | | |
| • Build in layers | 3 | 3 | 3 | 3 | 3 |
| • Design out waste | 3 | 3 | 3 | 3 | 3 |
| • Design for adaptability | 2 | 3 | 2 | 2 | 3 |
| • Design for disassembly | 3 | 2 | 2 | 1 | 2 |
| • Material selection | 2 | 2 | 3 | 2 | 3 |

Source: Authors

Appendix Table 6
Sources for Project Descriptions

- Netherlands:
  - Superlocal Play Publicatie (n.d.); Superlocal Indexation (n.d.); Burenboek (A)BCD (n.d.) downloadable from www.superlocal.eu
  - Fernandez, A. (2020). *Superplocal Flash Lecture*, Zuyd University of Applied Sciences
  - Ritzen et al. (2019)

- Belgium:
  - Paduart et al. (2018)
  - D12. Feasibility Report,
  - D13 Protyping + Feedback report,
  - D14 4 Pilots Built + Feedback Report downloadable from https://www.bamb2020.eu/topics/pilot-cases-in-bamb/retrofit-lab/

- Germany:
  - https://www.archtopia.com/karo-architekten-open-air-library/
  - https://inspiration.detail.de/open-air-library-in-magdeburg-103470.html
  - https://designlike.com/open-air-library-in-magdeburg-germany/
  - https://inhabitat.com/stunning-open-air-library-pops-up-in-east-germany/
  - https://creatingvibrantcommunities.com/shared/open-air-library-magdeburg-germany/
  - https://miesarch.com/work/2769
  - https://www.dezeen.com/2009/10/28/open-air-library-by-karo/
  - https://www.archdaily.com/39417/open-air-library-karo-architekten
- France:
  - CE100 (2015)
  - Gravagnuolo et al. (2019)
  - Derbal et al. (2017)
- Denmark:
  - https://lendager.com/en/architecture/upcycle-studios-en/
  - https://urbannext.net/upcycled-studios/
  - https://arcspace.com/feature/upcycle-studios/
  - https://www.dezeen.com/2019/04/16/upcycle-studios-townhouses-lendager-group-copenhagen-recycled-materials/
  - https://architizer.com/projects/upcycle-studios/
  - https://issuu.com/lendagertcw/docs/achangemakersguidetothefuture_2.udg

Appendix Table 7
Description of Environmental Indicators

| Indicator | Description |
|---|---|
| Reduction of GHG emissions | refers to $CO_2$ emissions from energy consumed in building material production as well as release of volatile organic compounds such as nitrogen oxides and sulphur dioxides in the manufacture and transport of construction materials that lead to ozone depletion |
| Savings on water use | pertains to reduction in water use in construction activity but also developing building products that enable building occupants to recycle or use alternative water sources such as rainwater runoff |
| Savings on energy consumption | relates to use of energy saving materials in the design and construction of buildings, improving building insulation and sealing performance, and enhancing the efficiency of heating, cooling, ventilation and lighting systems to reduce energy consumption |
| Embodied pollution | concerns the extent of energy, carbon and water used in the extraction, production, transport and use of construction materials, implying higher ecological consequences |
| Solid waste reduction | entails the extent that unwanted products from demolition, pre-construction, on-site construction and post construction are managed by sorting, reducing, reusing and recycling of waste materials including sorting before use as landfill |
| Product life cycle extension | relates to selection of materials with recycled content, high durability and resilience from shocks and high recycling or reuse potential in the design and pre-construction phase; building performance that minimizes need for maintenance and replacement; and maximizing recovery of materials at end of life considering as well ease of deconstruction or disassembly |
| Use of biotic materials | concerns the use of conventional bio-based materials made from plant and animal materials that are biodegradable and emerging bio-renewable materials that are extracted by industrial processes or produced from materials with biological origins but can be regrown |
| Reduction of exposure to human toxicity | involves segregation and safe disposal of hazardous and toxic construction waste upon demolition, replacement of existing toxic building materials such as asbestos-based materials, radioactive materials, lead plumbing, materials that release toxic fumes during a fire, and emit volatile organic compounds from paints and varnishes and impregnating agents (see Pacheco-Torgal et al., 2011) |
| Land use | Relates to land occupied by construction assets as well as embodied land use in non-renewable and renewable raw materials and energy over the building's value chain which affect soil sealing (when agricultural or other non-developed land is built upon), soil compaction (high loads caused by construction that reduces soil absorptive capacity), change of land use (ecological land which is built up causing biodiversity loss, soil degradation, and changes in hydrology) (see Häkkinen et al.,2013). |

Appendix Table 8. Environmental Impact of Selected EU Projects

| Indicators for Environmental Sustainability | Rank of Importance | **Superlocal, NL** | Score | **Circular Retrofit Lab, BE** | Score | **Open Air Library, DE** | Score | **Rehafutur House, FR** | Score | **Upcycle Studios, DK** | Score |
|---|---|---|---|---|---|---|---|---|---|---|---|
| • Reduction of GHG Emissions | 3 | Planned use of equipment (cranes, loaders) and labor to lower $CO_2$ emissions plus on-site construction to minimize transportation | 4 | Construction practices and plants are used to reduce emissions. Building materials used are prefabricated offsite to minimize onsite construction and reduce emission. | 3 | Designed as open air by residents to reduce emissions | 3 | Avoided use of 17 semi-trailers to carry debris to landfill and monitoring of $CO_2$ emissions during use phase of the building | 5 | The upcycled materials used to build the residential buildings help minimize emissions. Upcycle Studios reduces the total $CO_2$ emissions considerably over the life of the building. Upcycle Studios potentially reduces the total CO2 emissions over 50 years by about 60 percent | 5 |
| • Savings on Water Use | 3 | Less water used in technique 3 on upcycled concrete & bricks without wet connection. Worked with WML to develop closed water cycle using rainwater collected and purified to drinking water | 5 | Use of grey water or rain water in construction and operation of the building. Reversible Building solutions considers sustainable use of water | 3 | | 1 | | 1 | | 1 |
| • Savings on Energy Consumption | 3 | Lower diesel and energy costs of equipment by planning distance travelling from demolition to construction site | 4 | Construction practices and plants are used to reduce energy | 3 | The project with limited funds kept construction costs low and avoided the energy consumption of air-conditioning, by designing it as open air. | 4 | Reduction in heating demand to 34 kWh/m$^2$ from 104 kWh/m$^2$ which is maximum in region | 5 | Upcycle Studios saves 15 metric tons of $CO_2$ annually per dwelling using renewable heating through solar power combined with geothermal energy. With rooftop | 5 |

| | | | | | | | | | | | |
|---|---|---|---|---|---|---|---|---|---|---|---|
| | | | | | | | | | | solar panels the houses are potentially self-sufficient | |
| • Embodied energy and pollution | 3 | The investigated pilot dwelling consists of 3.35E02 GJ embodied energy, of which 65% is embodied in re-used materials, and 4.62E01 tons of embodied $CO_2$, of which 90% is embodied in re-used materials (Ritzen et al., 2019) | 5 | Sustainability of supply, durability, embodied carbon and embodied water were considered when specifying materials | 5 | Cut down on crates of pressed tree pulp by using shared resources such as book donations and beer crates | 3 | Hygrothermal behaviour (repeated wetting, drying, freezing and thawing of the fabric of a building leading to mold growth, condensation and loss of thermal performance), of the insulation materials were analyzed and quantified. Each façade is monitored to measure the thermal resistance of the insulation materials on-site (including the roof). Comfort parameters are also monitored; room temperature, relative air humidity and carbon dioxide levels. | 5 | Biodiversity of the neighborhood increases with the roof gardens; Each of Upcycle Studio's building components are analyzed in isolation, scrutinizing its whole-life energy performance against a strict set of parameters | 5 |
| • Solid Waste Reduction | 3 | Project recycled concrete and reused entire apartment hulls which were supplanted on construction site to build stand-alone | 5 | Building elements can be maintained, upgraded or replaced without creating excessive waste; identification of where on-site | 5 | Assemblage of 1,000 empty beer cartons donated by residents as building material | 5 | 350m$^3$ of rubble stored at rear of house was reused to level parking spaces and access paths | 5 | The wood for floors, walls and facades are produced from offcuts and surplus materials from Dinesen – these materials could have | 5 |

| | | | | | | | | | | | |
|---|---|---|---|---|---|---|---|---|---|---|---|
| | | houses to minimize waste. Materials of demolition were segregated per material for reuse. | | waste arises; use of materials with high recycled content | | | | | | ended up as landfill or incineration plant. | |
| • Product life cycle extension | 2 | Technique 2 developed bricks that are not glued but stacked on steel cables<br>Whole apartment hulls that were extracted and moved to construction site extended their product cycles. | 5 | Design ensures that building is flexible in use and adaptable to changes in use over its life span, allows building materials to be easily disassembled and can be reused and recycled at end of life. Life cycle of materials used estimated at:<br>Fibre-cement façade panels (outside finishing) (100 years)<br>Wooden battens (15-60 years)<br>Waterproof layer (30-60 years)<br>Insulation between wooden substructure (30-50 years)<br>Vapour barrier (sealing tape) (10 years) | 5 | Reuse of modernist façade of old department store Horten warehouse of city of Hamm without any structural change due to its being properly maintained | 5 | 2 marble fireplaces were moved to become ornamental features in public room; 62m$^2$ of spruce floor boards were dismantled and relayed after installing insulation; 18m2 of multicolored cement tiles were reused as features | 5 | About 75 percent of the windows come from abandoned buildings in North Jutland, Denmark. And 1400 tons of upcycle concrete are casted from very durable concrete waste from the construction of Copenhagen Metro. | 5 |
| • Use of biotic materials | 2 | The ground-level houses are being built | 3 | Use of fiber boards as wall materials | 5 | | 1 | Use of 8 bio-based materials for | 5 | | 1 |

| | | | | | | | | | | | |
|---|---|---|---|---|---|---|---|---|---|---|---|
| | | based on three construction principles:<br>1. Materials come from their own area; Those materials were recently released at the demolition of the adjacent flat and/or<br>2. Materials are bio-based; These raw materials grow in one natural way back and/or<br>3. It must be dismountable. But to what bio-based materials were used is not known. | | being lightweight and impact resistant and mineral wool as insulators | | | | interior insulation – both vegetal (i.e., wood fiber, flax fiber, hemp bricks, expanded cork) and animal-based (sheep wool) and recycled materials (cellulose, recycled textile and recycled glass) | | | |
| • Reduction of Exposure to Human Toxicity | 2 | | 1 | Wall materials meet acoustical and fire safety requirements | 2 | | 1 | Comfort and health parameters being monitored such as indoor air quality, relative air humidity, and $CO_2$ levels as well as hygrothermal properties of eco-materials | 5 | Waste materials used in Upcycle Studios are upcycled to nontoxic and certified materials. All hazardous and toxic waste from debris are sorted out from demolished buildings | 5 |
| • Land use | 2 | Thinning out housing stock because of declining population and increase green shell within fabric of the city | 4 | The current building standards and the respect of the heritage value of the site had to be taken into consideration. | 4 | Involved reuse of abandoned industrial site with fallow land and conversion to green public space | 5 | Preservation of mining houses in the site which was declared a UNESCO heritage site so external façade | 5 | Upcycle Studios has street spaces on either side and therefore not the same clear front/back situation as the rest of the district, balancing high and low, shadow and sun, housing and business. The site is | 4 |

IOP Conf. Series: Earth and Environmental Science **855** (2021) 012017 doi:10.1088/1755-1315/855/1/012017

| | | | | | | | | | | | |
|---|---|---|---|---|---|---|---|---|---|---|---|
| | | | | | | | | | | designed after shared economy principle balancing commonality and individuality with series of flexible spaces. Roof gardens add value to physical space. | |
| **Overall Score** | | | **95** | | **89** | | **72** | | **103** | | **93** |

Source: Authors compilations from sources listed on Appendix Table 4

IOP Conf. Series: Earth and Environmental Science **855** (2021) 012017 doi:10.1088/1755-1315/855/1/012017

Appendix Table 9
Description of Economic Indicators

| Indicator | Description |
|---|---|
| Employment generation | concerns potential to create new jobs and skills related to any of the circularity stages |
| Trade effects | relates to trade creating effects by way of imports and exports and potential to raise net exports by reducing imports of extractive and energy-intensive primary materials and export of recycled building materials |
| Business creation and multiplier effects on the value chain | pertains to new businesses created such as small and medium enterprises revolving around new building materials that result from any of the circularity stages and multiplier rounds of buying and selling on suppliers and users of these materials |
| Financial self-sufficiency | refers to extent the project is scalable and can be financed sustainably from operating profits |
| Consumption patterns | entails changes in consumer usage and buyer acceptance of building materials that are reused, recycled or refurbished |
| Savings on maintenance costs | involves reduction of building maintenance and operational costs due to savings on energy and water use |
| Higher residual values | concerns extension of product life cycles of building materials that raise residual values of the building or the materials themselves |
| Increase in innovation and new technologies and associated investments | relates to innovation effort and investment potential of new technologies and innovations arising from any of the circularity stages in the built environment |
| Fiscal effects | pertains to potential for not adversely affecting the budget position by reducing reliance on government or EU subsidies and potential for revised and progressive taxation systems to finance investment costs of the projects |

IOP Conf. Series: Earth and Environmental Science **855** (2021) 012017 doi:10.1088/1755-1315/855/1/012017

Appendix Table 10. Economic Impact of Selected EU Projects

| Indicators for Economic Feasibility | Rank of Importance | **Superlocal, NL** | Score | **Circular Retrofit Lab, BE** | Score | **Open Air Library, DE** | Score | **Rehafutur House, FR** | Score | **Upcycle Studios, DK** | Score |
|---|---|---|---|---|---|---|---|---|---|---|---|
| • Employment generation | 3 | Employment generated limited to demolition of buildings and construction of the houses | 3 | The project involves training architecture students about reversible solutions in circular economy which will potentially enhance future employment in the sector. Project also involves local actors such as local architects, contractors, producers and suppliers in project development and implementation, hence contributing to local employment. | 3 | The project is located in an abandoned industrial site in a district with high unemployment and vacancy rates of 80 percent but employment effects are not known, although houses surrounding the site have been renovated. | 2 | The developers found that in reusing materials, the majority of the budget goes towards labor costs rather than in the purchase of new, virgin materials – which is a positive for a region in need of employment. | 4 | Upcycled building materials are locally sourced, creating local jobs | 5 |
| • Trade effects | 2 | Importation of machinery from China for molding the brick mortars | 3 | Project designed against backdrop of Belgium being an importer of building materials and need to reduce this reliance on outside suppliers | 4 | | 1 | The region is classified as fuel poor so the project will reduce fuel imports | 5 | The products are harvested and processed locally so no imports involved although there is potential for exports of upcycled materials | 5 |
| • Business creation and value chain multiplier effects | 3 | Potential business created for brick technique if commercialized; houses made are bought by local housing corporation who leases them out | 4 | Reversible building solutions for e.g. wall partitions, can be developed in the future with a larger application in mind, namely as a building systems that can be interchanged over the | 5 | A reading café and stage was added among residents to host public plays, youth concerts and other communal gatherings | 4 | Value creation and value chain effects of the eco-materials used as insulation. Coordination of different tradespeople to achieve specific | 5 | Designed townhouses as series of flexible spaces that can be used as offices, rental accommodation, workshops and other user-defined activities. Collaboration with wood manufacturer | 5 |

| | | | | | | | | | | | |
|---|---|---|---|---|---|---|---|---|---|---|---|
| | | | | total university campus. | | | | training on the worksite with all the teams was undertaken on this issue. Multiplier effects on 70,000 mining houses in the area | | Dinesen for surplus and off cuts of wood. Lendager Group has developed upcycled materials designed for implementation on industrial scale | |
| • Financial self-sufficiency | 3 | Potential self-sufficiency from rental income from completed houses for the housing corporation and commercialization of mortar brick technique | 4 | Potential self-financing from use of available materials and commercialization of insulation techniques used in partition walls that can easily be dismounted. | 3 | The project relied primarily on donations by the community of 1,000 beer crates and books. The community took over management and maintenance of the library | 4 | The Rehafutur House was converted into office spaces which would generate rental income for the landlord | 4 | Commercial project, built on market conditions and designed for scale. Also, the landfill tax system of Danish government was restructured to finance reuse and recycling of waste. Financed by real estate developers. | 5 |
| • Consumption patterns | 3 | Possible change in consumer reaction to recycled building materials | 2 | The living lab acts as a demonstrator of how the existing modules' transformation potential can be enhanced. The project applies kit of parts modular building design related to prefabricated components where suppliers were invited to provide take back and end-of-life scenario for products. There is a reception desk and information center to show the used techniques. It will raise awareness and | 5 | | 1 | Tradespeople used to demolishing floorboards in other projects and this project changed their mentality | 4 | Finding multiple alternative uses for innovative materials and utilizing efficiency dynamics and architectural aesthetics to increase the affordability and viability of these new products for consumers | 5 |

| | | | | | | | | | | | |
|---|---|---|---|---|---|---|---|---|---|---|---|
| | | | | foster circular thinking within the future generation of local construction stakeholders. | | | | | | | |
| • Savings on maintenance costs | 2 | Closed water loop system will reduce maintenance costs | 4 | Project will improve the quality and comfort of the buildings, since they can be adapted and maintained more easily, without the need for destructive transformation works. The project improves maintenance savings by designing for assembly and disassembly. Selecting durable materials is important to lengthen the lifespan of the building elements and because the materials and building elements must withstand being reused, which includes the wear-and-tear of frequent transport and intensive use of construction products | 5 | Being designed as public space and open air, the project cut down on maintenance costs although later, there were signs of vandalism but residents took care of library as if it were a public toy. | 3 | The project was designed to make the building efficient during the use phase so that it has less environmental impact but is also less expensive to run. The house used to burn off 1,000 liters of fuel each year; after renovation, the insulation is such that the heating demand is very low at 34 kWh/m2 a year | 5 | Upcycle Studios are exclusively sustainable housing with minimal or no operating costs. The mission of the project was to construct the first residential area built by upcycled and local waste materials without compromising on aesthetics, quality or price. Upcycle Studios saves 15 tons of CO2 annually per unit using renewable heating using solar power combined with geothermal energy. Shared economy principles potentially reduces maintenance costs of users. | 5 |
| • Higher residual values | 2 | Higher residual values for mortar detachable bricks and houses that were developed from whole demolished studios | 5 | The financial impact will be evaluated of transformability in building valuation and throughout the total building life cycle. In this respect, the team together with the | 5 | The community reused the façade of the warehouse which was properly maintained from the 1960s without any structural change during | 5 | Energy efficiency of eco-materials used together with thermal monitoring endures higher residual values. The project intended to keep | 5 | Upcycled building materials were designed to be sustainable both from an environmental perspective, an operation perspective, an indoor climate | 5 |

IOP Conf. Series: Earth and Environmental Science **855** (2021) 012017 doi:10.1088/1755-1315/855/1/012017

| | | | | | | | | | | | |
|---|---|---|---|---|---|---|---|---|---|---|---|
| | | | | industry partners implemented efficient operational solutions, such as the use of dry, robust and reversible technical solutions and the use of materials able to endure multiple reuses without being damaged. | | construction except for recoloring of aluminum modules | | high heritage value of the building (floorboard, fireplaces, tiles) | | perspective, a biodiversity perspective, raising their residual values. | |
| • Increase in investments in innovation and new technologies | 3 | Innovation and new technology developed in the techniques 2 and 3 involving mortar bricks and upcycled concrete rubble | 5 | For investors, the future flexibility of buildings may create new business opportunities, as project creates more spatial and functional opportunities. The project tested on two cases of buildings as infrastructure and buildings as investments. Depending on their expected rate of change in the floor plan, three different types of walls were defined, analyzed, constructed and transformed: walls with (1) a high rate of change, (2) a high degree of flexibility for the integration of technical infrastructure and 3) a low rate of change. The CRL pilot project applied a step- | 5 | | 1 | A recycled material called *metisse* was used for the insulation. This regional material is made out of old clothes (mainly cotton). Lot of innovation went into testing thermal capacity of the eco-materials. | 5 | Materials were processed so they both appear as beautiful and sustainable in quality and function as well as aesthetically compared to traditional building materials | 5 |

| | | | | | | | | | | | |
|---|---|---|---|---|---|---|---|---|---|---|---|
| | | | | by-step innovation strategy based on products available on the market. This strategy thus incrementally improves products that are already technically and commercially viable. | | | | | | | |
| • Fiscal effects | 2 | Funding from EU subsidy | 2 | Funding from EU subsidy | 2 | Relied on public donations but partly funded by federal government to raise required money for construction | 4 | Funding from EU subsidy | 2 | No reliance on subsidies but commercially financed | 5 |
| **Overall Score** | | | **82** | | **95** | | **62** | | **100** | | **115** |

Source: Source: Authors compilations from sources listed on Appendix Table 4

Appendix Table 11
Description of Social Indicators

| Indicator | Description |
|---|---|
| Stakeholder participation | refers to the extent of consultation, workshops, meeting, and participation of different stakeholder groups affected by circular buildings |
| Community engagement | relates to commitment and actual engagement and ownership of affected communities in the construction and upkeep of the project |
| Preservation of cultural heritage | involves reconnecting, reminding, preserving cultural values, historical memories and traditions of the project including architectural landmarks |
| Development of social capital | concerns promotion of networks among people who live and work in particular society to enable it to function effectively |
| Regional inequality | entails extent the project addresses regional disparities in income, wealth and human development opportunities |
| Formation of human capital | relates to the extent the project enhances human capital by way of education, upskilling, literacy, and health |
| Demographic transition and change | refers to the extent the project responds to demographic changes such as declining and ageing population, outmigration, and urbanization |
| Social cohesion | concerns the promotion of local communities, sense of community spirt and belongingness. |

Appendix Table 12. Social Impact of Selected EU Projects

| Indicators for Economic Feasibility | Rank of Importance | **Superlocal, NL** | Score | **Circular Retrofit Lab, BE** | Score | **Open Air Library, DE** | Score | **Rehafutur House, FR** | Score | **Upcycle Studios, DK** | Score |
|---|---|---|---|---|---|---|---|---|---|---|---|
| • Stakeholder participation | 3 | In behalf of the Urban Innovation Action application for the project, no less than twelve partners are intensively cooperating: housing association HEEMwonen, the municipality of Kerkrade, IBA Parkstad, City Region of Parkstad Limburg, demolition firm Dusseldorp Infra, Sloop en Milieutechniek B.V., VolkerWessels Construction and Real Estate Development South (VW VolkerWessels Construction and Real Estate Development South (VW), Waterboard Company Limburg (WBL), Limburg Drinking Water Company (WML), the municipalities of Landgraaf and Brunssum, Zuyd | 5 | For the **Circular Retrofit Lab**, a steering group was set up to check at any time if the initial aims are being respected during the development of the CRL solutions, and to evaluate the feasibility of the planning and budget. The group consists of the building owner (VUB), the TRANSFORM research team (VUB), the architect (Kaderstudio), IBGE (as the Work Package 4 leader), the general construction coordinator (Groep Van Roey) and the engineering office (MK engineering). Groep van Roey, a general building provider. Furthermore, a group of stakeholders, including industrial partners, has been put together for this | 5 | The project was planned right from the beginning as a social sculpture. The design and the functions were planned in a very close and open participation process. During the participation process several citizens demanded to construct the facade out of recycling material | 5 | The management team were very collaborative when it came to working towards circularity, especially since there was a desire to increase skills amongst building professionals to influence future projects. With this demonstration site, the cluster Ekwation engage in partnerships with regional enterprises, laboratories, schools, universities and education centers, stakeholders and policy-makers in order to offer them a complete tool to exhibit, test and understand the process of renovating the ancient housing stock in the | 5 | | 1 |

| | | | | | | | | | | | |
|---|---|---|---|---|---|---|---|---|---|---|---|
| | | University of Applied Sciences and Association of Demolition Contractors (VERAS). The knowledge the partners attain during the implementation of the SCE project is open-source, the acquired skills are therefore available for use in other international projects. | | specific Pilot Project, enabling the stakeholders to work together on the development of innovative dynamic and circular solutions | | | | region (especially mining housing). CAPEM brought together the expertise of 11 partner organisations to improve the production, distribution and use of eco-materials. | | | |
| • Community engagement | 3 | Residents and neighbors can contribute ideas about the future of the neighborhood through a residents' panel. Residents and students were consulted for ideas about the project. | 5 | Industrial stakeholders, like manufacturers, were involved from the design phase, but also during the progress of the works. This allowed co-developing the systems and making intermediate improvements. Three working groups were formed: external systems, internal systems and service. The university organized several round tables with industry stakeholders where design solutions were debated and improved, as well as hands-on workshops | 5 | The library began as an assemblage of 1,000 empty beer cartons pulled together by residents. The shelves of the temporary library were filled by the residents with book donations. With limited resources, the residents have brought into being a building that is innovative in both form and function and which, thanks to their involvement in the project, precisely responds to their demands. The citizen participation that is revitalizing the social fabric does not end | 5 | The construction workers were very proud to work on the project, with continuous visits and communication around the building site. Those who set the insulation materials were used to eco-materials but enjoyed working with the flax fiber enough that they have made themselves a tool to cut it. | 4 | | 1 |

IOP Conf. Series: Earth and Environmental Science **855** (2021) 012017 doi:10.1088/1755-1315/855/1/012017

| | | | | | | | | | | | |
|---|---|---|---|---|---|---|---|---|---|---|---|
| | | | | with students where solutions were tested. | | with conceiving a building since, now the professionals who acted as advisers have gone, the residents have taken over its management. | | | | | |
| • Preservation of cultural heritage | 2 | After almost 50 years in high-rise blocks, it is time for an evaluation to see if the apartment blocks served their purpose well (structurally, architecturally, socially), their value during their lifespan, partly from the residents' view. In order to explore these issues, a book of neighbors is currently being compiled together with the last remaining residents. This book of neighbors gives a more detailed illustration of the matrix 'past- present- future' versus 'individual- community- location' of the residents and the buildings. The project also envisions to create links to the | 5 | The development of the transformable façades encountered major challenges. The current building standards and the respect of the heritage value of the site had to be taken into consideration. A case was made to keep part of these student housing units for their architectural specificity using Willy Van Der Meeren concept of prefabricated concrete modules were installed in the 1970's to cope with the increasing demand for student housing on site. The flexibility of this modular system developed in Switzerland allowed a random layout and generated a sequence of qualitative green urban spaces. | 4 | Remembrance, history and narratives provided the background for the "re-occupation" of the abandoned expanse. An old empty shop was used as base for a temporary library and camp for a building workshop. The architectural highlight of the project is the re-use of the modernist facade of the old HORTEN warehouse of the City of Hamm [built in 1966], which has been knocked down in 2007. The warehouses of the entrepreneur Horten which have been built in numerous German cities represent *sustainability of* | 5 | Given the building's significant heritage value, it was important for the team to reuse as much as possible whilst ensuring the longevity of the site – so circularity was a strong element from early on. Interior insulation compulsory as the building is on the UNESCO list of world heritage sites. Due to its heritage value, the renovation of this building could not affect the external façade, meaning that retrofitting could only involve interiors, | 5 | By reusing the walls from the abandoned dwellings as new facade elements, the project not only saves $CO_2$ and virgin materials, but also gets a new building with history and character with 75 percent of the windows coming from abandoned buildings in North Jutland, Denmark and 1400 tons of upcycle concrete casted from very durable concrete waste from the construction of Copenhagen Metro. | 3 |

IOP Conf. Series: Earth and Environmental Science **855** (2021) 012017 doi:10.1088/1755-1315/855/1/012017

| | | | | | | | | | | | |
|---|---|---|---|---|---|---|---|---|---|---|---|
| | | mining past of the area. | | | | *signs* of the postwar period | | | | | |
| • Development of social capital | 3 | In a declining population, residents are a unique form of capital. The residents have strong ties with the area. Many have lived here their whole lives. Cohesion is vital to a district, particularly one with a declining population. The current residents are being mobilized to preserve and develop the qualities present in the area. The project s want to recycle their knowledge of and vision on the area and, if possible, keep residents in the area. | 4 | In order to build this demonstrator lab, VUB and IBGE are working together with pioneers in the building sector, like product developers, entrepreneurs and enterprises that support this circular demonstrator lab. Their support consists of integrating existing (dynamic) solutions in the lab, joint development of new building solutions, funding through material donations or material leasing, etc. The Circular Retrofit Lab has developed partnerships with a large group of industrial partners. In a first phase, a temporary lab allowed exploring different building solutions, mainly wall assemblies, that will be used in the final demonstrator lab | 5 | The grassy plaza features a reading cafe and a stage that hosts elementary school theater plays, public readings, concerts-gigs of local youth bands and other cultural events. | 5 | Potential networks created between trades people and material suppliers on insulation properties of eco-materials used. The project creates opportunities to connect farmers living in the communes around the mining town and light industry artisans and textile makers on use of eco-fibers for insulation | 5 | Upcycle Studios is designed as a new model for shared living, which seeks to find a balance between commonality and individuality, as 'the community thrives best if there is space for the individual.' This taps into various aspects of the 'sharing economy', by prioritizing access over ownership and providing generous spaces for shared living and private accommodation. The result is a series of flexible spaces that can be used as offices, rental accommodation, workshops or a multitude of other user-defined activities, with each townhouse able to be divided and shared in different ways. Much like their approach to material use, Lendager Group emphasize the economic incentive of this 'sharing', with its potential benefit for | 5 |

IOP Conf. Series: Earth and Environmental Science **855** (2021) 012017 doi:10.1088/1755-1315/855/1/012017

| | | | | | | | | | | | |
|---|---|---|---|---|---|---|---|---|---|---|---|
| | | | | | | | | | | both 'owner and consumer'. | |
| • Regional inequality | 3 | | 1 | | 1 | The southeast of Magdeburg belongs to these urban areas in Eastern Germany which are characterized by shrinkage, abandoned industrial plants and fallow land. A post-industrial city landscape with high unemployment and figures of vacancy up to 80%. Like many other suburbs of East Germany, the Salbke district is deeply immersed in a state of physical and social decadence. Unemployment figures are high and public facilities are scarce. | 5 | The Rehafutur Engineer's house project responds to the challenge that France faces in renovating existing building stock to high energy efficiency standards. Rehafutur aims to offer solutions for the renovation of the typical mining housing in the region which have specific high energy requirements as the region is affected by fuel poverty. Demonstrating circular solutions at Rehafutur has a huge multiplier effect as there are approximately 70,000 mining houses still occupied in Northern France. | 5 | | 1 |
| • Formation of human capital (education and health) | 3 | The playbook of the project presents many collaborations with students from elementary, high | 5 | The project will show how an existing structure can get more potential for the future in order to be easily | 5 | The residents which take care themselves for a reading-café as well as for the Open-Air-Library call it a | 5 | Training sessions were held on the work site on airtightness | 5 | Waste materials used in Upcycle Studios are upcycled to nontoxic and certified materials | 5 |

IOP Conf. Series: Earth and Environmental Science **855** (2021) 012017 doi:10.1088/1755-1315/855/1/012017

| | | | | | | | | | | | |
|---|---|---|---|---|---|---|---|---|---|---|---|
| | | school, university and technical schools in eliciting designs for the project including asoects related to circularity and sustainability, finance, architecture and technical specifications of materials used for the project. | | transformed into different functions (e.g. a dissemination space, a co-working space, an eco-guesthouse in the middle of the campus). The CRL will demonstrate how this change can be anticipated by supporting other functions every 6 months on the first floor level. On the ground level a dissemination space will be situated in which the results of the BAMB project will be shown to a large audience, sustainable lectures can be given, the technical solutions developed with the industrial stakeholders can be observed, etc. During Summer School, these solutions were tested by a group of enthusiastic students. | | "library of confidence": There is no registration needed and there is no control. A book can be taken whenever wanted, but should be returned voluntarily or replaced with another one. The shelves are never closed - the library is opened for 24 hours a day. This contributes to increasing literacy and education of residents. | | with all trades people to make this intricate system work, so collaboration was key to the success to the project. | | | |
| • Demographic transition and change | 2 | The project is based in a sustainable shrinkage area in Stadsregio Parkstad | 5 | | 1 | In the old center closed-down businesses abound along with | 5 | The very rural village changed into a town, with | 5 | Three million people will have completed their move from rural to urban areas. The | 5 |

IOP Conf. Series: Earth and Environmental Science **855** (2021) 012017 doi:10.1088/1755-1315/855/1/012017

| | | | | | | | | | | | |
|---|---|---|---|---|---|---|---|---|---|---|---|
| | | Limburg in the South of the Netherlands. In the next 30 years, the Parkstad Limburg region's population will shrink by 27% due to population ageing and young people moving to urban agglomeration cities such as Amsterdam and Rotterdam. This implies less housing accommodation will be demanded in the following decades. High-rise apartment buildings, which were mainly built in the 1960s when housing shortage was a primary issue, no longer satisfy requirements and inhabitants' needs. Three vacant high-rise apartment buildings in the project-area in the city of Kerkrade contain valuable materials, qualities and former social structures. Demolition of these buildings will irreversibly impair these values. The | | | | unoccupied housing, vacant lots and abandoned factories that populate the post-industrial landscape. One of the vacant lots was the triangular space defined by the intersection of the main street with Blumenberger Strasse, which remained vacant after the old district library was demolished. | | many foreign workers, especially from Poland, swelling the workforce. A former mining site (*Écopole*) has been preserved and now hosts many cultural, economic and environmental activities which are the symbol of new developments for the commune.<br>In recent years, population left the mining towns seeking work elsewhere, about 1000 homes have been demolished in the recent past.<br>The mines of the Nord-Pas de Calais region have become the 38th French site on the list of UNESCO World Heritage.<br>As of 2008, Loos-en-Gohelle was a town whose economy was dominated by light industry, textiles and workshops for | | increased demand for housing means that an estimated 60 percent of the new urban areas needed in 2030 have not yet been built. The entrepreneurial spirit has increased in recent years in the metropolitan area. For more and more people, the day is not clean cut into distinct categories as it used to be and the dream of the good life has become more about living and working with a flexible schedule rather than the well-known "9-17". Concepts such as flexibility, symbiosis, exchange and resource efficiency are high on the agenda for many Danes, which goes to show that alternative ways of thinking of life and social structure are emerging. | |

| | | | | | | | | | | | |
|---|---|---|---|---|---|---|---|---|---|---|---|
| | | objective of the project is to circularly reuse these values within the project area while boosting the local economy and creating a high-quality and desirable urban environment. The former social structures will be recovered by actively stimulating former inhabitants to take residence in the area again. The project serves as a best example for approximately 1250 similar housing accommodations in the Parkstad Limburg region. | | | | | | artisans and professionals, but it also still retained a strong rural connection with thirty farms within the commune. The Rehafutur project converts one of the mining houses into office spaces. | | | |
| • Social cohesion | 3 | Some neighbors lived next to each other for years. They had good social contacts. If the flats were demolished, those qualities would be lost. With demolition, not only the tangible value, such as the building and materials, but also invisible values, such as social structures between | 4 | The project provides opportunities for actors in the supply chain for co-creation / co-making processes. The client, design team, suppliers and builders can help build this trust. This, however, requires time and willingness to adapt the procurement process. The CRL involved contractors, | 4 | Open to the public twenty-four hours a day, the library is managed by the residents themselves who, without any checks or monitoring, freely borrow and return books. Although, as if testifying to the harsh surroundings, the new library has been vandalized on | 5 | Coordination of different tradespeople to achieve specific training on the worksite with all the teams was undertaken on this issue. | 4 | The Upcycle Studios is designed socially by fostering neighborly relations and increasing sharing economy potentials – all of which adds a value to the physical space. The core focus for the development of Upcycle Studios is based on the Sharing Economy. Communities are developing around the | 3 |

| | | | | | | | | | | |
|---|---|---|---|---|---|---|---|---|---|---|
| | | residents and the ideas about life in and around the flats, disappears. The wish is therefore to maintain the qualities of the existing flats when building new homes and to use materials. In line with circularity principles, former inhabitants will be invited back into the area. These potential tenants will be strongly involved in the co-design, operation and monitoring of the new collaborative economy services and facilities (such as a shared mobility platform and a social services center) within the area. | | manufacturers, designers, and researchers upstream, in an intensive collaborative design process with very good circular results. Once the construction has started, more and more manufacturers and construction contractors from small to big companies became interested in the "reuse" potential of materials. | | occasion, it is a fully functioning concern. | | | | concept of sharing resources, which has the benefit of providing economic gain for both owner and consumer. | |
| **Overall Score** | | | **92** | | **85** | | **110** | | **104** | | **64** |

Source: Authors compilations from sources listed on Appendix Table 4

## REFERENCES

Allenby BR 1992 Achieving sustainable development through industrial ecology. *International Environmental Affairs* **4** (1) pp. 56-68.

Bal M, Bryde D, Fearon, D and Ochieng E 2013 Stakeholder Engagement: Achieving sustainability in the Construction Sector *Sustainability* 2013 **6**, pp. 695-710 doi:10.3390/su5020695.

Bayern Innovativ 2019 *The circular economy in Southern Germany* Market Study Final report Retrieved on 10 December 2020 from www.nederlandwereldwijd.nl

Benyus J 2002 *Biomimicry: innovation inspired by natu*re (2nd ed.) New York: Harper Perennial.

Best A, Duin L and Chelminska M 2018 *Macroeconomic and Societal Impacts of Mainstreaming the Circular Economy* Retrieved on 15 December 2020 from http://circular-impacts.eu/deliverables.

Bloomsma F and Brennan G 2017 The emergence of circular economy: a new framing around prolonging resource productivity *Journal of Industrial Ecology* https:// doi.org/10.1111/jiec. 12603.

Boulding K 1966 The economics of the coming spaceship earth. In Jarett H ed., *Environmental quality in a growing economy* pp. 3-14 Baltimore, Md Resources for the Future/Johns Hopkins University Press.

Brand S 1994 *How buildings learn – what happens after they're built* London: Phoenix, p.13.

Breen C and Vangsbo P 2018 Municipality-led circular economy case studies EIT Climate-KIC, Retrieved on 5 December 2020 from https://circulareconomy.europa.eu/platform/en/knowledge/municipality-led-circular-economy-case-studies

Bringezu S 2003 Industrial ecology and material flow analysis In Bourg D and Erkman S eds. *Perspectives on industrial ecology* Sheffield, UK: Greenleaf.

CE100 2015 *Circularity in the Built Environment Case Studies: A compilation of case studies from CE100*, April

Chertow M 2000 Industrial symbiosis: Literature and taxonomy *Annual Review of Energy and Environment*, **25**(1), pp. 313-37

Cheshire D 2016 *Building Revolutions: Applying the Circular Economy to the Built Environment.* Riba Publishing ISBN 10: 1859466451.

Daly HE 1992 Allocation, distribution and scale: towards an economics that is efficient, just and sustainable *Ecological Economics* **6**, pp. 185-93.

Das S, Bhowmick M., Chattopadhyay, SK and Basak S. 2015 Application of biomimicry in textiles. *Current Science* **109**: pp. 893-901

De Angelis R 2018 *Business Models in the Circular Economy*. Palgrave Macmillan. ISBN 978-3-319-75126-9.

Deloitte 2015 *Construction and Demolition Waste Management in Denmark.* September, Retrieved on 10 December 2020 from https://ec.europa.eu/environment/waste/studies/deliverables/CDW_Denmark_Factsheet_Final.pdf

De Groene Zaak and Ethica 2015 *Boosting Circular Design for a Circular Economy* Joint Report funded by EU Climate KIC Pioneers into Practice Programs.

Derbal R, Lucas, A, Brachelet F, Defer D and Antczak E 2017 Rehafutur Insitu instrumentation for a comprehensive building analysis *Energy Procedia* **139**, pp.301-06 Elsevier.

Dul J 2015 Necessary condition analysis : Logic and methodology of "necessary but not sufficient" causality. Organizational Research Methods **19** (1), pp. 10-52.

EMF and McKinsey 2012 Towards the circular economy Economic and business rationale for an accelerated transition, Retrieved 8 December 2020, from https://www.ellenmacarthurfoundation.org/assets/downloads/publications/Ellen-MacArthur-Foundation-Towards-the-Circular-Economy-vol.1.pdf

EMF 2017 *Cities in the circular economy: An initial exploration,* Retrieved on 5 December 2020 from https://www.ellenmacarthurfoundation.org/publications/cities-in-the-circular-economy-aninitial-exploration

ESPON 2020 Stocktaking and assessment of typologies of Urban Circular Collaborative Economy initiatives *ESPON SHARING Final Report* 01/04/2020.

EU Commission 2015 *Closing the loop: An EU Action plan for the circular economy* Communication from the Commission to the European Parliament, the Council, the European Economic and Social Committee and the Committee of the Regions COM(2015) 614 final Brussels, 2/12/2015

Fernandez A 2020 *Superlocal: concrete recycling techniques developed in the project* Flash lecture given in the Rescue Minor Sustainability Course at Zuyd University of Applied Sciences, 5 October 2020.

Fontaine F and Dewulf B eds. 2018 *Feuille de Route des Acteurs de la Construction a Bruxelles vers une Economie Circulaire,* Retrieved on 10 December 2020 from https://www.circulareconomy.brussels/decouvrez-la-feuille-de-route-des-acteurs-de-la-construction-vers-une-economie-circulaire/

Fontaine F and Dewulf B eds. 2018 *The Construction Sector in Brussels: Findings and Perspectives towards a Circular Economy*, Retrieved on 10 December 2020 from https://www.circulareconomy.brussels/wp-content/uploads/2017/05/be_prec_construction_sector_EN.pdf.

Frosch R and Gallopoulos N 1989 Strategies for Manufacturing. *Scientific American* **261**, pp. 94-102.

Ghisellini P, Ripa M. and Ulgiati S. 2018 Exploring environmental and economic costs and benefits of a circular economy approach to the construction and demolition sector: A literature review. *Journal of Cleaner Production* **178**, Elsevier, pp. 618-43.

Grant D, Trautrims A. and Wong CY 2017 *Sustainable Logistics and Supply Chain Management* (2nd Ed.) London: Kogan Page Ltd. ISBN 9780749478278.

Gravagnuolo A, De Angelis R and Iodice S 2019 Circular economy strategies in the Historic Built Environment: Cultural heritage adaptive reuse *Proc. of the STS Conference*, Graz, 2019, pp 121-44.

Green DW, Lee JM and Jung HS 2015 Marine structural biomaterials in medical biomimicry *Tissue Engineering Part B Reviews* **21** (5), pp.438-50.

Habib MK 2011 Biomimetics: innovations and robotics *International Journal of Mechatronics and Manufacturing Systems* **4** (2), pp. 113-34.

Häkkinen T, Helin T, Antuña C, Supper S, Schiopu N and Nibel S 2013 Land use as an aspect of sustainable building. *International Journal of Sustainable Land Use and Urban Planning* **1** (1): pp.21-41.

Haupt M and Hellweg S 2019 Measuring the environmental sustainability of a circular economy *Environmental and Sustainability Indicators* **1-2** 10005, September Science Direct

Hawken P, Lovins A and Lovins L 2000 *Natural capitalism; The next industrial revolution* London: Earthscan

Holland Circular Hotspot 2018 *Circular Construction Economy Transition Agenda 2018*, Retrieved on 15 December 2020 from https://hollandcircularhotspot.nl/wp-content/uploads/2019/09/Circular-Construction-Economy.pdf.

Hysa E, Kruja A, Ur Rehman N and Laurenti R 2020 Circular economy innovation and environmental sustainability impact on economic growth: An integrated model for sustainable development *Sustainability 2020*, **12**, 4831, doi:10.3390/su12124831

Inter-American Development Bank (IADB) 2018 *Social impact assessment: integrating social issues in development projects*, IDB Series on Environmental and Social Risk and Opportunity.

Kennedy C, Pincetl S and Bunje P 2007 The study of urban metabolism and its applications to urban planning and design *Environmental Pollution* xxx, pp. 1-9, doi:10.1016/j.envpol.2010.10.022

Kennedy S 2007 Biomimicry.bimimetics: general principles and practical examples *The Science Creative Quarterly,* Retrieved on 10 December 2020 from https://www.scq.ubc.ca/biomimicrybimimetics-general-principles-and-practical-examples/

Kirchherr J, Reike D and Hekkert M 2017 Conceptualizing the circular economy: An analysis of 114 definitions *Resources, Conservation & Recycling* **127**, pp. 221-32. Elsevier.

Lah O 2016 Circular Economy Policies and Strategies of Germany. In Anbumozhi V and Kim J eds., *Towards a Circular Economy: Corporate Management and Policy Pathways* ERIA Research Project Report, 2014-44, Jakarta: ERIA, pp.59-74.

Lifset R and Boons F 2012 Industrial ecology: Business management in a material world. In Bansal P and Hoffman A eds.*The Oxford handbook of business and natural environment*, pp. 311-26. Oxford University Press.

Linton J, Klassen R and Jayaramman V 2007 Sustainable supply chains; An introduction *Journal of Operations Management* **25**, pp 1075-82.

Lovins A, Lovins I and Hawken P 1999 A road map for natural capitalism *Harvard Business Review* **77** (May-June), pp. 145-58.

Lyle JT 1994 *Regenerative design for sustainable development* New York: John Wiley & Sons

Mäler KG 1974 *Environmental economics: a theoretical inquiry* Baltimore: Johns Hopkins University Press.

Marshall A and Lozeva S 2009 Questioning the theory and practice of biomimicry *International Journal of Design & Nature & Ecodynamics* **4**(1), pp.1-10.

McDonough W and Braungart M 2002 *Cradle to cradle: Remaking the way we make things* New York: North Point Press

Migliore M, Talamo C and Paganin G 2020 *Strategies for Circular Economy and Cross-sectoral Exchanges for Sustainable Building Products* Springer Tracts in Civil Engineering ISBN 978-3-030-30317-4.

Ministry of Ecological and Solidarity Transition 2018 *Circular economy Roadmap for France: 50 measures for a 100% circular economy*, Retrieved on 12 December 2020 from https://www.ecologique-solidaire-gouv.fr/sites/default/files/FREC%20anglais.pdf.

Ministry of Environment and Food and Ministry of Industry, Business and Financial Affairs 2018 *Strategy for Circular Economy* Danish Government, Retrieved on 22 December 2020 from https://circulareconomy.europa.eu/platform/sites/default/files/eng_mfvm_cirkulaer_oekonomi_as5_uk_final_web.pdf

Mossberg M 2018 *Evaluation of Circular Economy design principles by material Flow Analysis*. Chalmers University of Technology Masteral Thesis. Gothenburg, Sweden.

Odu, G.O. 2019 Weighting methods for multi-criteria decision making technique. *Journal of Applied Sciences and Environmental Management* **23** (8), pp. 1449-1457, August.

OECD 2019 *Building Models for the Circular Economy: Opportunities and Challenges for Policy* OECD Publishing, Paris, https://doi.org/10.1787/g2g9dd62-en.

Ortiz O, Castells F and Sonnemann G 2009 Sustainability in the construction industry: A review of recent developments based on LCA *Construction and Building Materials* **23**, pp. 28-39 Science Direct.

Pacheco-Torgal F and Jalali S 2011 Toxicity of building materials *International Journal of Sustainable Engineering* iFirst article, pp. 1-7 Taylor & Francis Group.

Padilla-Rivera A, Russo-Garrido S and Merveille N 2020 Addressing the social aspects of a circular economy: a systematic literature review *Sustainability 2020*, **12**, 7912, pp. 1-17.

Paduart A, Elsen S and De Temmerman N 2018 *Circular Retrofit Lab; A Brussels' renovation experiment on how to redesign existing buildings into circular buildings* Powerpoint presented at the WASCON Conference 6 June 2018.

Pauli G 2010 *The blue economy: 10 years, 100 innovations, 100 million jobs* Taos, NM: Paradigm Publications.

Pearce DW and Turner KT 1990 *Economics of natural resources and the environment* Baltimore: Johns Hopkins University Press.

Persson U and Olander S 2004 Methods to estimate stakeholder views of sustainability for construction projects *The 21st Conference on Passive and Low Energy Architecture* **19-22** September 2004 Eindhoven, Netherlands.

Pötting J and Hanemaaijer A eds. 2018 *Circular economy: what we want to know and can measure. Framework and baseline assessment for monitoring the progress of the circular economy in the Netherlands* Policy Report PBL Netherlands Environmental Assessment Agency.

Ritzen M, van Oorschot J, Cammans M, Segers M, Wieland T, Scheer P, Creugers B, Abujidi N 2019 Circular (de)construction in the Superlocal project *IOP Conference Series: Earth and Environmental Science* 225. doi:10.1088/1755-1315/225/1/012048

Rizos V, Tuokko K and Behrens A 2017 The circular economy: A review of definitions, processes and impact *Center for European Policy Studies (CEPS) Research Report* No. 2017/8, April 2017.

Schut E, Crielaard M and Mesman M 2015 *Circular economy in the Dutch construction sector* Final Report. Rijkswaterstaat Ministry of Infrastructure and the Environment and the National Institute of Public Health and the Environment (RIVM).

Sillanpää M and Ncibi C 2019 *The Circular Economy: Case Studies about the Transition to the Linear Economy* London: Elsevier/Academic Press. ISBN 978-0-12-815267-6.

Singh A, Sharma B, Gaurav N and Singh N 2016 Environmental impact assessment as a tool to achieve the sustainable development *IMPACT International Journal of Research in Applied, Natural and Social Sciences,* Vol. **4**, Issue 8, pp. 143-56.

Stahel WR 1986 Product life as a variable: the notion of utilization. *Science and Public Policy* **13** (4), pp.185-93.

Tebbatt Adams K, Osmani M, Thorpe T and Thornbach J 2017 Circular economy in construction: current awareness, challenges and enablers *Proc. of the Institution of Civil Engineers Paper 1600011.*

Tietenberg T 1984 *Environmental and natural resource economics* Illinois, Scott: Foresman & Company.

United Nations Environment Program (UNEP) 2009 *Guidelines for social life cycle assessments of products* ISBN: 978-92-807-3021-0

Van Vliet MM 2018 *Disassembling the steps towards Building Circularity: Reinventing the Building Disassembly assessment method in the Building Circularity indicator* Eindhoven University of Technology Masteral Thesis.

Wahlström M, Bergmans J, Teittinen T, Bacher J, Smeets A and Paduart A 2020 Construction and Demolition Waste: challenges and opportunities in a circular economy *Eionet Report –* ETC/WMGE 2020/1 European Environmental Agency, European Topic Center on Waste and Materials in a Green Economy.

Wells P and Seitz M 2005 Business models and closed loop chains: A typology *Supply Chain Management: An International Journal* **10**, pp. 249-51.

Windapo A and Ogunsanmi O 2013 Construction sector views of sustainable building materials *Proc. of the Institution of Civil Engineers Paper* 1300011 ICE Publishing.

Yijun J, Ying H and Xuhong S 2011 Applying circular economy theory in Environmental Impact Assessment *2010 International Conference on Biology, Environment and Chemistry* IPCBEE 1, Singapore: IACSIT Press

Yilmaz M and Bakis A 2015 Sustainability in Construction Sector *Procedia – Social and Behavioral Sciences* **195**, pp. 2253-62 Science Direct.

Zabihi H, Habib F and Mirsaeedie L 2012 Sustainability in building and construction: Revising definitions and concepts *International Journal of Emerging Sciences* **2**(4), pp.570-78.

Zavadskas EK, Saparauskas J and Antucheviciene J 2018 Sustainability in Construction Engineering. *Sustainability* 2018 **10**, 2236, doi:10.3390/su10072236.

Zhang X, Wu Y, Shen L and Skitmore M 2014 A prototype system dynamic model for assessing the sustainability of construction projects *International Journal of Project Management* **32**, pp.66-76 Science Direct.

IOP Conf. Series: Earth and Environmental Science **855** (2021) 012018 doi:10.1088/1755-1315/855/1/012018

# A critical review on circular and regenerative urban areas

**L Gejer**[1] **and V Gomes da Silva**[1]

[1] School of Civil Engineering, Architecture and Urbanism, University of Campinas - UNICAMP, Brazil | vangomes@unicamp.br

**Abstract.** Circular economy (CE) has gained relevance as a new economic-environmental paradigm. Despite their key role in this model, managing cities towards CE has taken different meanings, approaches, concepts and methods. Therefore, this study aims to clarify circularity approaches to urban areas, by identifying main trends and exploring potential organization into a framework for policymakers and urban managers. We first conducted a systematic literature review (SLR) to understand limits and divergences when spatially expressing circularity. Four approaches covering different possibilities regarding circularity in urban areas stood out: (i) specific flows within a circular city; (ii) flows integration for resource looping; (iii) planning the transition from linear to circular cities; and (iv) concepts of circular or regenerative urban areas. We then hypothesized that one of these concepts, the Cradle to Cradle (C2C) approach, embraces the multiplicity of quantitative and qualitative requisites needed for developing circular urban areas. Our contribution, in the second part of the paper, organizes the requisites and indicators raised during the SLR according to the C2C principles into a draft framework to enable optimization and integration of different flows with human activities to various urban and socioeconomic contexts.

## 1. Introduction

Urban centers stand as key elements to the transition to a circular economy (CE), which has gained relevance as a new economic-environmental paradigm to overcome environmental issues by decoupling resource use and economic growth. Although cities occupy only 2% of the Earth's surface, they consume 80% of all energy generated, produce 75% of carbon emissions and represent almost 80% of global resource use and waste release [1]. Closing metabolic cycles within urban areas not only tackles environmental issues, but also creates positive conditions for regenerating ecological systems, optimizing social and health conditions and maximizing regional economic opportunities, by integrating urban environments with their supporting areas. However, transiting from linear to circular cities urges many challenges, given the multiple urban layers, actors and sectors and locally-bound cultural, political and social background.

A clear concept for circularity in urban areas may allow urban managers to replicate and transfer criteria and requisites to different scenarios. Despite the debate held over the past decade, conceptualization of CE is still being discussed, and somewhat confusing: Kirchherr *et al.* (2017) found over 114 definitions in their review. This lack of consensus worsens when applying CE approach to cities. Cities are complex systems in which interactions between human beings and the environment are more intense than those observed for products and businesses. Furthermore, many of the existing CE frameworks focus on micro-level activities (industrial products) and meso-level (buildings), whilst studies in macro-level (cities) are still missing [3]. Designing such multi-dimensional circular cities requires connection of multiple spatial practices and frameworks which are currently disconnected [4].

Therefore, this study aims at clarifying circularity approaches to urban areas, by identifying main trends and approaches, and at exploring their organization into a draft framework for policymakers and urban managers. We applied the Systematic Literature Review (SLR) method to critically analyze CE frameworks applied to cities and to understand the complexities and goals to achieve circularity into urban regions.

## 2. Methods

Firstly, we carried out a Systematic Literature Review (SLR), based on [5], to understand the current trends on theoretical frameworks (Figure 1 and Appendix 1). Four research questions were posed: "How has circularity been applied to the built environment?", "How has the approach been used in medium-scale urban planning?", "What are the key trends for circularity in urban areas being discussed in the academic environment?", and "What are the requisites being used and how are they implemented?".

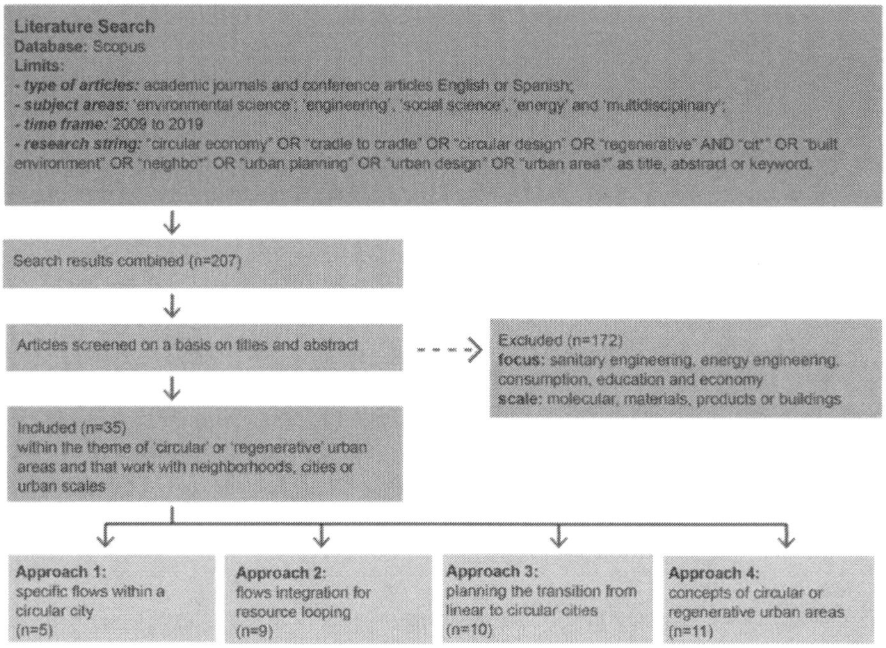


**Figure 1.** Systematic Literature Review Protocol

After title and abstract filtering, 35 articles were analyzed according to their understanding of circularity applied to urban areas. We then observed the locations of the researchers and case studies, looking up at how the scholars have worked with circularity and regeneration criteria, requisites, indicators and ways of monitoring them. We categorized four complementary approaches on circularity in urban areas found in the sample (Appendix 2). For in-depth understanding of the approaches, we complemented the sample using the 'snowball' technique, which presented complementary bibliographic references.

Subsequently, we grouped categories, requisites and indicators for circular urban areas raised in the literature sample to understand the state of the art of circular and regenerative cities agenda, and to fulfill the gaps revealed by the SLR. Based on our preliminary findings, we hypothesized that Cradle to Cradle (C2C) approach and principles would embrace the multiplicity of quantitative and qualitative requisites needed for developing tools for creating circular urban areas. To verify this hypothesis, we devised an in-progress C2C-driven framework, which is being further developed for testing in practical cases.

## 3. Results and discussion

### *3.1. Bibliometric information*

The number of published articles has increased substantially in recent years. The SLR shows that circularity in urban areas has been poorly discussed from 2009 to 2016, with an average of one article per year. This began to change in 2017 when the debate was heated up and raised exponentially after 2019.

European leadership in the field is evident: Europe concentrated most of the publications (87%) and case studies (92%); with only two cases set in emerging economies (United Arab Emirates and Cuba). Sustainability and CE policies in Europe empowered over the last decade probably explain the debate concentration. In 2015, the European Union determined a 'Circular Economy Action Plan', as a comprehensive body of legislative and non-legislative actions aiming the transition from a linear to a circular model [6]; for which the EU has allocated more than ten billions of Euros between 2016 and 2020 [7]. Also, a new 'European Green Deal' was set in 2019 to determine a new growth strategy for Europe to tackle climate and environmental challenges and to decouple economic growth from resource use. All these measures aim at Europe's economy to fit for a *'green future, strengthen competitiveness and circularity while protecting the environment'* [7]. This enlightens not only the increased number of European articles on the theme, but also of case studies in cities e.g. Amsterdam [8,9], Glasgow [10] and London [11], and the regions of Lombardy, Catalonia and Gelderland, among others [12].

### *3.2. Circularity approaches for urban areas*

We grouped the SLR sample texts into the four most frequent approaches for circularity in urban areas (Figure 2). These views are complementary and emphasize interdisciplinarity, complexity and the need for multi-dimensional approaches and solutions to enable replication into diverse social, cultural and economic contexts.

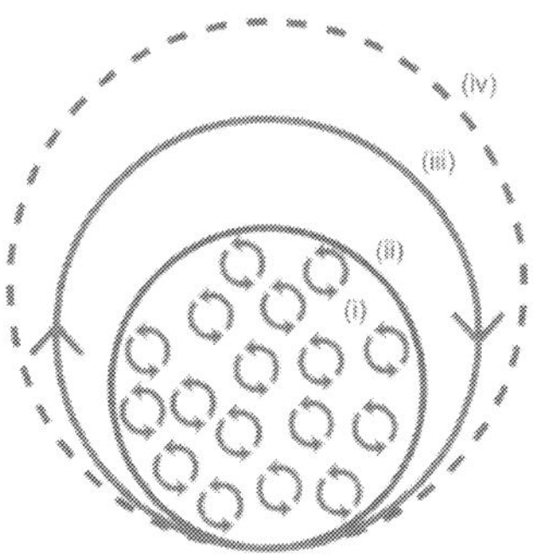


**Figure 2.** The four approaches for CE in urban areas found in the SLR: (i) **specific flows** within a circular city; (ii) **flows integration for resource looping**; (iii) **concepts** of circular or regenerative urban area; and (iv) **planning the transition** from linear to circular cities.

**(i) Specific flows within a circular city**

The first approach covers the debate about selected flows that comprise the complexity of an urban area. Most of these texts investigates technologies, management and planning of specific flows, expressing criteria, requisites and indicators for looping actions. These articles and flows connect and complement each other, and bring the discussion of closing cycles of flows such as urban waste management [32], biogas production [33], food stream and waste food [33–36], building stock [37–39], water and energy [40], and land use [39].

**(ii) Flows integration for resource looping**

This group of articles studies the management of flows within urban metabolic systems. Cities are seen as complex systems, and they are consumers and producers of materials, natural resources and food. Therefore, cities have inputs of resources and outputs of emissions and wastes. Two papers work on the Nexus solution, which is an ecocycle model for cities implemented in Sweden and China. In the first article, the author identifies three action types for looping actions: reuse, recycle and recovery and applies them into five resource categories: materials, infrastructure, water, energy and land [28]. These categories differ from the second article, in which the urban structure is characterized according to three subsystems: food, energy and water. The subsystems are applied to the Urban Circular Economy Calculator, a tool developed for calculating impacts and supporting decisions for urban management [29,30]. A third article looks into the Eurostat method for Material Flow Analysis, proposing complementary requisites and indicators such as local sourcing of primary and secondary resources [31].

**(iii) Planning the transition from linear to circular cities**
Research on CE has primarily focused on technical and economic strategies for closing resource loops, with new technologies of production and business models [2]. However, when expanding CE concept to a spatial perspective, infrastructure and land use also becomes crucial as it materializes the spatial connection between flows, people and places. Cities must therefore overcome limitations in current governance and planning to create and develop such infrastructure [40]. This group of articles explores requisites for tackling cities' transition and planning in a multi-actor and multi-sector sphere, and contribute to accelerate CE implementation in metropolitan areas, through social and cultural involvement of different sectors and stakeholders.

Complexity and comprehensiveness of transitioning towards circularity in cities call for a multi-faceted governance approach and an integrated political and legal guiding framework, citizens awareness, and technological developments [41]. As current spatial planning systems have multiple system failures, the transition should focus on flexible demand-driven and participatory processes involving various actors. Besides, spatial areas can be redeveloped for beneficial growth, by integrating social structures, focusing on citizen empowerment, and integrating a long-term vision. The demand-driven connects diverse functions of a city and will be able to cope with uncertainties [18].

**(iv) Concepts of circular or regenerative urban area**
The fourth approach debates concepts for circular or regenerative urban areas by optimizing existing frameworks. For that, some of the authors adapt existing CE frameworks developed for businesses [13,14], or for products design [15–18]. Other scholars propose the application of the UN's sustainability agenda [19] or suggest to optimize existing eco-cities models [4] or urban ecology principles [20].

*ReSOLVE Framework*
According to the Ellen MacArthur Foundation (EMF), CE fosters system effectiveness designing a society where waste should not exist while preserving and enhancing natural capital through finite stocks control and renewable resource flows balance [21]. The EMF has developed the ReSOLVE framework to optimize businesses for a CE, which consists of six principles: Regenerate, Share, Optimize, Loop, Virtualize, Exchange [22]. Two articles from the SLR looked at ways to adapt ReSOLVE to urban environments [13,14]. In the first article, Williams (2019a) states that although ReSOLVE is the most comprehensive and well-succeeded CE framework for businesses, it has strong limitations when applied to cities. The author identifies key components that should be added when applying CE into cities: complexity of urban ecosystems, consumption, land, infrastructure, adaptation, scale and localization. Similarly, Prendeville *et al.* (2018) see gaps for transferring CE frameworks for businesses to cities' and suggest top-down and bottom-up changes to be applied for each ReSOLVE's principle. Top-down changes are institution-driven such as strategies and policies created by local government. And bottom-up changes are pushed by social innovations such as activities from NGOs, communities and businesses.

*Cradle to Cradle (C2C)*
The *Cradle to Cradle* paradigm proposes an approach for designing products for circular economy and sets as one of the theoretical bases for the EMF's framework for circular economy [23]. It comprehends three design principles (1) Waste is food: eliminating waste by designing for healthy distinct biological and technical cycles; (2) Use of clean and renewable energy; and (3) Celebrate diversity: supporting biodiversity, socio-cultural and conceptual diversity [24].

In 2012, a C2C certification program was launched for industrial products, validating the framework as a measure of continuous optimization for CE. The program has been applied to thousands of products, which are assessed for environmental and social performance across five categories: material health, material reutilization, renewable energy and carbon management, water stewardship and social fairness [25]. Thus, C2C has been extensively applied to micro scale (materials and products) [3]. On the meso scale, employing C2C to buildings means to transform them in material banks adding integration and circularity of water, energy and nutrient systems [26,27]. Whilst C2C has been applied to several products and buildings, it is still to be scaled up for wider application to the built environment [17]. In the SLR sample, three articles explored C2C for urban environment, studying its quality dimensions [15,17] or the design and planning beneficial areas [18].

*Urban metabolism for spatial practices (Vandrebroek's approach)*
Vandrebroek's approach organizes urban metabolism spatial practices differentiating them as objectivist or constructivist worldview. Objectivist practices focus on performance and efficiency, whether the constructivism concentrates on potential, acting from within the intervention space. This framework outlines circularity drivers, mixing planning and design approaches with resource management. It envisions how people relate to the ecosystems they inhabit and combines technocratic and emancipatory approaches to draw cities' management and politics. It also distinguishes two CE research levels: *application of CE*, by adopting tools, metrics, instruments and indicators; and *qualitative* research, addressing norms, values and concepts. Lastly, urban landscape design is seen as the core of circular urban areas, set as a hub that connects diverse needed disciplinary fields [4].

*Low-entropy city*
Pelorosso et al. (2017) affirms that CE is based on nature's principles and functions and, when applied to cities, regards the thermodynamics of open systems. The authors explore the *low-entropy cities* concept considering the role of Urban Green Infrastructure (UGI) in reducing cities' entropy. The UGI components capture unused energy and reuse wastes coming from other human activities transforming them into benefits for the population or environment. These components enhance cities' socio-ecological and structural complexity by reducing internal entropy and by adding and optimizing functional elements and synapses among those elements, while minimizing wastes. Besides, cities should be autopoietic systems, maintaining their autonomy while remaining open to interact and compensate for the inevitable losses due to the second law of thermodynamics [20].

*3.3. A circularity framework for cities*
From all the concepts which emerged in the SLR, discussed in the approach (iv), the *Cradle to Cradle* - or C2C – framework outstands, for being based on design principles and criteria, and recognized as effective for evaluating and optimizing industrial products, materials and buildings for CE; and whose design principles and criteria are potentially upscalable to urban levels. Hence, we hypothesized that C2C principles can embrace the multiplicity of quantitative and qualitative requisites needed for developing circular urban areas.

To advance in our hypothesis verification, we regrouped the requisites and indicators presented in the SLR into the three C2C principles and adapted categories from its buildings' design application (Figure 3). Starting from this preliminary framework, we will be able to further complete it with indicators, strategies and patterns to propose an integrated vision for circular and regenerative urban areas. The main challenges are to accommodate consideration of local urban characteristics whilst being flexible enough to be transferred or adapted to other contexts, to manage interactions with other locations or regions, to encompass diverse social structures focusing on citizen empowerment, and to develop an integrated long-term vision incorporating multiple actors and nature. To achieve this, analyses of urban circular systems and corresponding data gathering are fundamental to calibrate practical metrics and tools.

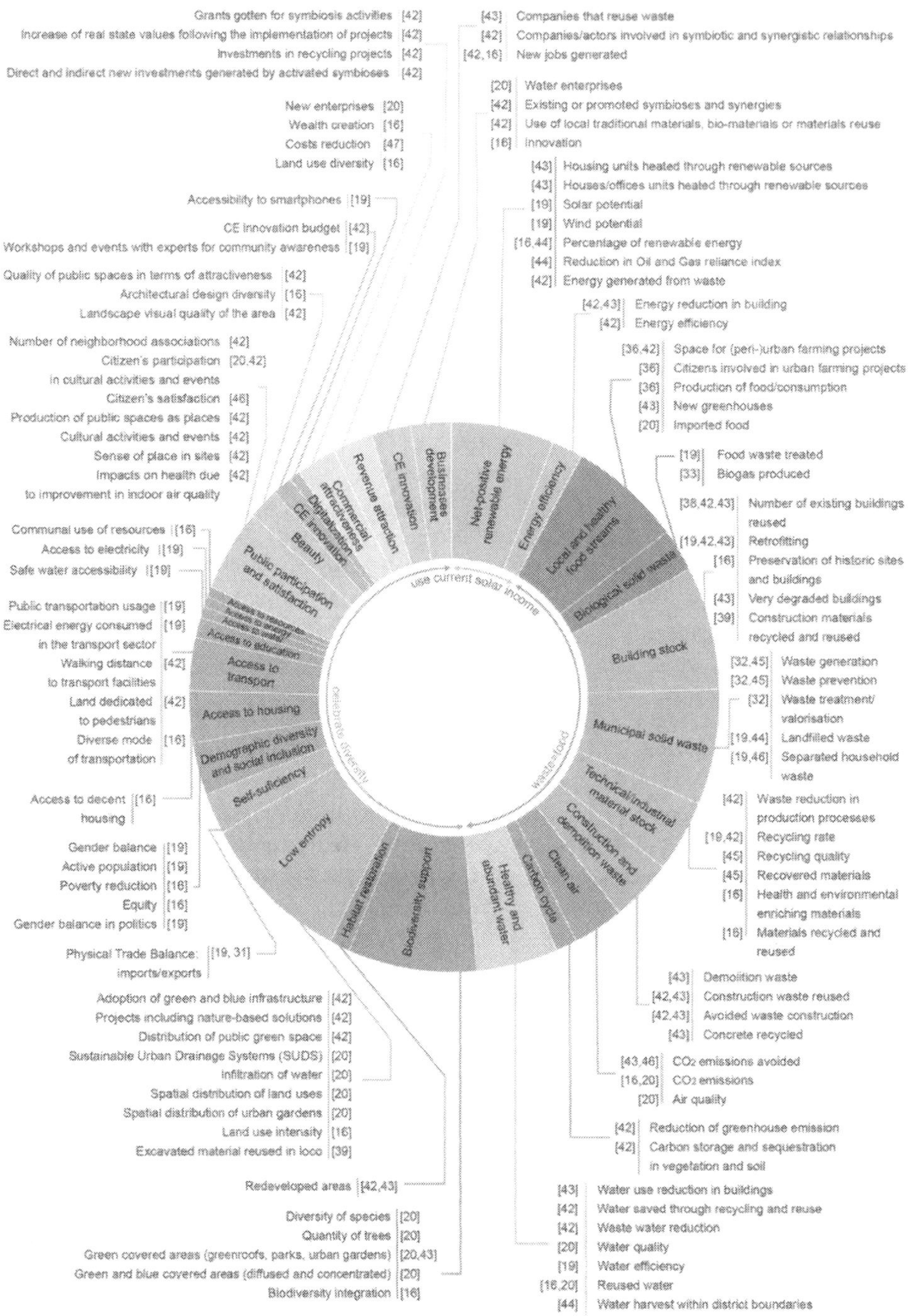


**Figure 3**. C2C-inspired framework for circularity in urban areas. Category/requisite slice represents the weighted contribution (by number of mentions and discussion depth) in SLR.

## 4. Final remarks and outlook

CE frameworks have already been applied to products and businesses, but they lack societal and behavioral factors, and a future-oriented, multidisciplinary and participatory approach fundamental for their implementation in urban areas. Translating from such perspectives into urban areas is challenging, as it crosses multiple - currently disconnected - spatial practice fields.

Such integration requires a flexible design perspective, replicable and transferable demands to encompass different spatial and socioeconomic contexts, from urban areas facing rapid urbanization throughout the global South to cities that are becoming centers of intense re-urbanization in the global North. However, the SLR showed that the discussion is still concentrated in Europe. Indeed, European regulation sets examples for driving and accelerating CE in its several practice fields, but such concentration fails to acknowledge complexities inherent to other socioeconomic contexts. The global North and South have shown varied barriers, potentials and abilities for implementing CE, which aggravates when the concept is applied to the urban scale.

Besides, CE conceptualization and its application to urban areas remains confusing. Our Systematic Literature Review revealed substantial differences in the approaches. The frameworks studied are either too technical, incomplete, or incipient, usually lacking indicators comprehensiveness and applicability, and pragmatic ways of evaluating and replicating to a set of scenarios. The SLR also underlined the benefits of standardizing criteria, requisites and indicators that enable urban areas worldwide to monitor success in their transition towards circularity and regeneration. In this sense, adopting C2C principles and categories seems a fruitful vehicle to embrace urban complexity.

The preliminary framework herein presented is a first step for developing a framework that can be used by decision makers to look at urban areas as complex ecosystems connecting resource flows to human activities when planning, designing and transitioning towards circular and regenerative urban environments. Additional work is required to support such task though. The framework should still consider examples of urban areas already applying CE, additional frameworks and testing in real case studies to understand the main challenges for its implementation and optimization. To this end, learning and testing circularity aspects in the global South will bring more data and new requisites and parameters to consider, to be then compared to the patterns functioning in the North, in the quest for replicable or adaptable models that best fit the context of study.

## References

[1] UN-Habitat 2011 *Cities and climate change: global report on human settlemans* (London: Earthscan)
[2] Kirchherr J, Reike D and Hekkert M 2017 *Resour Conserv Recycl.* **127** 221–32.
[3] Pomponi F and Moncaster A 2016 *J Clean Prod* 10 1016
[4] Marin J and De Meulder B 2018 *Sustain.* 2018 10 1310
[5] Kitchenham B 2004 *Procedures for Performing Systematic Reviews* 2004 (Keele: Keele Univ.)
[6] European Commission 2015 Closing the loop - an EU action plan for the circular economy (Brussels: European Comission)
[7] European Commission 2020 *The EU circular economy action plan* (Brussels: European Commission)
[8] Circle Economy, TNO and Fabric 2015 *Amsterdam Circular: a vision and roadmap for the city and region* (Amsterdam: Gemeente Amsterdam)
[9] Gemeente Amsterdam 2020 *Amsterdam Circular 2020-2025 Strategy* (Amsterdam: Gemeente Amsterdam)
[10] Glasgow Chamber of Commerce, Zero Waste Scotland, Glasgow City Council and Circle Economy 2016 *Circular Glasgow: A vision and action plan for the city of Glasgow* (Amsterdam: Circle Economy)
[11] LWARB 2017 *London's circular economy route map* (London: LWARB)
[12] European Union. *European regions toward Circular Economy* [Internet] available from: https://www.interregeurope.eu/circe/ acesssed: 2020 Dec 1
[13] Williams J 2019 *Urban Stud.* **56** 2746–62.
[14] Prendeville S, Cherim E and Bocken N 2018 *Environ Innov Soc Transitions* **26** 171–94

[15] Ankrah NA, Manu E and Booth C 2015 *Energy Procedia* **83** 31–40
[16] Baffour Awuah K G and Booth C A 2014 *WIT Trans Ecol Environ* **191** 111–23.
[17] Booth C A *et al.* 2012 *WIT Trans Ecol Environ.* **155** 515–26.
[18] Lodder M, Rotmans J and Braungart M 2014 *WIT Trans Ecol Environ* **191** 151–63.
[19] de Ferreira A C, Fuso-Nerini F 2019 *Sustain.* 11 1813
[20] Pelorosso R, Gobattoni F, Leone A 2017 *Landsc Urban Plan* **168** 22–30.
[21] Ellen MacArthur Foundation 2015 *Growth within: a circular economy vision for a competitive Europe* (London: Ellen MacArthur Found.)
[22] Ellen MacArthur Foundation 2015 *Delivering the circular economy: a toolkit for policymakers* (London: Ellen MacArthur Found.)
[23] Ellen MacArthur Foundation. Circular Economy Schools of Thoughts [Internet] Available from: https://www.ellenmacarthurfoundation.org/circular-economy/concept/schools-of-thought [acesssed: 2020 Dec 1]
[24] McDonough W, Braungart M 2002 *Cradle to Cradle: Remaking the Way We Make Things* (New York: North Point Press)
[25] Cradle to Cradle Products Innovation Institute [Internet] Available from: http://www.c2ccertified.org [assessed 2020 Aug 16]
[26] Mulhall D, Braungart M 2010 *Criterios 'Cradle to Cradle' para el entorno construido*. Ekonomias Rev vasca Econ. **75** 182–93.
[27] Mulhall D, Braungart M and Hansen K 2013 *How to plan a big beneficial footprint: guide to innovation tools for cradle to cradle® - inspired value in building developments* (Rotterdam: Erasmus University)
[28] Williams J. 2019 *Sustain.* **11** 423
[29] Xue J, Liu G, Casazza M and Ulgiati S 2018 *Energy* **164** 475–95.
[30] Yang Z, Liu G and Parthernope S. Urban Circular Economy Calculator (UCEC) [Internet]. Available from: http://ucec.umsoft.cn/introduce.html [assessed 2020 Aug 2].
[31] Voskamp I M, Stremke S, Spiller M, Perrotti D, van der Hoek J P and Rijnaarts H H M 2017 *J Ind Ecol.* **21** 887–902.
[32] Zeller V, Towa E, Degrez M, Achten W M J 2019 *Waste Manag.* **83** 83–94.
[33] Loizia P, Neofytou N, and Zorpas A A 2019 *Environ Sci Pollut Res.* **26** 14766–73.
[34] Fassio F, Minotti B 2019 *Sustain.* 2019 11 6078
[35] Piezer K, Petit-Boix A, Sanjuan-Delmás D, Briese E, Celik I, Rieradevall J, Gabarrell X, Josa A and Apul D 2019 *Sci Total Environ* **651** 1495–504
[36] Säumel I, Reddy S E and Wachtel T 2019 *Sustain.* **11** 972
[37] Kyrö R, Jylhä T and Peltokorpi A 2019 *Facilities* **37** 75–90.
[38] De Medici S, Riganti P and Viola S 2018 *Sustain.* **10** 4305
[39] Cross M 2017 *Waste Resour Manag.* **170** 3–14.
[40] Giezen M 2018 *Sustain.***10** 3487
[41] Obersteg A *et al.* 2019 *Urban Plan.* **4** 19–31.
[42] Girard L F and Nocca F 2019 *Sustain.* **11** 6253
[43] Gravagnuolo A, Angrisano M, Girard L F 2019 *Sustain.* **11** 3512
[44] Keough N and Ghitter G 2019 *Sustain Sci* **15** 203–17
[45] Bolger K, Doyon 2019 *A Eur Plan Stud.* **27** 2184–205.
[46] Longato D, Lucertini G, Fontana M D and Musco F 2019 *Sustain.* **11** 2101
[47] van der Hoek J P, Struker A and de Danschutter J E M 2017 *Urban Water J.* **14** 61–8.

IOP Conf. Series: Earth and Environmental Science **855** (2021) 012019 doi:10.1088/1755-1315/855/1/012019

# Policy instruments for circular built environment implementation: A systematic literature review.

**Felipe Bucci Ancapi**[1]

[1]Faculty of Architecture and the Built Environment, Delft University of Technology. Address: Julianalaan 134, 2628 BL, Delft, the Netherlands.

**Abstract.** The built environment (BE) is of fundamental importance in the transition towards circular economy (CE), for it concentrates major consuming and polluting human activities. CE in the BE research has rapidly increased in recent years. However, aspects concerning its policymaking and implementation, governance, and management are acknowledged to be widely over-looked. Such context may jeopardize effective implementation of circular built environments (CBE). In this article, I conduct a systematic literature review to characterize the relation between circular built environments and the policy instruments suggested for its implementation. Results show that only 7% of publications address policy and instruments for CBE implementation. Yet, identified publications seem to cover a wide variety of policy instruments according to prevailing classifications. Finally, operationalized concepts in publications mostly relate to technological aspects of CBE implementation, which calls for increasing research efforts over systemic challenges in governance, and policy integration and coherence.

## 1. Introduction

Integrating circular economy (CE) strategies into the built environment (BE) has been pointed out as crucial for sustainable urban transitions [1], due to BE's profiles as a major global resource consumer and polluter human activity [2-4]. At different scales, a variety of frameworks and methods are used to measure cities' performance in terms of flows of materials and energy – e.g., urban metabolism, material flow analysis and input/output analysis – and emissions – e.g., lifecycle assessment. Although these measurements are essential to support the management of sustainable built environments [5, 6], predominant narrow perspective on economic and environmental performance [7, 8] may not be sufficient to bring about the CE [2] since, for instance, political, social and behavioral aspects are normally over-looked [1, 9].

Policies for transitioning towards CBE require systemic understanding of BE's constituent parts, their interdependence, and its connection to a wider context of ecological crisis driven by the ever-increasing consumption of energy and resources. However, there is a lack of a comprehensive examination of the policy implications in the transition towards CBE – i.e., the role of governments and of policy, both public and private. Whether current research of CBE address the needed policy settings to make them emerge remain an under-revised aspect. Such a preliminary insight follows the work of Munaro et al. [7], who conducted systematic review of CE in BE research as for 2019. Their results highlight the need for studying the boundaries of CE, and, specifically, the challenges in the governance and management of CBE and their transitions.

The aim of this research is to provide a review of policies for CBE implementation. I do so by answering (1) how many publications elaborate about policy requirements for CBE, (2) what kind of policy instruments are mentioned, and (3) what concepts are operationalized in the selected publication.

## 2. Environmental policy instruments

Following Huppes and Simonis [10] environmental policy instruments 'link policy development and decision-making to policy implementation' (p. 239), by influencing citizens' and businesses' behavior [11]. What is more, policy instruments are the way policies' visions and goals are operationalized to trigger their desired effects. Such instruments are proven to be essential to operationalize the goals of policy. Echoing this need, the OECD [12] created a world database for policy instruments relevant to the environment and natural resources management. The Policy Instruments for the Environment (PINE) gathers over 3400 instruments around the world.

Several policy instruments are available for the transition towards sustainable BE construction and management. Kibert [13] categorized the ones used for the construction sector in the United States into 5 sub-groups, namely: regulatory instruments, economic instruments, information tools, voluntary tools, and research and development tools. It is worth noting that this kind of categorization is not exclusive to the BE, but common among environmental policy set-ups, as it can be found in Bouwma et al. [11]. However, other classifications also exist, as it is the case of the one proposed by Huppes and Simonis [10], which depends on the kind of actors' relations (political-administrative, regulatory and social instruments), mechanisms (prohibiting, prescriptive, option-creating, economic, cultural, structural and procedural) and objects (single objects or classes of objects) of the policy itself. Likewise, OECD's PINE database divides them into taxes, fees and charges, tradable permits, deposit-refund systems, subsidies, and voluntary approaches. The preference of economic instruments over more traditional binding regulations follows a trend in international policy-making as past command-and-control interventions lose their effectiveness in new global, dynamic contexts that privilege consensus building over strict top-down regulations [10].

## 3. Methods

For this systematic literature review, the Preferred Reporting Items for Systematic Reviews and Meta-Analyses (PRISMA) protocol was used. PRISMA is the result of an analysis about available methods and tools for systematic literature review in medical studies. It provides a checklist of 27 steps and a flow diagram to summarize the process of study selection in terms of identification, screening, eligibility, and inclusion. Likewise, the checklist is divided in 7 main parts, namely: Title, Abstract, Introduction, Methods, Results, Discussion, and Funding [14], although more steps can be added if needed. Both the checklist and flow diagram enable a rigorous review that can be checked and replicated by others. However, PRISMA does not ensure the quality of a systematic review since study selection can still be biased. Because of its reporting meticulosity, PRISMA is increasingly being used in social science and qualitative research – i.e., [15], [16] and [17].

### *3.1. Eligibility criteria*

Firstly, from the literature search we will only consider published articles, reviews, books, and book chapters available in the selected online databases. Secondly, the period 1990-2020 was selected because it ensures that eligible early developments in the BE in China, Japan, United Kingdom, Germany, and European countries in general, as CE frontrunners are taken into account [7, 18]. This decision seeks to include only those actions strictly aligned with CE research and policy, avoiding those including so-called circular strategies [19] – i.e. reduce, recycle – in previous BE research and policy without a clear CE framework (i.e. publications based on waste or environmental management). Thirdly, eligible manuscripts must be written in English. Fourthly, the words 'circular*', 'built environment', and 'polic*', 'govern*', plan*' or 'manag*' must be included either in the text's title, abstract and/or keywords. I acknowledge that this selection criteria may lead to the discrimination of valuable articles, reviews, proceedings, and books; however, it ensures that only manuscripts explicitly linked to the field of CE in the BE are covered, thus reducing possible bias in the selection process.

### *3.2. Information sources*

Three strategies were used to identify eligible publications. First, I searched 2 online databases, namely: Web of Science and Scopus, to ensure a wide pool of scientific inputs in our literature search. Secondly, to add a more specialized scope to this review, I decided to search publications of 5 top journals – in terms of number of publications – in the field of CE in the BE. Journal of Cleaner Production; Sustainability; Resources, Conservation and Recycling; Materials; and, Construction and Building Materials were selected as they concentrate approximately 30% of scientific articles in the field in 2019 [7]. Thirdly, I explored Google Books looking for books and chapters matching the above-mentioned search criteria. The last search for these three strategies was October 26, 2020. The information flow is shown in Figure 1.

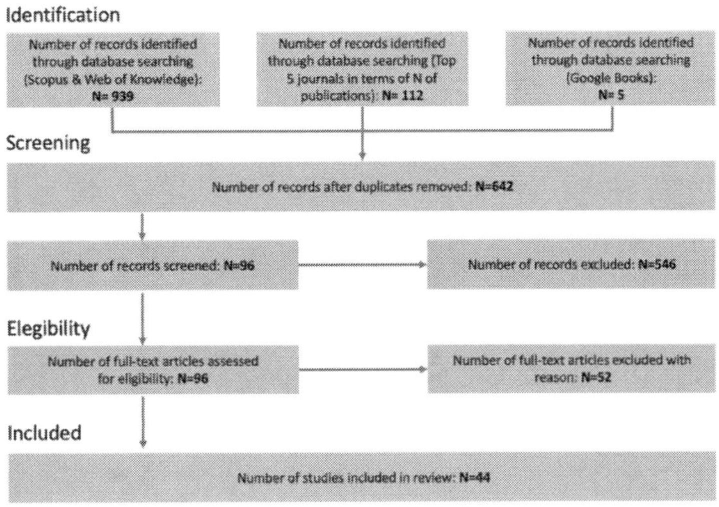


Figure 1. Information flow for final selection of studies included in review, based on the PRISMA protocol. Source: author.

### *3.3. Search and data collection process*

For searching into the two different online databases, I conducted the following search strategy: I searched for the phrases 'circular economy' AND '('built environment' OR 'construct*')'. In Web of Knowledge, the field 'topic', which searches authors, abstracts and keywords was selected. In Scopus, I selected the field 'Article title, Abstract, Keywords'. Only articles, reviews and book chapters were included in the search. For Google Books, I searched for 'circular economy' AND '('built environment' OR 'construct*')'. The resulting findings were exported as RIS, CSV and Plain text files containing full information. They were stored and grouped in EndNotes X9 and visualized using VOSviewer. Finally, using the search engine of EndNotes X9, four groups were created to contain those publications that include 'circular', 'built environment, and (1) 'polic', (2) 'manag', (3) 'govern' or (4) 'plan' in the text's title, abstract or keywords.

## 4. Results

### *4.1. How many publications elaborate about policy requirements for CBE?*

A total of 44 articles, reviews and book chapters met the selection criteria and were included for analysis (see Table 2). 29 (65%) publications correspond to articles, 9 (20%) to book chapters and 6 (14%) to reviews. In terms of journals with most contributions, Journal of Cleaner Production provided 10 publications, followed by Sustainability with 6, and Journal of Resource Conservation and Recycling with 4.

Four sub-groups were created to contain publications having one or more of the following words in their titles, abstracts, and keywords: polic*, manag*, govern* and plan*, as they can be considered representative of policy related aspects to the BE. My selection of 44 articles equals 7% of the total 642 publications after duplicates were removed. As result, 16 publications were grouped under "policy", 21 publications were grouped under "management", 7 under "governance", and 13 corresponded under "planning". That is to say, firstly, that most publications do not focus on the way frameworks and approach should be operationalized and monitored along policy implementation, but, often, they just mention policy recommendations and further research directions. Secondly, that such a final selection was only possible by extending the search through multiple queries, otherwise selected publication would have accounted for a smaller sample. Therefore, it is worth noting that, although the above-mentioned words were present in the selected publications' titles, abstracts and/or keywords, only a few of the publications are based on discussions explicitly related to policy aspects of BE in their transition towards circular ones. Figure 2 shows keyword co-occurrence for the final selection. The size of the nodes shows how often a keyword occurred, while the width of the links shows co-occurrence among the nodes. Arguably, none of the keywords relate explicitly to mainstream terms in policy discussion, denoting that, although policy and related concepts were present, they were mostly part of final policy recommendations and further research agendas. A few exceptions argue explicitly about policy aspects of CE in the BE – i.e., [20], [21].

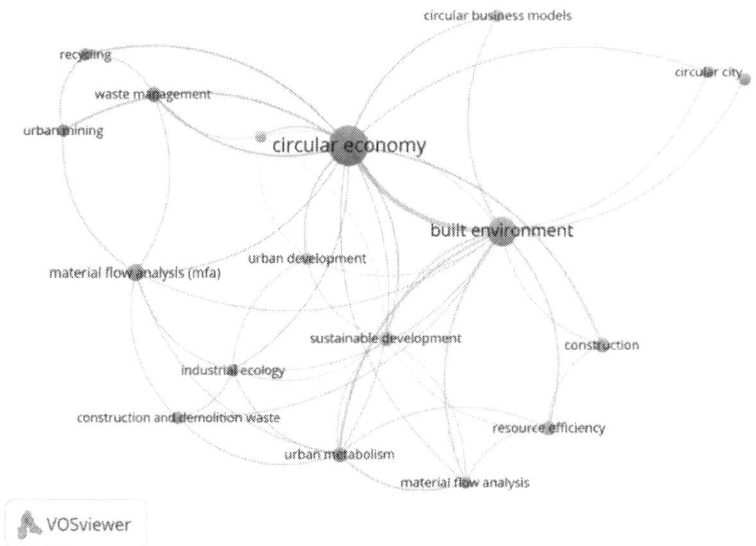


Figure 2. Keywords co-occurrence for final literature selection, based on Scopus database (37 publications available). Made with VOSviewer, co-occurrence threshold = 2. Source: author.

### 4.2. *What kind of policy instruments are mentioned?*

Following Kibert [13] categories for policy instruments, I classified the selected publications according to whether they mention, implicitly and explicitly, policy instruments for the implementation of CBE. Findings are shown in Table 1. Explicitly mentioned policy instruments from implicitly mentioned ones were differentiated through content analysis and the level of detail provided in the publication. For instance, a need for sharing resource information along a circular supply chain was considered as *implicitly* talking about regulatory instruments, while the need for databases to support the implementation of material passports in the construction sector was identified as *explicitly* talking about regulatory instruments. Table 1 is followed by explicitly mentioned policy instruments in highlighted articles.

| Regulatory | | Economic | | Information | | Voluntary | | R&D | |
|---|---|---|---|---|---|---|---|---|---|
| Exp. | Imp. | Exp. | Imp. | Exp. | Imp. | Exp. | Imp. | Exp. | Imp. |
| 15 | 12 | 11 | 13 | 9 | 6 | 5 | 7 | 15 | 12 |

Table 1. Number of publications referring to policy instruments, explicitly or implicitly, in selection. Source: author.

### *4.3. What concepts are operationalized in the selected publications?*

The policy instruments identified in the previous section resulted from the operationalization of specific concepts in articles, reviews, and books. Therefore, knowing which concepts are discussed when policy for CBE implementation is mentioned or proposed provide an account of the main research directions to date (specifically related to policy, see Munaro [7] for a general overview about CE in the BE research). To answer this research question, I reviewed the methodological section of each publication and identified the main concepts that were operationalized for sake of each research. A reduced number of publications did not have a methodological section. For instance, some present or discuss new approaches to design or construction, for which methods were not required. In such cases I reviewed the abstracts and introductions to identify the main concepts under discussion. Results are presented in Table 2.

| Ref. Number & Author | Year | Policy instruments | | | | | Operationalized concept(s) |
|---|---|---|---|---|---|---|---|
| | | Reg | Eco | Inf | Vol | R&D | |
| [20 (Al Hosni)] | 2020 | E | E | 0 | 0 | E | Circular Economy, Built Environment |
| [22 (Arora)] | 2019 | 0 | 0 | 0 | 0 | E | Urban Mining (inflows, stocks and outflows) |
| [23 (Arora)] | 2020 | I | 0 | 0 | I | 0 | Urban Mining (inflows, stocks and outflows) |
| [24 (Attia)] | 2018 | 0 | 0 | 0 | 0 | 0 | Positive Impact Buildings |
| [25 (Bolger)] | 2019 | I | I | 0 | ! | I | Strategic Planning, Circular Economy |
| [26 (Cai)] | 2019 | E | E | 0 | 0 | E | Material Reuse |
| [27 (Cerreta)] | 2020 | 0 | 0 | 0 | 0 | 0 | Landscape Regeneration |
| [28 (Cross)] | 2017 | 0 | 0 | 0 | 0 | 0 | Recycle and Reuse of Materials |
| [29 (Desing)] | 2019 | 0 | 0 | 0 | 0 | 0 | Renewable Energy Potential |
| [30 (Edike)] | 2020 | 0 | 0 | 0 | 0 | 0 | Eco-bricks |
| [31 (Eray)] | 2019 | I | I | I | I | I | Adaptive Reuse, Interface Management |
| [32 (Gallego-Schmid)] | 2020 | E | E | E | E | E | Circular Economy, GHG mitigation |
| [33 (Gassner)] | 2020 | 0 | 0 | 0 | 0 | 0 | Urban Mining (inflows, stocks and outflows) |
| [34 (Geldermans)] | 2019 | E | I | E | I | I | Circular Building Design and Health/Well-being |
| [35 (Ghaffar)] | 2020 | E | E | E | E | E | Resource Recovery |
| [36 (Gravagnuolo)] | 2019 | I | I | I | I | I | Circular City Implementation |
| [37 (Heesbeen)] | 2020 | 0 | ! | 0 | 0 | I | Circular Business Models |
| [38 (Heisel)] | 2020 | E | E | 0 | 0 | E | Material Documentation |
| [39 (Joensuu)] | 2020 | E | E | E | E | E | Circular Economy, Built Environment |
| [40 (Katriniaris)] | 2018 | 0 | 0 | 0 | 0 | 0 | Cradle to cradle, Regenerative Design |
| [41 (Lanau)] | 2020 | E | E | 0 | 0 | I | Urban Mining (inflows, stocks and outflows) |
| [42 (Lanau)] | 2019 | 0 | 0 | E | 0 | E | Urban Mining (inflows, stocks and outflows) |
| [43 (Laurenti)] | 2018 | I | I | 0 | 0 | I | Sustainable Physical Resource Management |
| [44 (Liaros)] | 2019 | 0 | 0 | 0 | 0 | 0 | Regenerative Development |
| [45 (Lowe)] | 2005 | 0 | 0 | 0 | 0 | 0 | Sustainable Economic Practices |
| [46 (Mangialardo)] | 2018 | 0 | 0 | 0 | 0 | 0 | Circular Construction |
| [47 (Marcellus-Zamora)] | 2020 | 0 | I | 0 | 0 | I | Urban Mining (inflows, stocks and outflows) |
| [48 (Marinova)] | 2020 | E | 0 | E | 0 | E | Urban Mining (inflows, stocks and outflows) |
| [7 (Munaro)] | 2020 | E | E | E | E | E | Circular Economy, Built Environment |
| [49 (Ness)] | 2017 | 0 | I | 0 | 0 | 0 | Resource-Efficient Built Environment |
| [50 (Noll)] | 2019 | E | E | 0 | 0 | E | Material Flows |
| [51 (Omwoma)] | 2017 | I | 0 | 0 | 0 | 0 | Technological Tools for Sustainable Development |
| [52 (Romero Perez)] | 2020 | 0 | 0 | 0 | 0 | 0 | Urban Mining (inflows, stocks and outflows) |
| [53 (Schiller)] | 2017 | I | 0 | 0 | 0 | E | Urban Mining (inflows, stocks and outflows) |
| [54 (Talamo)] | 2020 | I | I | I | 0 | I | Building Re-manufacturing and Reuse |
| [55 (Tingley)] | 2017 | E | E | E | I | E | Material (steel) Reuse |
| [56 (Tucci)] | 2018 | I | E | 0 | 0 | E | Resilience in the Built Environment |
| [57 (van der Leer)] | 2018 | I | 0 | 0 | 0 | 0 | CE Integration into Urban Planning |
| [58 (Wuyts)] | 2019 | E | E | 0 | 0 | 0 | Short-lived Buildings |
| [59 (Wuyts)] | 2020 | I | 0 | 0 | 0 | 0 | Urban Mining (inflows, stocks and outflows) |
| [21 (Ness)] | 2019 | E | E | I | 0 | E | Policy and Economic Instruments |

| | | | | | | | |
|---|---|---|---|---|---|---|---|
| [60 (Densley Tingley)] | 2018 | I | I | E | I | I | Embodied Carbon in Buildings |
| [61 (Chesire)] | 2016 | 0 | I | I | 0 | I | Circular Economy, Built Environment |
| [62 (Giorgi)] | 2020 | E | E | I | E | E | Regeneration of Building Stocks and Policy |

Table 2. Operationalized concepts and identified policy instruments in selected publications (E = explicitly mentioned, I = implicitly mentioned, 0 = does not mention). Source: author.

## 5. Discussion and conclusions

Some years ago Pomponi and Moncaster [2] pointed out that *"the initiatives themed around CE in the built environment however demonstrated little interdisciplinarity underpinning the complexity of such transition"* (p. 717). Likewise, more recently, Munaro et al. [7] stated that a *"systematic regulation and policy system, with better interactions among governmental institutions, policymakers, communities, and manufacturing industries"* (p. 15) was still required. Therefore, a systemic view of the CBE transition is still lacking, governance being its Achilles' heel. However, scholars such as Hartley et al. [63] have started to identified necessary policy changes to trigger a more coherent CE transition, although from a more general, regional-based perspective; one that do not address the specific challenges of the BE.

The intention of this research was to characterize the relation between CE in the BE and policies for its implementation. I did so by examining which publications in the field touch upon concepts such as policy, govern, plan and management for CBE. In numbers, to date a reduced quantity of articles, reviews, and book chapters (n= 44) somewhat elaborate about policies needed to transition towards CBE. What is more, most of publications mention policy, but they do as policy recommendations or further research agendas. In terms of policy instruments suggested for such transition, a variety of perspectives are covered –for instance, norms to include material passports, taxation for both secondary resources incentive and unsustainable resources avoidance, circular guidelines for stakeholders in the building sector, sector-wide agreements to accelerate the circular transition, and public investment in research and technology for material recovery and reuse. Finally, when connecting such policy instruments to the operationalized concepts in the selected publications, great attention has been given to technical aspects such as urban mining, material flows and material reuse. Meanwhile, publications with a strict focus on policy and policy instruments – e.g., policymaking, implementation– were scarcely found. It can be argued that the thinking of policy instrumentation is influenced by the thinking in technical solutions, which leads to a focus on policy instruments in support of those particular solutions, while leaving out the more obvious instruments when looking purely at how actors in the BE are influenced by current policies and regulations.

Sustainable transition research [64] may serve to organize future research around CE in the BE and systematic policymaking and -implementation, since its framework is based on lacking/required concepts such as complexity. So far, policymaking and -implementation tend to be associated with governments and State interventions. However, the indispensable role of private policy is increasingly being recognized. Perhaps, one of the main attributes of sustainable transition research is to recognize the limited but still essential role of government in complex decision-making arenas. The latter may help to legitimize long-term intervention for change [65]. Therefore, for sound CE in the BE policy, it is required not only to think about adequate policy, but new complexity-driven ways of governance from which new policies can emerge.

## Acknowledgments

I want to acknowledge dr. Ellen van Bueren and dr. Karel Van den Berghe for their guidance and constant contribution to this (still in progress) research. Likewise, I want to thank the sponsorship of Agencia Nacional de Investigación y Desarrollo (ANID), Chile.

## 6. References

1. Schröder, P., A. Lemille, and P. Desmond, *Making the circular economy work for human development.* Resources, Conservation and Recycling, 2020. **156**.
2. Pomponi, F. and A. Moncaster, *Circular economy for the built environment: A research framework.* Journal of Cleaner Production, 2017. **143**: p. 710-718.
3. van Bueren, E., *Greening governance: an evolutionary approach to policy making for a sustainable built environment.* Delft Center for Sustainable Urban Areas. Vol. 330. 2009, Delft: Delft University Press.
4. Ness, D.A. and K. Xing, *Toward a Resource-Efficient Built Environment: A Literature Review and Conceptual Model.* Journal of Industrial Ecology, 2017. **21**(3): p. 572-592.
5. Kaviti Musango, J.K., P. Currie, and B. Robinson, *Urban metabolism for resource efficient cities: from theory to implementation.* 2017, UN Environment: Paris.
6. Lucertini, G. and F. Musco, *Circular Urban Metabolism Framework.* One Earth, 2020. **2**(2): p. 138-142.
7. Munaro, M.R., S.F. Tavares, and L. Bragança, *Towards circular and more sustainable buildings: A systematic literature review on the circular economy in the built environment.* Journal of Cleaner Production, 2020. **260**.
8. Kirchherr, J., D. Reike, and M. Hekkert, *Conceptualizing the circular economy: An analysis of 114 definitions.* Resources, Conservation and Recycling, 2017. **127**: p. 221-232.
9. Korhonen, J., A. Honkasalo, and J. Seppälä, *Circular Economy: The Concept and its Limitations.* Ecological Economics, 2018. **143**: p. 37-46.
10. Huppes, G. and U. Simonis, *Environmental Policy Instruments*, in *Principles of Environmental Science*. 2009, Springer.
11. Bouwma, I., et al., *Policy instruments and modes of governance in environmental policies of the European Union.* 2015, Wageningen University & Research: Wageningen.
12. OECD, *Policy INstruments for the Environment.* 2017, Organisation for Economic Co-operation and Development (OECD).
13. Kibert, C.J., *Policy instruments for a sustainable built environment.* Journal of Land Use & Environmental Law, 2002. **17**(2): p. 379-394.
14. Moher, D., et al., *Preferred reporting items for systematic reviews and meta-analyses: the PRISMA statement.* Annals of Internal Medicine, 2009. **151**(4): p. 264-269.
15. De Vries, H., V. Bekkers, and L. Tummers, *Innovation in the public sector: a systematic review and future research agenda.* Public Administration, 2015. **94**(1): p. 146-166.
16. Sadick, A. and I. Kamardeen, *Enhancing employees' performance and well-being with nature exposure embedded office workplace design.* Journal of Building Engineering, 2020. **32**: p. 11.
17. Huijbregts, R., B. George, and V. Bekkers, *Public values assessment as a practice: integration of evidence and research agenda.* Public Management Review, 2021: p. 1-20.
18. Geissdoerfer, M., et al., *The Circular Economy – A new sustainability paradigm?* Journal of Cleaner Production, 2017. **143**: p. 757-768.
19. Potting, J., et al., *Circular Economy: measuring innovation in the product chain.* 2017, Netherlands Environmental assessment Agency (PBL).
20. Al Hosni, I.S., O. Amoudi, and N. Callaghan, *An exploratory study on challenges of circular economy in the built environment in Oman.* Proceedings of the Institution of Civil Engineers-Management Procurement and Law, 2020. **173**(3): p. 104-113.
21. Ness, D., *The Impact of Overbuilding on People and the Planet.* 2019: Cambridge Scholars Publisher.
22. Arora, M., et al., *Residential building material stocks and component-level circularity: The case of Singapore.* Journal of Cleaner Production, 2019. **216**: p. 239-248.
23. Arora, M., et al., *Buildings and the circular economy: Estimating urban mining, recovery and reuse potential of building components.* Resources Conservation and Recycling, 2020. **154**.
24. Attia, S., *Definitions and Paradigm Shift.* Regenerative and Positive Impact Architecture: Learning from Case Studies, 2018: p. 13-31.

25. Bolger, K. and A. Doyon, *Circular cities: exploring local government strategies to facilitate a circular economy*. European Planning Studies, 2019. **27**(11): p. 2184-2205.
26. Cai, G. and D. Waldmann, *A material and component bank to facilitate material recycling and component reuse for a sustainable construction: concept and preliminary study*. Clean Technologies and Environmental Policy, 2019. **21**(10): p. 2015-2032.
27. Cerreta, M., et al., *Operationalizing the Circular City Model for Naples' City-Port: A Hybrid Development Strategy*. Sustainability, 2020. **12**(7).
28. Cross, M., *Wallasea Island Wild Coast Project, UK: circular economy in the built environment*. Proceedings of the Institution of Civil Engineers-Waste and Resource Management, 2017. **170**(1): p. 3-14.
29. Desing, H., et al., *Powering a Sustainable and Circular Economy-An Engineering Approach to Estimating Renewable Energy Potentials within Earth System Boundaries*. Energies, 2019. **12**(24).
30. Edike, U.E., O.J. Ameh, and M.O. Dada, *Production and optimization of eco-bricks*. Journal of Cleaner Production, 2020. **266**: p. 121640.
31. Eray, E., B. Sanchez, and C. Haas, *Usage of Interface Management System in Adaptive Reuse of Buildings*. Buildings, 2019. **9**(5).
32. Gallego-Schmid, A., et al., *Links between circular economy and climate change mitigation in the built environment*. Journal of Cleaner Production, 2020. **260**: p. 121115.
33. Gassner, A., J. Lederer, and J. Fellner, *Material stock development of the transport sector in the city of Vienna*. Journal of Industrial Ecology, 2020.
34. Geldermans, B., M. Tenpierik, and P. Luscuere, *Human Health and Well-Being in Relation to Circular and Flexible Infill Design: Assessment Criteria on the Operational Level*. Sustainability, 2019. **11**(7).
35. Ghaffar, S.H., M. Burman, and N. Braimah, *Pathways to circular construction: An integrated management of construction and demolition waste for resource recovery*. Journal of Cleaner Production, 2020. **244**.
36. Gravagnuolo, A., M. Angrisano, and L. Fusco Girard, *Circular Economy Strategies in Eight Historic Port Cities: Criteria and Indicators Towards a Circular City Assessment Framework*. Sustainability, 2019. **11**(13): p. 3512.
37. Heesbeen, C. and A. Prieto, *Archetypical CBMs in Construction and a Translation to Industrialized Manufacture*. Sustainability, 2020. **12**(4).
38. Heisel, F. and S. Rau-Oberhuber, *Calculation and evaluation of circularity indicators for the built environment using the case studies of UMAR and Madaster*. Journal of Cleaner Production, 2020. **243**.
39. Joensuu, T., H. Edelman, and A. Saari, *Circular economy practices in the built environment*. Journal of Cleaner Production, 2020. **276**: p. 124215.
40. Katriniaris, A., *Cradle to cradle regeneration design: from circular economy to sustainable construction*. WIT Transactions on Ecology and the Environment, 2019. **217**: p. 9.
41. Lanau, M. and G. Liu, *Developing an Urban Resource Cadaster for Circular Economy: A Case of Odense, Denmark*. Environmental Science & Technology, 2020. **54**(7): p. 4675-4685.
42. Lanau, M., et al., *Taking Stock of Built Environment Stock Studies: Progress and Prospects*. Environmental Science & Technology, 2019. **53**(15): p. 8499-8515.
43. Laurenti, R., et al., *The Socio-Economic Embeddedness of the Circular Economy: An Integrative Framework*. Sustainability, 2018. **10**(7).
44. Liaros, S., *Implementing a new human settlement theory: Strategic planning for a network of regenerative villages*. Smart and Sustainable Built Environment, 2019. **9**(3): p. 258-271.
45. Lowe, E., *Economic solutions*, in *Environmental Solutions*. 2005, Elsevier Inc. p. 61-114.
46. Mangialardo, A. and E. Micelli, *Rethinking the Construction Industry Under the Circular Economy: Principles and Case Studies*. Smart and Sustainable Planning for Cities and Regions, Sspcr 2017, 2018: p. 333-344.

47. Marcellus-Zamora, K.A., P.M. Gallagher, and S. Spatari, *Can Public Construction and Demolition Data Describe Trends in Building Material Recycling? Observations From Philadelphia.* Frontiers in Built Environment, 2020. **6**: p. 7.
48. Marinova, S., et al., *Global construction materials database and stock analysis of residential buildings between 1970-2050.* Journal of Cleaner Production, 2020. **247**.
49. Ness, D. and K. Xing, *Toward a Resource-Efficient Built Environment: A Literature Review and Conceptual Model.* Journal of Industrial Ecology, 2017. **21**(3): p. 572-592.
50. Noll, D., et al., *The expansion of the built environment, waste generation and EU recycling targets on Samothraki, Greece: An island's dilemma.* Resources Conservation and Recycling, 2019. **150**.
51. Omwoma, S., et al., *Technological tools for sustainable development in developing countries: The example of Africa, a review.* Sustainable Chemistry and Pharmacy, 2017. **6**: p. 67-81.
52. Romero Perez , A., C.M. Rose, and J.A. Stegemann, *Quantification of material stocks in existing buildings using secondary data—A case study for timber in a London Borough.* Resources, Conservation & Recycling: X, 2020. **5**: p. 100027.
53. Schiller, G., K. Gruhler, and R. Ortlepp, *Continuous Material Flow Analysis Approach for Bulk Nonmetallic Mineral Building Materials Applied to the German Building Sector.* Journal of Industrial Ecology, 2017. **21**(3): p. 673-688.
54. Talamo, C., et al., *Re-NetTA. re-manufacturing networks for tertiary architectures*, in *Research for Development*. 2020, Springer. p. 303-314.
55. Tingley, D.D., S. Cooper, and J. Cullen, *Understanding and overcoming the barriers to structural steel reuse, a UK perspective.* Journal of Cleaner Production, 2017. **148**: p. 642-652.
56. Tucci, F., *Resilience and green economies for the future of architecture and the built environment.* Techne-Journal of Technology for Architecture and Environment, 2018. **15**: p. 153-164.
57. van der Leer, J., A. van Timmeren, and A. Wandl, *Social-Ecological-Technical systems in urban planning for a circular economy: an opportunity for horizontal integration.* Architectural Science Review, 2018. **61**(5): p. 298-304.
58. Wuyts, W., et al., *Extending or ending the life of residential buildings in Japan: A social circular economy approach to the problem of short-lived constructions.* Journal of Cleaner Production, 2019. **231**: p. 660-670.
59. Wuyts, W., et al., *Understanding and Managing Vacant Houses in Support of a Material Stock-Type Society-The Case of Kitakyushu, Japan.* Sustainability, 2020. **12**(13).
60. Densley Tingley, D., J. Giesekam, and S. Cooper-Searle, *Applying Circular Economic Principles to Reduce Embodied Carbon*, in *Embodied Carbon in Buildings*, F. Pomponi, C. De Wolf, and A. Moncaster, Editors. 2018, Springer.
61. Cheshire, D., *Building revolutions*. 2016: RIBA Publishing.
62. Giorgi, S., M. Lavagna, and A. Campioli, *Circular economy and regeneration of building stock: policy improvements, stakeholder networking and life cycle tools*, in *Research for Development*. 2020, Springer. p. 291-301.
63. Hartley, K., R. van Santen, and J. Kirchherr, *Policies for transitioning towards a circular economy: Expectations from the European Union (EU).* Resources, Conservation and Recycling, 2020. **155**.
64. Köhler, J., et al., *An agenda for sustainability transitions research: State of the art and future directions.* Environmental Innovation and Societal Transitions, 2019. **31**: p. 1-32.
65. Rotmans, J., R. Kemp, and M. van Asselt, *More evolution than revolution: transition management in public policy.* Foresight, 2001. **3**(1): p. 15-31.

IOP Conf. Series: Earth and Environmental Science **855** (2021) 012020 doi:10.1088/1755-1315/855/1/012020

# Creating an enabling environment for accelerating condominium energy retrofitting: case studies in Grenoble (FR) and Brussels (BE)

**S Monfils[1] and A van Zeijl-Rozema[2]**

[1] Department of Sciences and Environmental Management, University of Liege, 185 Avenue de Longwy, Arlon, Belgium

[2] Maastricht Sustainability Institute, School of Business and Economics, University of Maastricht, PO Box 616, 6200 MD Maastricht, Netherlands

**Abstract.** Energy retrofitting in condominiums involves collective decision making, the complexity of which is believed to be one of the main barriers to retrofitting projects. A strong heterogeneity among the occupants in terms of age, education, income, or occupancy status, as well as varying interests and perceptions, can make it difficult to gather people around and agree on a common project. In this paper, we present the results of two in depth case studies about condominium renovation projects that took place in Grenoble (France) and Brussels (Belgium). Those in-depth case studies investigate the elements that are needed to create an enabling environment stimulating the acceleration of energy retrofitting of condominiums, and took place within the Interreg NWE "ACE Retrofitting" project. The enabling environment surrounding a successful retrofit is shown in these case studies to be more than just providing process facilitation or a financing scheme. It takes time to put an enabling environment in place. For different locations across Europe, different enabling environments are needed. Clearly, this requires giving more attention to condominiums at the local, regional, national and European levels. Among other crucial element to look for, social acceptance is at the heart of decision-making in retrofitting projects. This study concludes with propositions and policy recommendations, at the different levels of the social-ecological model.

## 1. Introduction: ACE Retrofitting Project Summary

A majority of buildings in North-West Europe are still energy-inefficient and in need of deep renovation. Even though local governments are tackling climate change, the building sector remains a weak point. If action in this field is not reinforced, local governments will be locked in a high carbon future. Condominiums (tenements and flatted buildings) are an important area. By directing efforts towards co-owned residential buildings, a large number of private owners at once can benefit from more energy efficient homes, reduced energy bills and carbon footprint (50-70%) [1].

Through the Interreg NWE funded ACE Retrofitting project local governments learn to act as facilitators between co-owners and building experts. Energy Cities, the European city network, along with six local authorities from five different countries, have designed a method and governance framework to help overcome the legal, human and financial barriers that are currently hindering energy retrofitting of condominiums [1].

Part of the project consisted of the analysis of a great number of case studies, with the intent to shed light onto success stories and stimulate potential candidates to condominium retrofit. The objectives of Maastricht University's Sustainability Institute were to develop an understanding of the enabling environment that led to a retrofitting project of specific condominiums. The overall research question addressed was what are the crucial elements of an enabling environment that allowed the energy retrofit of condominiums?

## 2. Theoretical framework

Energy retrofit in condominiums involves collective decision making, the complexity of which is believed to be one of the main barriers to retrofitting as there is a strong heterogeneity in the occupants in terms of age, education, income, or occupancy status. Varying interests and perceptions can make it difficult to gather people around a common project. Whereas some owners are willing to, and pushing for, the energy retrofit of their condominium, others can block the decision by voting against the project. When looking at ways to accelerate retrofitting, one can focus activities on the owners directly, but also on creating an environment that is conducive for owners to retrofit. This paper focuses on the latter by investigating two case studies where successful retrofitting of condominiums took place. In order to paint a picture of the enabling environment in both case studies, owners and other stakeholders were asked what helped them to start and implement the retrofitting project in their condominium [2-4].

### 2.1. *Enabling Environment.*

An enabling environment is generally understood as a set of interrelated conditions that impacts the potential to bring about sustained and effective change; in this case, impacts on the capacity of co-owners to engage in energy retrofitting. According to the literature [2-3], four main pillars are involved in creating enabling environments:

1. **Policy**, a generic term regrouping political will and support through policies, strategies and governments' capacity to engage, such as the creation of energy efficiency requirements [5] or the creation of mandatory multi-annual plans for renovation, ensuring that co-owners adopt a long-term vision toward retrofitting [6-7].
2. **Finance**, through the development of supportive financial regulations and mechanisms (e.g. tax credits, tax reduction/exemption, public loans/grants, green public procurement) and the mobilization of funds for implementation [8-13]. Among the solutions specific to energy retrofit, literature suggests the creation of financial incentives and attractive financing programs to deal with the initial cost of energy retrofitting such as zero interest financing [5-6] or clean energy assessment districts.
3. **Capacity and internal organisation**, to ensure sufficient investment in human (condominiums) capacity and skills development. For instance, there should be a dedicated budget and staff for the retrofit, and technical training opportunities for the energy referent [9-10, 12-13]. A prerequisite for enabling environment could also be the identification of clear management roles and responsibilities (condominiums' structure and role delegation among co-owners) across institutions and agencies (condominiums) (e.g. data and information, creation of planning tools and management guidelines) [9-10, 12-15]. Capacity and skills development could also mean providing a free (= subsidized by public authorities) consultant who guides the condominiums with the procedures [6].
4. **Socio-cultural acceptance**, through the presence of social capital and trust (within the condominium, towards the service providers and towards the intervention) [9], which involves communication practices, information networks [8] and third-party services [6-7]. The intervention should match the users' perceptions, preferences and commitments [10]. For example, changing paradigm from energy retrofit to global amelioration of buildings would help the decision to retrofit by taking into account co-owners' priorities and interests [7]. Also, in order to maintain positive perceptions, information regarding energy-efficiency interventions should be made clear and easily available [16], with measures including minimal paper work requirements and information services that help owners in choosing contractors [5]. Additionally, renovation professionals, technologies and contractors must be reliable and capable of meeting participants' expectations to increase social capital and trust.

*2.2. The Social-Ecological Model.*

Creating an enabling environment implies interacting with a variety of actors and external factors (legislation, finances, etc.), that must be in place in order to accelerate the energy retrofit uptake. In order to understand the interrelations of personal and external factors, the social-ecological model (SEM) can be used. This model is based on ecological systems theory, more precisely on Bronfenbrenner's ecological framework for human development [17]. The term "ecological" refers to the societal environment which influences the way human beings develop and behave. This model, therefore, is a theory-based framework or set of theoretical principles for understanding the intricate dynamic interactions between individual and environmental factors (at interpersonal, organisational, community and societal levels) that determine behaviour [18]. This is consistent with systems thinking, where a system is understood to be a structure of interacting, interrelated of interdependent factors, activities, actors and settings that influence or are influenced by a specific situation [19]. The effectiveness of planning behavioural change intervention can be increased when using a systems perspective to assess the needs and strengths of a population; to comprehend a problem and its causes; to create a group of stakeholders to design, carry out and diffuse an intervention; and to choose the most efficient influence points to address a certain problem [2]. An approach of looking at environmental agents (decision makers or role actors) at every ecological level allows for multiple influences both within and across levels to enhance behavioural change. Interventions using SEM centre their attention on agents in positions to exert control over aspects of the environment [20].

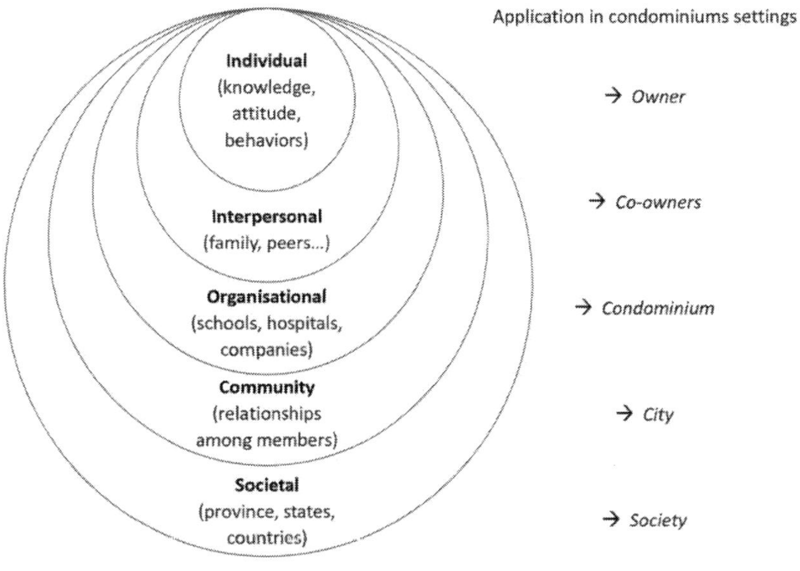


**Figure 1**. The Social-Ecological Model, of the individual in relation to its environment, and its application in condominium settings [2], [3].

Five scale levels can be identified (Fig. 1). Translated towards the condominium setting they could be interpreted as follows:

- The individual level is composed of individual home-owners and other dwellers of the condominiums.
- The interpersonal level is represented by the community of co-owners of the condominium.
- The organisational level can be seen in different bodies, for instance the property managers ("syndics", in French), as well as the condominium board, and internal temporary commissions overseeing the retrofit such as the one in the Grenoble case study (see below).
- The community level of the SEM is represented by officials of the municipality and the municipality's services.
- The societal level of the SEM is composed of the higher administrative levels in society (such as legislators).

This framework (the four pillars of the enabling environment coupled with the five scale levels of the social-ecological model) has been used to analyse the findings collected by means of focus groups in Grenoble and Brussels.

## 3. Methodology

*3.1 Focus groups.*
In order to investigate the enabling environment surrounding successful energy retrofits, next to some eighty small case studies, several in-depth case studies were conducted within the Interreg NWE "ACE-retrofitting" project. Two of those in-depth case studies are discussed in this paper: Grenoble and Brussels. Focus Groups have been used, as a means to gather people's opinions, ideas, and beliefs on a certain topic [21]. Although surveys or questionnaires could have been useful too, they cannot capture the depth of a person's thoughts, feelings or understanding, whereas focus groups allow to gather more useful information in a shorter period of time, which is what was needed here. In this interactive context, people are encouraged to discuss thoughts freely with other participants, which typically generate ideas and can provide a wealth of information through open ended, broad, and qualitative responses [22].

*3.2 Selection of case studies.*
Case studies were selected on the basis of:

- Their condominium status: these buildings needed to be privately-owned by multiple owners. The presence of renters inside the building was a welcome condition, but not an obligation. There were a priori no limitations on the number of apartments or owners (as long as it was more than one), or on the size of the building.
- Their location, depending on the list of countries and regions forming a part of the Interreg North-West Europe territory (Austria, Belgium, France, Germany, Ireland, Luxemburg, Netherlands, Switzerland, UK). Condominiums which had already been studied in the ACE Retrofitting project were ruled out.
- Their renovation status: the condominium board had to have reached a decision on starting a deep retrofit project, and on the types of works to implement. The works had to be started but not necessarily finished. The renovation project needed to involve some "deep retrofit" works – works which brought enough inconvenience to the condominium to imply difficulties in the decision-making process (such as multiple interventions, change of windows, insulation of facades, or works on collectively-owned systems).
- The retrofit project could have been initiated by the board of the condominium, or by the existence of a larger (European, national, regional or local) renovation program. A diversity in the incentive dynamic was welcomed, to facilitate replicability analysis.
- Several types of stakeholders had to be invited to share their experience: the decision makers (which involved members of the condominium board, and eventually other co-owners), the technical assistance team (architect and possibly others), the building manager, representatives of the financial stakeholders (banks, insurances), the governance bodies, and potential other stakeholders involved in the project (NGOs, associations, researchers…).

*3.3 Data Collection.*
The data collection of the "in-depth" case studies in Grenoble and Brussels took place during a two-and-a-half-hour meeting for the Grenoble case study and a two-hour online meeting for the Brussels case study in February and March 2020, using a semi-structured group interview approach. The data were recorded and notes were taken by the researchers. One researcher was leading the discussion, while the other kept track of the topics to be discussed. Participants were asked to fill out a small questionnaire about their background. The participants were selected based on their role in the retrofit process and on an objective to cover at least the first four levels of the SEM. The societal level (fifth SEM level) was represented by desk research into legislation and policies. The interview recordings were fully transcribed and translated into English.

IOP Conf. Series: Earth and Environmental Science **855** (2021) 012020 doi:10.1088/1755-1315/855/1/012020

The interviews were organised in three sequences, where the different success (and failure) factors creating the enabling environment "before retrofit", during retrofit" and "after retrofit" were discussed, based on interview guidelines that were prepared (Fig. 2).

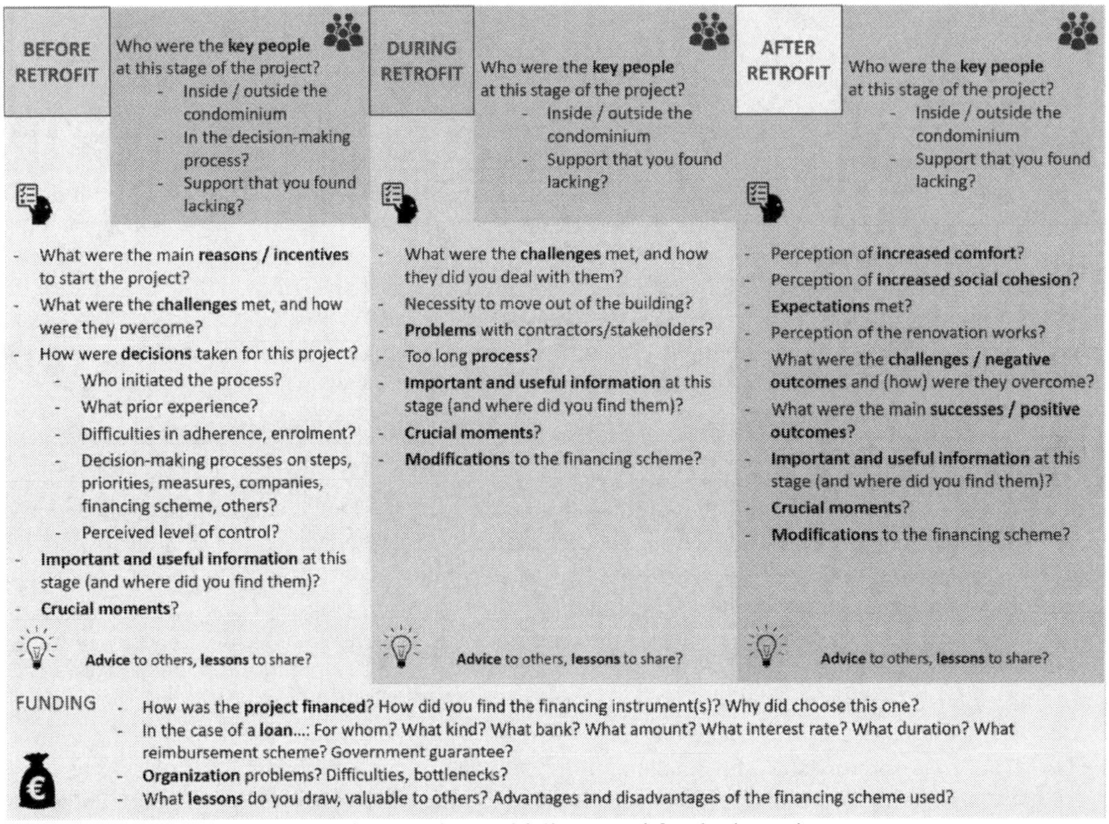


**Figure 2**. Interview guidelines used for the interviews

*3.4 Data analysis*

Data analysis was done without software, using content analysis and grounded theory to identify patterns and connections. Content analysis was used when the researchers had some ideas about hypotheses or themes that might emerge, using a color-coding system to identify text about the different themes, grouping together ideas and gathering evidence about views on each theme. Grounded theory was used when themes that were not foreseen emerged from the discussion. Data was then summarized to link research findings to hypotheses or research aim and objectives. Noteworthy quotations were used in order to highlight major themes within findings and possible contradictions. Six subcategories of themes were created to simplify the analysis, based on grounded principles rather than a specific theory:

- Finance: related to financial issues such as financial aids, initial cost and return on investments, impacting the decision to retrofit.
- Legislation: related to energy legislation impacting co-owners in their decision to retrofit.
- Communication: related to communication issues affecting the decision to retrofit.
- Condominium rules: related to condominium rules and legislation indirectly impacting co-owners in their decision to retrofit, such as majority rules.
- Social: related to issues with other co-owners impacting the decision to retrofit.
- Technical: related to problems with the technicity of the energy retrofit process affecting the decision to retrofit.

Those themes are regrouped around the four pillars described in section 2.1, in the findings, in a review of the most salient details for both case studies, linked to the different levels of the social ecological model (SEM).

## 4. Case studies

*Grenoble, France*

The first focus group gathered in Grenoble on the 13th of February 2020 in Grenoble, to discuss the energy retrofit of a condominium called "Le Manival", located in Meylan, near Grenoble. Important facts about the building and its dwellers are listed in Table 1 below.

The retrofit of the Manival Condominium has been realised within the MURMUR project, a support and incentive system for condominiums and owners of individual houses, implemented since 2010 in the Metropole of Grenoble. This operation is led in partnership with the French State, notably through financial support from the French Environment and Energy Management Agency, and actors from the banking, municipalities and building sectors. MURMUR2 saw 15 million euros invested to renovate 5000 dwellings.

*Brussels, Belgium*

The focus group in Brussels gathered on a virtual platform on the 23rd of March 2020, to discuss the energy retrofit of a condominium called "Expo I", located in Jette, near Brussels. Important facts about the building and its dwellers are listed in Table 1 below. This retrofit project was not part of a national, regional or city plan, it has been implemented because the syndic took charge of the condominium in 2017 with a plan, a pre-established course of action that they apply to all the condominium they manage. By renegotiating operating contracts, working on improving the systems, and sometimes installing renewable electricity production systems (PV panels or co-generation units), they generate savings (and sometimes income) that can be used later to finance renovation works on the envelope.

**Table 1.** Facts about the case studies.

| | Grenoble, France | Brussels, Belgium |
|---|---|---|
| Name of the condominium | Le Manival | Expo I |
| Year of construction | 1972 | 1967 |
| Nb of buildings | 4 | 2 |
| Nb of apartments | 144 | 130 |
| Average area of apts | 82m$^2$ | 85m$^2$ |
| % of rented apts | 33% | 40% |
| Repartition of costs for energy consumption | heating: based on dwelling areas (quotas); DHW: based on individual meters recordings | 25% divided based on dwellings areas (quotas), 75% distributed based on calorimeters readings (for heating) or individual meters (for DHW) |
| Condo decision to start process | 2015 (contacts) / 2017 (vote) | May 2017 (vote) |
| Start of retrofit project | 2018 | October 2019 |
| Vote process | Majority of the present and represented at the general assembly (article 24 of the Law of 1965) | Qualified majority of more than 75% of the co-owners present or represented (law applicable in 2018) during the General Assembly |

IOP Conf. Series: Earth and Environmental Science **855** (2021) 012020 doi:10.1088/1755-1315/855/1/012020

| | | |
|---|---|---|
| Co-ownership | Facades, roof, technical and aesthetic choices, common spaces | Facades, roof, technical and aesthetic choices, common spaces |
| Single ownership | Windows (obligation to replace by new windows with max U-value of 1,7W/m²K) | Windows (most of it done prior to facade works) |
| Main problems before retrofit | Energy consumption (too high), aesthetics needed to evolve | Water infiltrations, thermal bridges, water tightness problems (balconies, passageways), deterioration of concrete |
| Facades composition (before retrofit) | Concrete slab, roughcast cover | structural skeleton of reinforced concrete beams and columns, filling in light aerated concrete blocks, French stone imitation cladding |
| Facades renovation | External insulation by 110mm of resolic foam (Rth = 5m²K/W) + treatment of potential thermal bridges (parapet walls, vertical walls separating balconies…) | Facades without balconies: external insulation by 100mm of polyisocyanurate, new covering in cement panels; facades with balconies: external insulation by 120mm of expanded polystyrene, roughcast cover; insulation of floors and ceiling of balconies to reduce thermal bridges |
| Roof composition (before retrofit) | Concrete slab | Concrete slab; in 2016, a renovation included waterproofing renovation and insulation by 120mm of polyisocyanurate (note: financed by telecom provider who rents the roof for their antennas) |
| Roof renovation | External insulation by 140mm of polyurethane (Rth = 6m²K/W) | |
| Floor composition (before retrofit) | Concrete slab | Concrete slab |
| Floor renovation | Insulation by 160mm of mineral wool (Rth = 4m²K/W) | not in this project |
| Systems (heating / DHW) | Replaced in 2015 (condensing boilers) - not part of this project | 2 co-generation units placed in 2018 (43,5kWth / 20kWe each) (note: electricity produced is sold to generate income and finance renovation works) |
| Ventilation | Installed during retrofit (completely mechanical, with heat recovery, hygro-adjustable) | not present before works, not implemented in this project |
| Additional (non energy-related) works | None | security: new railings on balconies; salubrity (punctual damages from water leakages); concrete repair and water tightness of balconies and passageways; collective water network retrofit |

| | | |
|---|---|---|
| Total costs | 3,250,000€ | 2,574,000€ |
| Repartition of costs for retrofit projects | Based on dwellings areas (quotas) | Based on dwellings areas (quotas) |
| Financing the retrofit | Individual savings + collective or individual loans | 400,000€ from condominium savings + individual savings + collective or individual loans |
| Expected outcomes | 40% in energy consumption reduction | 20% in energy consumption reduction; aesthetics improvement; increase in real estate values |

*Participants to the focus groups*
The Grenoble focus group consisted of 9 people (apart from the researchers), 2 of whom were females. The group was composed of

- People from different age groups: 1 in the (15, 24) group (a trainee at the Local Agency for Energy and Climate, organiser of the meeting), 4 in the (35, 44) group (the syndic and people from the community and society levels of the SEM, see below), 1 in the (55,64) group (the architect), 3 over 65 years old (the co-owners and members of the condominium board)
- People with different educational backgrounds: 1 with a high school degree (a co-owner), 1 with a bachelor degree (the architect), 6 with a master degree, 1 with a PhD (a co-owner)

The Brussels focus group consisted of 6 people (apart from the researchers), all males. The group was composed of

- People from different age groups: 4 in the (45, 54) group (the syndic, the architect and people from the community level of the SEM, see below), 2 in the (55,64) group (the co-owner and president of the condominium board, and a technical expert)
- People with different educational backgrounds: 2 with a high school degree (the co-owner and the technical expert), the 4 others with a master degree.

It was important to gather, in those focus groups, representatives of the different levels of the SEM:

- The individual level (owners, renters): in Grenoble, 3 home-owners were present; 1 in Brussels (the president of the condominium board).
- The interpersonal level (co-owners): represented by the same people in those focus groups. The difference lies in the fact that in this level, they represent the important interpersonal relations between co-owners of a same building.
- The organisational level regrouped different actors: the property manager ("syndic", in French; either an insider or an outsider from the condominium, (s)he is in charge of managing the condominium, board meetings, budgets, maintenance plans…) and the condominium board (elected body, composed of some co-owners, which takes decisions but is accountable to the others co-owners in the General Assembly, manages the condominium with the "syndic", deals with maintenance issues). In Grenoble, the 3 members of the condominium board were also members of its "Commission travaux", an internal board of co-owners who were in charge of following the renovation project, mandated by and accountable to the condominium board and the General Assembly. The Figure 3 hereunder displays organisational Venn diagrams of condominiums, depending on whether the property manager is (left) or is not (right) a co-owner and therefore a member of the General Assembly of the condominium.
  Both focus groups also accounted for the presence of the architects of the retrofit projects, who can also be considered part of the organisational level (although their organisational skills applied more to the project than the condominium structure). In Brussels, he was backed up by an additional representative of the organisational level, a technical expert with many experiences dealing with condominiums and helping them reach decisions in retrofit.

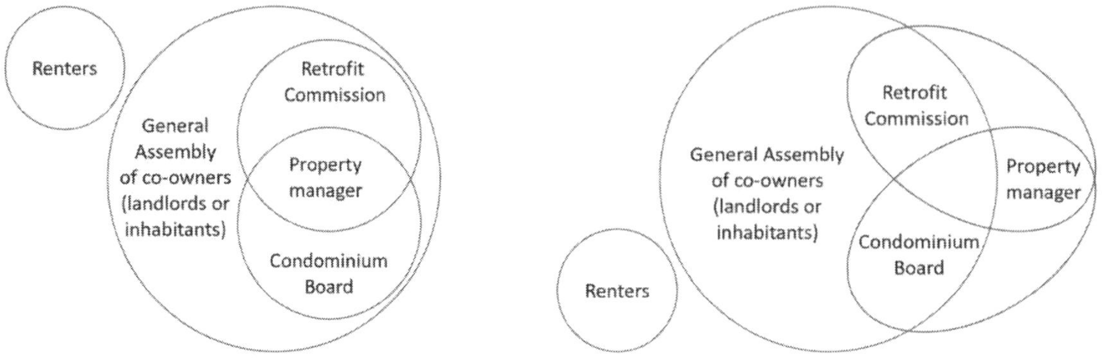


**Figure 3**. organisational Venn diagrams of condominiums, depending on whether the property manager is (left) or is not (right) a co-owner of the condominium

- The community level: In Grenoble, this level would be represented by the Local Energy and Climate Agency (ALEC), an agency describing itself as "a local tool, a decision-making aid, a place for discussion and advice on energy, for all consumers in the Grenoble Metropole (Grenoble city and surrounding municipalities)", which mission is to "contribute locally to the energy transition [...] by taking part in the development and implementation of public policies, stimulating innovative actions and forging partnerships, and providing everyone with personalized advice and support" [23]. Another Grenoble representative of the community level of the SEM was SoliHa ("Solidaires pour l'habitat" - Solidarity for housing), an "associative movement", an important actor of accompanied rehabilitation which missions include the "provision of solutions for decent and affordable housing for all" [24]. In Brussels, representatives of the community level of the SEM were a representative of the "Fédérale Assurance" finance and insurance body, which developed a loan solution for co-owners' associations destined to finance renovation works on condominiums; and a representative of the Atradius insurance group, acting as the insurer guaranteeing the financing scheme.
- Finally, at the societal level of the SEM: In Grenoble, one member represented the "Grenoble Alpes Metropole" administration and governance body [25], a structure that exercises major influence on a territory regrouping 49 municipalities centred around Grenoble, on domains such as waste management, mobility, economic development, water supply and treatment, equipment management, tourism, urban planning, environment protection, urban policies, energy transition, housing policies and energy supply. They are at the policy origin of the MURMUR program and participate in financing the ALEC and SoliHa structures. For Brussels, this level was looked into via desk research.

## 5. Findings

Hereunder is a review of the most salient details for both case studies, linked to the different levels of the social ecological model (SEM), and the four pillars described in section 2.1.

### *5.1. Policies*

Policies are enabling factors that are implemented at the higher level (community and society), but have direct impact at lower levels (organisational and interpersonal). As legislation is very specific to countries, we have separated our findings per case study.

*Grenoble:*

At the French **society level,** we find several important laws that influenced retrofitting in condominiums, such as:

- The Law of 1965 which states that decisions on such important works as a condominium's building renovations have to be voted by the majority of the present and represented at the general assembly.

- The Law of 2010 requiring energy suppliers to realise energy savings, which gave life to energy saving certificates. Since then, those energy suppliers became very interested in buying back energy savings made by others, which in turn allowed financial success to renovation projects, and a revolving fund at the Grenoble Alpes Metropole.
- The "Warsmann" law[1], which created for condominiums the possibility to secure a collective loan (not a mortgage) without collateral.

At the **community level** of the SEM, local policies to enforce global plans targeting ambitious environmental objectives and to provide significant budgets to make them possible, were important too. Such policies have been implemented by the Grenoble Alpes Metropole for the past 15 years, based on the Territorial Climate Plan or the Air Energy Climate Plan[2].

*Brussels:*
At the Belgian **society level,** we find, for example, the recently revised articles 577-2 to 577-9 of the Belgian Civil Code[3], regulating the functioning of condominiums, granting them a legal status (under conditions), requiring the constitution of a reserve fund, lowering the needed majority for important renovation works from 75% to 66%, and leaving abstentions out of the ballot in case a vote is needed at the General Assembly. Another important policy at the regional level is the one granting financial incentives to condominiums implementing energy retrofits.

Other European and national legislation should be mentioned here, not because they enable the decision of retrofitting in itself, but because they require from those who decide to do so, to aim for excellence. They are by essence more stringent as they impose a threshold in quality (for the materials), efficiency (for heating systems, for example) or quantity of insulation (for example) to respect; as such, they might appear to be a hindrance rather than an enabling factor. However, it must be acknowledged that their objective is to enable the outcome to be qualitative.[4]

In general, legal requirements and civil responsibility can be considered fitting levers to convince the General Assembly of co-owners (GA). The president of the Brussels condominium board, for example, insisted that convincing the GA had been possible by referring to obligations, civil responsibility and willingness to avoid problems, should any accidents happen in the pre-retrofit state of disrepair (with façade cladding falling on pedestrians, for example).

For both case studies we can note that specific "Policies" can also be found at lower levels of the SEM, at the **organisational level** specifically. They are not laws per se, but legal documents regulating the functioning of a particular condominium. Such are the basic condominium acts and statutes, the condominium rules and internal regulations. Such tools have been mentioned by the syndic, for example when describing the manner to share the overall costs of the condominium between co-owners, using this as an argument to convince all co-owners that they have a financial interest in deep-retrofitting the whole building.

---

[1] Decree 2012-1078 of 24th September 2012, taken in application of the article 2 of the law 2011-525 of the 17th of May 2011

[2] The Big Boulevard policy, the Local Program for Housing, and of course the MURMUR program.

[3] revised in 2018 (which came into force on the 1st of January 2019)

[4] For instance: The EPB legislation, which (at least for now) does not impose global energy retrofits, but minimum requirements for thermal insulation (only for walls submitted to important retrofit, other than a fresh layer of paint) or systems' efficiencies; or the Belgian and European standards imposed on materials to be used in renovation.

*5.2. Financing scheme*

The crucial availability of interesting financing schemes is coming from various levels: the **society** (through laws), **community** (through incentive mechanisms) and **organisational** (through banks) levels of the SEM. Condominiums (at the **interpersonal** level) are a special target group for financing bodies as they involve socio-economic diversity.

*Loans:*

At le Manival in Grenoble, a collective loan was used: the COPRO100. At the EXPO I condominium in Brussels, a collective loan was used, made available by the Fédérale Assurance Bank.

Such loans have a tremendous influence in alleviating the concerns many of those vulnerable households might face when having to find the money to finance their part of the project. In these cases, they were available to anyone, without any discrimination based on age, medical history or existing financial debts. There is no life or income insurance involved, which is also interesting for older people. By permitting vulnerable (the old and/or sick, those who cannot afford / be granted individual loans…) people in the condominium to join, they allow for some kind of solidarity in sharing the risks. This solidarity has nevertheless understandable limits, represented by the absence of any solidarity in payments (i.e. should there be a bad payer inside the condominium, the other co-owners, at personal levels, are not asked to compensate; the insurance intervenes, pays, and will turn against the bad payer).

*Subsidies*

The subsidies and incentives (whether collective or individual) were also an important factor at le Manival in Grenoble, in the way they decrease the amount to be financed by co-owners, more so for vulnerable people. By taking on the often-heavy administrative burden at the **organisational** level, SoliHa (financed by the community and society levels) helped to secure these subsidies.

In Brussels, the financial incentives at the **organisational** level were mentioned, also. The main problem was that they were only made available after completion of the works. Therefore, they were not factored in budgeting the renovation works, but were deposited by the syndic in the condominium reserve fund to encourage posterior works. There might be some obstacles in threshold effects and delays in payments (they often are paid after the works are done), but these "monetary gifts" that were kept out of the project budget influenced the co-owners positively.

*Grenoble*

At the **society and community** levels some additional financial factors were important to success at le Manival:

- The controlled rent mechanism, where landlords agree to make their newly-retrofitted property available at an affordable rent to people with modest resources, for a minimum period of 9 years, in return for a subsidy financing part of the works. This mechanism, however, is not used very often, according to the Local Agency for Energy and Climate, because owners tend to consider that the threshold imposed on those controlled rents is too low. The subsidy however is, as often, interesting to many owners who believe the mechanism would be a better incentive if those thresholds were adjusted.
- The significant tax deduction granted by the national authority on the gross property income of all owners who invested in the energy retrofit of their apartment (or the building) (**society** level).
- The Energy Savings Certificates mechanism granted to co-owners, who can sell their energy savings (which are given a market value) to the Metropole. The **individual** level is therefore helping the **community** and **society** levels to reach regional energy savings objectives, that is then used to generate a revolving fund for the Metropole which is used to subsidise more retrofit works.
- The ECOCITE governmental plan, financing the monitoring phase after retrofit, which appealed to the co-owners as a possible way to check the outcome of the operation, and enter a comparison contest with their neighbours who already have retrofitted their condominium (**individual and interpersonal** level).

*Brussels*
For Brussels too, some specific mechanism was an important part of the enabling environment:

- The renovation project in Brussels was helped by is the mechanism of "Green Certificates", a support system for the production of green electricity, governed by regional decrees in Belgium. When an installation allowing the production of green electricity (such as photovoltaic panels, or a cogeneration unit) is certified, the holder (who can be a private owner at the **individual** level, or a condominium at the **interpersonal** level of the SEM) is granted green certificates depending on the installation potential to save $CO_2$ emissions. Those certificates are given a monetary value by the market, and can therefore insure a revenue for the holder for 10 years. They can be "sold", each year, to electricity suppliers, who are required to meet obligations in green production. They are, therefore, interested to buy those Green Certificates from private "green electricity producers", and ultimately pass on the cost of their quota return obligation to all of their end customers.
- The earlier renovation of the roof is an example of another mechanism: this particular retrofit had been financed by the investment of a telecommunication provider in exchange for the placement of antennas. This scheme belongs more to the **organisational** level of the SEM: the roof was actually rented out to three telecommunication providers, who paid 9 years' worth of rent in one instalment, 168,000€. An additional regional financial incentive of 90,000€ helped the General Assembly to vote for a roof renovation that would virtually cost nothing to them. After the 9-years period, the telecom providers will have to pay the rent as they would have normally done.
- Finally, the insurance scheme must be acknowledged, because it is an important factor that enabled the decision-making in this project. The Atradius insurance [26-27] is provided by an insurance company that can be situated at the **community** level of the SEM. It covers all the co-owners of the condominium (not only the ones participating in the collective loan) against those who do not meet their obligations in the payment of their part of the project. It has the virtue of comforting all co-owners that they will not be held responsible for defaulters. It simplifies the work of the syndic, who does not have to worry about legal actions and recovery procedures; and reassures the bank on the reimbursement of the collective loan.

*5.3. Building capacity*
Enabling the renovation of condominiums also implies the necessity to use or build capacity, and to learn about adequate internal organisation. At the heart of this consideration is the ecosystem of key stakeholders, which can be divided in three main groups, depending on the level of the SEM in which they act.

At the **community** level of the SEM are the professionals acting as enablers, helping the co-owners to define the technical and financial aspects of the project:

- In Grenoble, an active and willing (local or regional) government was involved, which not only shows political motivation, but also offers to condominiums the (technical and financial) services they need to reach these objectives.
- The financing bodies, represented by the banks and experts such as SoliHa and the ANAH in France, and the Fédérale Assurance and Atradius Insurance groups in Belgium, who helped owners to get incentives and round up the financial files. The ALEC confirms that it is crucial to offer co-owners the possibility to be in contact with skilled, competent people, who have a close look at the budgets.

At the **organisational** level of the SEM are those with skills to answer owners' questions and complaints:

- The technical assistance team, mainly represented here by:
  - The ALEC in Grenoble, a motivated and skilled team of experts paid by the administration, which goal is to motivate condominiums to renovate. They propose participatory workshops, contests, visits, promotion, newsletters, tests, testimonials…

  - The architect or project manager, who took charge of the technical project, offered solutions and alternatives, followed the works, managed the workers, organised weekly meetings with the "renovation committee" of the condominium, informed on progress and delays, and generally acted as a crucial guide, federating people.
  - In Brussels, the architect was seconded by a "condominium expert", who saw his work particularly praised during the Brussels meeting, for his communication skills notably. It is important to stress this, as this profile is not always found in renovation projects; however, it seems undeniable that he acted as a great enabler in the implementation of this project, and took upon himself to overcome a number of usual obstacles in retrofits (e.g. convincing owners, finding skilled professionals, assessing technical solutions, negotiating with contractors…).
- The syndic or property manager, who acted as facilitator between various stakeholders and who was a motor to mobilize and motivate the co-owners. He was their advocate, administrative referent, and the project's orchestra conductor.
- At this level can also be considered the concierge, mentioned in the French case study. The architect highlighted the very important role the concierge had to play, which "should not be forgotten", even qualifying them as "fundamental".

At the **individual and interpersonal** levels, is the team of "insiders", such as the condominium board. In Brussels, the president of the board was specifically praised; in Grenoble, it was the presence (as is sometimes the case in large condominiums) of a specific "renovation committee", which represents the clients of the project (they are "more important than the syndic, more important than the architect", according to the syndic). They were the essential link between the co-owners whose interests they represent, and the professionals who have the answers and explanations they seek. They have the fundamental role to help in convincing the General Assembly of co-owners, but also to manage the Assembly, counter the opponents' arguments, keep the debates civil and constructive, and finally reach an agreement among the members of the Assembly in time for the vote to start a retrofit project.

*5.4. Socio-cultural acceptance*

Early engagement inside the condominium meant that the much-needed socio-cultural acceptance at the lower levels (**individual, interpersonal**) of the SEM tended to be increased by the following elements:

- The guidance by skilled, motivated, and dynamic professionals. Their presence on site was also a plus for elders, which favoured social inclusivity. They encouraged dialogue, communication, transparency of information, and gave reliable answers to the co-owners' questions.
- Although the experts were keen on maintaining a project dynamic, they valued the necessity to give time to owners, to let them appropriate the project, ponder the consequences, ask questions, get information, assimilate, understand… The difficulty lies in finding the right balance with the well-known problems triggered by a too-long process. In Brussels, the experts considered that it was crucial for the co-owners to feel involved, as what is presented to the GA must be seen as their own project, not a project by outsiders (the syndic, or the architect) imposed on them.
- The renovation project remained a voluntary decision, a process which did not feel forced on any of the owners. This tended to reinforce the sense of ownership.
- Examples of other condominiums that already did a retrofit, served as inspiration to these new candidates who felt that it was possible, and had access to their experience and the guarantee that positive outcomes will come after the difficult retrofit period. It is an existing knowledge on which all can lean. In some cases, these examples can go as far as proposing technical answers to particular problems that might arise. It has the benefit to inspire the architect, and to reassure the co-owners that solutions exist and are affordable.

*Grenoble: opportunity effect and snowball effect*

In Grenoble, the MURMUR programme illustrates the "opportunity effect" that drives many condominiums in or around the Metropole to register and implement projects like this one. The opportunity is indeed given to them, to take advantage of the professional frame and coaching; of the skills and competences of the support teams that are offered, freely, by the administration; of the financial schemes and technical support that are so important to set up and "tie up" a complete and realistic project.

A programme like MURMUR, implemented for many years, has become a success factor in time, having already overcome many difficulties and dealt with the problems that could have made it a failure in its early stages. Proof is given by the numbers mentioned by the ALEC on the votes: nowadays, around 90% of condominiums entering the program vote "for" the attribution of the project to an architect, and around 70%[5] of condominiums which go through the entire process, vote positively for the work package at the final general assembly. This percentage decreases to 45%, however, if we consider the condominiums which have given up[6] during the process.

The MURMUR programme observed a "snowball effect", where exemplar projects become inspiration for others to engage in the process. Co-owners and the architect of the French case study confirmed that they themselves started by seeking information among condominiums that had gone through the programme, and that they had visitors from other interested condominiums since they started the works. (interaction of **interpersonal, organisational** and **community** levels).

## 6. Discussion

What do these findings tell us about the enabling environment at various scale levels? It seems to be rather complex: projects seemingly can only be successful if all actors are mobilized, motivated and competent, and are able to mobilize a "team spirit" to work as a squad, to "speak the same language" (as one French co-owner said). The project must remain collective, and all actors have to be aware that it is a win-win-win situation, that they all want to work for the same objective, which is to retrofit the building.

But if we take a look at the enabling environment for condominium retrofitting in more detail, we can recognise several important elements that might help to design an enabling environment (Fig 4). The organisational level seems to be pivotal to the success of a retrofit project. We saw that the professionals must be the right experts, at the right place, and show "people skills" in the way they handle co-owners and assemblies. They have to be patient, mobilised, flexible, qualified, neutral, independent but kind teachers. The technical team should be composed of people who do not shy away from projects involving complex structures of "clients", and should therefore be seen as crucial guides, whose role is also to unite people. They must be competent, as they have to give the adequate answer to the questions that arise among co-owners, and must show reactivity when they inevitably hit an obstacle or encounter a challenge. They also need to be aware of the opportunities that exist at the community and societal level (subsidies, legislation, financing options), provided there are any.

The syndic is a very important link between the organisational level and the interpersonal level. (S)he needs to show all the qualities that are asked from professionals (patience, flexibility, qualifications, neutrality, teaching skills), but is allowed (even encouraged) to be mobilised for, and show loyalty to, the condominium's interests. They are pointed out as an essential piece of the puzzle everywhere, when they are active and push the condominium board to act.

---

[5] This number comes directly from the ALEC, who warns that they should be decreased by the end of the MURMUR 2 project, for several reasons: first votes have mainly happened in condominiums that were highly motivated and had already made progress in the process; work costs were lower at the beginning of the MURMUR2 program due to the slow economy recovery in the construction sector, and have increased with the workmanship workload; finally, the Covid19 sanitary crisis is expected to lower the number of positive votes in high-budget projects.

[6] Reasons for withdrawals from the programme may be diverse and due to different partners, such as the syndic (being a hindrance more than a help), the condominium board (losing motivation), or the architect (driving the project to another work program due to apparent complexity of the MURMUR program)

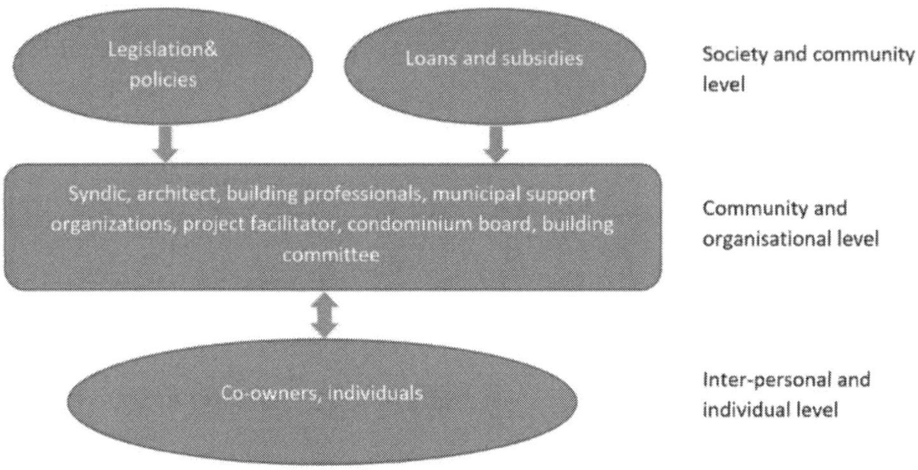


**Figure 4**. Linking findings to the Social-Ecological Model.

Only a condominium (interpersonal level) that trusts its support team will ultimately vote for the works to be implemented. This tends to confirm the need for an efficient, frequent and transparent communication between all stakeholders. Professionals also need to be comprehensive about the necessity, for the co-owners, to have time to digest crucial information and appropriate the project. All actors must understand that they need to let time do its wonders, do not try to rush things, and let people absorb information and take ownership of the project. Time must be taken to explain the situation, to encourage more frequent dialogue, and to "tie" the project sufficiently to present it to the General Assembly.

Financing the renovation is a key aspect that could either win or lose the vote. Although transparency is essential on that aspect too, it might be wise to wait for the project to be sufficiently defined and developed before numbers are presented in General Assembly. A first phase of interaction between the professionals and the "renovation committee" (or condominium board) is necessary to settle on details, negotiate and finalize the project.

It is a difficult balance to find for the team of professionals. If the condominium feels abandoned (by a lack of response, for example), confused (by a project that does not feel technically mastered, or by financial numbers that change too often), or pressured (forced into a too-tight schedule, for example), bad press and distrust towards their professionalism and skills can rapidly spread, and create feelings of dissatisfaction, uncertainty, insecurity, indecision… Most likely, General Assemblies would then either vote against projects, postpone the decisions or, in the worst-case scenario, vote for a simple façade restoration without insulation.

Although the French case study ended on the impression that everything went smoothly, everyone did their job, and no consequent obstacle was met, the case study of Brussels highlighted some missing pieces of the puzzle, which have to be formulated into recommendations to the different levels of the SEM in order for enabling environments to improve:

- Several participants to the meeting mentioned that some obstacles could be lifted by implementing more legal obligations. The syndic, in this case, lamented that a law, such as the one that exists in the Flemish Region of Belgium, does not exist to impose some retrofitting works to residential buildings. This obligation has the virtue to close all debates rapidly, when it comes to the decision to retrofit.
- Also, nearly all participants (in Brussels) wished for more public investments, in energy, time and money. The co-owner and president of the Expo I condominium board wished for support on real-estate taxes, renovation incentives, and legal requirements. The architect and the condominium expert, acknowledging the necessity to renovate an ever-aging existing stock, mentioned that the biggest challenge is convincing the large number of stakeholders who have to be involved, and that a lot of political actors need to go beyond the political discourse, take action on the field, and hand out bigger means to achieve their goals.

## 7. Conclusion

The enabling environment surrounding a successful retrofit is shown in these case studies to be more than just providing process facilitation or a financing scheme. It is a combination of engaged co-owners, an active building manager (syndic), outgoing and patient building professionals, process facilitation, local, regional and national policies or laws, proper financing mechanisms, subsidies, communication and, in this case, prior examples. Tentatively we would like to state that the organisational level is incredibly important. However, without the proper legislation (both as stick and carrot) or financial incentives, the organisational level will not be very effective. Also, the interpersonal and individual levels need to be receptive and organised in order for the information to flow freely between the various levels.

The case studies show that it takes time to put an enabling environment in place because of its complexity. For different locations across Europe, different enabling environments are needed. Clearly, this requires giving more attention to condominiums at the local, regional, national and European levels.

## 8. Acknowledgements

Financial support for this research has been provided by Interreg North Western Europe, and the Province of Limburg (the Netherlands). This research paper is based on thesis work by Armandina Ramirez Tamez, Celine Sottovia and Ilian de Meersman. We wish to thank them for their hard work! Furthermore, we are grateful for the invaluable contributions of the stakeholders who shared their thoughts and experiences with their retrofitting project. We would like to thank especially Gaetan Bogotto of the ALEC offices in Grenoble and Vincent Spruytte of Managimm in Brussels.

## 9. Bibliography

[1] Energy Cities. (2020). ACE-Retrofitting: *Accelerating Condominium Energy Retrofitting - Project Overview.*
Retrieved 2020-11-06, from https://www.nweurope.eu/projects/project-search/accelerating-condominium-energy-retrofitting-ace-retrofitting/

[2] Ramirez Tamez, A. (2018). *Understanding behavior change in energy retrofitting of condominiums.* Master thesis, Maastricht University.

[3] Sottovia, C. (2018). *Accelerating Energy Retrofitting in Condominiums in Liège and Paris via behavioral change: recommendations to the Interreg Program ACE-Retrofitting.* Master thesis, Maastricht University.

[4] De Meersman, I. (2020). *Creating an enabling environment to support co-owners towards deep energy retrofit of condominiums in the Region of Brussels-Capital.* Master thesis, Maastricht University.

[5] Living Cities, I. f. S. C. (2009). *Scaling up building energy retrofitting in U.S. cities. A resource guide for local leaders.* New York, Living Cities: 79.

[6] Liverzay, T., Teissier, O., and Togo, G. (2016). *Rénovation en copropriété : analyse des blocages et leviers. Quelles stratégies pour une massification?* Paris, Agence Parisienne du Climat: 53.

[7] Plan Bâtiment Durable (2018). *Plan de rénovation énergétique des bâtiments. Regards et propositions du Plan Bâtiment Durable : 50 propositions issues de la consultation.* Paris, Ministère de la Transition écologique et solidaire: 26.

[8] World Bank (2003). *Enabling Environments for Civic Engagements in PRSP Countries.* Washington, DC. **82**.

[9] Brinkerhoff, D. W. (2004). *The Enabling Environment for Implementing the Millennium Development Goals: Government Actions to Support NGOs. The Role of NGOs in Implementing the Millennium Development Goals.* Washington DC: 18.

[10] Lüthi, C., Morel, A., Tilley, E., and Ulrich, L. (2011). *Community-Led Urban Environmental Sanitation Planning (CLUES).* Dübendorf, Switzerland, Swiss Federal Institute of Aquatic Science and Technology (Eawag)

[11] Charabi, Y. and Al-Badi, A. (2014). *Creating An Enabling Environment for Renewable Energy Application in the Sultanate of Oman.* International Journal of Green Energy **12**.

[12] Climate Investment Funds (2014). Enabling Renewable Energy Investment in Low-Income Countries, Climate Investment Funds: 2.
[13] Sustainable Energy for all (2017). *Enabling policies: Which African and Asian countries have an enabling environment for investment in energy access?*, Sustainable energy for all.
[14] Konold, M. (2015). *Creating an enabling environment for project development.* Platts Caribbean Energy Conference. Worldwatch Institure.
[15] USAID. (2020). *Enabling Environmental Principles.*
Retrieved 2020-11-27, from https://www.usaid.gov/powerafrica/enablingenvironmentreforms.
[16] Crosbie, T. and Baker, K. (2010). *Energy-efficiency interventions in housing: learning from the inhabitants.* Building Research & Information **38**(1): 70-79.
[17] Bronfenbrenner, U. (1977). *Toward an experimental ecology of human development.* American Psychologist, 32, 513–529.
[18] Centers for Disease Control and Prevention. (2020). "The social-ecological model: a framework for prevention."
Retrieved 2020-02-19 from https://www.cdc.gov/ViolencePrevention/pdf/SEM_Framewrk-a.pdf.
[19] Bronfenbrenner, U. (1989). *Ecological systems theory.* In Vasta, Ross (ed.). *Annals of Child Development: Vol. 6.* London, UK: Jessica Kingsley Publishers. pp. 187–249
[20] Bartholomew, L. K., Markham, C. M., Ruiter, R. A. C., Fernandez, M. E., Kok, G., and Parcel, G. S. (2016). *Planning health promotion programs: an intervention mapping approach.* New York City, USA, John Wiley & Sons Inc.
[21] Copley Focus Center. (2012). *Why are focus groups used.*
Retrieved 19/02, 2020, from https://copleyfocus.com/focus_info/why-are-focus-groups-used/.
[22] Adams, A. and Cox, A. (2008). *Questionnaires, in-depth interviews and focus groups.* Research Methods for Human-Computer Interaction.
[23] Agence Locale pour l'Energie et le Climat. (2020). Retrieved 09/11/2020, 2020, from http://www.alec-grenoble.org/4189-presentation-alec.htm
[24] SoliHa. (2019). *SoliHa, Solidaires pour l'Habitat: qui sommes-nous?* Retrieved 2020-11-29, from https://www.soliha.fr/mouvement-soliha/
[25] Grenoble Alpes Metropole. (2020). Retrieved 09/11/2020, 2020, from https://www.lametro.fr/.
[26] Atradius ICP. (2016). *L'assurance qui couvre la copropriété contre le risque de non-paiement des charges dues par un ou plusieurs copropriétaires.* Retrieved 2020-11-29 from https://atradiusicp.com/index.htm
[27] Syndicat National des Propriétaires et Copropriétaires. (2019). *Atradius ICP - Assurances aux copropriétés.* Retrieved 2020-11-29 from https://www.snpc-nems.be/fr/content/atradius-icp-assurances-aux-copropri%C3%A9t%C3%A9s

# Review of archetype development strategies for energy assessment at urban scale

**G M Colleto[1] and V Gomes[1]**

[1] School of Civil Engineering and Architecture and Urbanism, University of Campinas, Brazil, vangomes@unicamp.br

**Abstract.** Understanding the built environment's impacts is essential to support strategic planning and policy design for sustainable development now, and in the future. Modelling individual buildings and infrastructure at high level of detail is resource intensive. Thus, urban scale analyses demand simplifications that balance level of detail and scope broadness. For combining simplified modelling and extended scope, classification by archetypes emerge as a promising methodological approach to extend assessment scope beyond energy use simulation. Archetypes that include life cycle assessment (LCA) parameters can support circularity challenges diagnosis, mapping and predictions, strategies to close material and energy loops and their monitoring within urban built environments. We hypothesized that, upon limited complementation, operational and pre/post construction (embodied) datasets coupled with building grouping techniques satisfactorily represent the built stock to support cradle to grave LCA of built environments. Studies on the use of archetypes for this purpose are scarce, so this article reports findings of a Systematic Literature Review (SLR) on archetypes-based approaches for energy assessment that could inspire application in LCA studies at urban scale. The SLR highlighted a lack of methodological consensus, and that data availability seems to be the major limitation for archetype creation in most studies, which rarely present in-depth information on their development. The few investigations providing consistent methodological procedures actually detail the initial classification process. Transposing the archetype strategy from energy assessment to LCA at urban scale faces practical limitations. Several databases support operational energy studies, but the same does not apply to LCA. Urban building energy models typically overlook infrastructure. Also, statistical results depend directly on data input quality and data availability may compromise the quality of variable selection.

## 1. Introduction

Globally, the construction sector concentrates a large portion of GHG emissions and is responsible for about 40% of final energy consumption [1]. Considering the growth projected for cities over the next decades, achieving established energy use goals become indispensable to mitigate urban environmental impacts. Estimating and understanding these impacts are essential to assist in performance assessment and monitoring and to support strategic planning and energy and environmental policy design for sustainable development now, and in the future. However, if we consider the city as a set of individual buildings, the high level of detail in the analysis of segregated structures requires the intensive use of financial, temporal, technological and human resources. Thus, the complexity at the urban scale demands strategies to simplify the built stock modelling, to balance level of detail and scope broadness. The archetype-based approach, for instance, abstracts existing urban tissue into a number of representative buildings.

Construction archetypes are theoretical average buildings that represent specific groups of buildings aggregated by similar characteristics, defined according to the scope of the study at hand. For combining simplified modelling and extended scope, the classification by archetypes is commonly used in research on new technologies [2] and energy efficiency policy design [3].

So far, their use has increased yet been focused on operational energy studies at the urban scale, but building archetypes emerge as a promising methodological approach to extend assessment scope beyond energy use. Internationally recognized projects, such as EPISCOPE and TABULA [23] projects, use archetypes as a way of simplifying the built stock. The robustness of theirs results proves the viability of this technique for energy assessment in the urban context. Thus, we aim to deeply understand the development process of energy archetypes in different studies to apply them simultaneously in Life Cycle Assessments on an urban scale, in order to expand the results of energy analyses for assessments. For this paper, we hypothesized that, upon limited complementation, operational and embodied datasets coupled with building grouping techniques would satisfactorily represent the built stock to support integrated life cycle simulations.

As few studies have been published on the use of archetypes to support life cycle assessment (LCA) of built stocks, this article reports findings of a Systematic Literature Review that retrieves and discusses the most relevant studies on archetypes strategies and methodological approaches used for energy assessment that could inspire application in Life Cycle Assessment studies at urban scale.

## 2. Method

A Systematic Literature Review (SLR) was carried out to identify publication trends on the subject, and the main published contributions. The research questions fuelling the first review round were: "Which methods are used to develop building archetypes for energy assessments at urban scale?" and "Are there similar experiences for life cycle assessments?"

Based on the Procedures for Performing Systematic Review [22], we adopted the following protocol:

### *2.1. Selection criteria*

- Database: Scopus
- Limits: i) document type: journal articles; ii) fields: title, abstract, keywords; iii) language: English; iv) time range: all years (until ago/2020)
- Keywords: i) Archetypes and Building and Energy; ii) Archetypes and Building and LCA

### *2.2. Exclusion criteria and paper grouping:*

Paper exclusion - by title and by abstract - took effect only for those clearly non-relevant for the study at hand. Articles were then grouped according with the nature of information provided, to focus on those actually contributing to future methodological design (archetype definition or classification). The resulting sample supported our quest to answer five specific research questions through this SLR:

- How to define the ideal set of input variables to be used in the grouping algorithms?
- Which variables standardization methods are the most appropriate?
- Which grouping techniques should be employed and how to define the building sample?
- How to define the optimal number of clusters?
- How to evaluate the clusterization quality?

## 3. Results and Discussion

### *3.1. Bibliometric analysis*

The search output (Figure 1) retrieved publication lists corresponding to the strings including LCA (7 hits) and energy (168 hits).

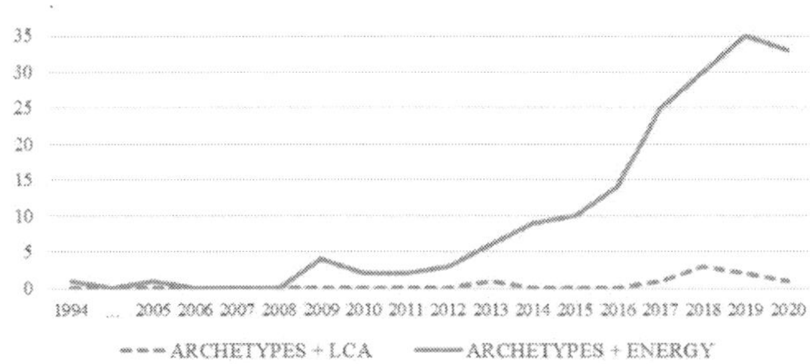


**Figure 1.** Chronological publications distribution retrieved with the search strings used (articles/year)

After 27 papers were excluded by title and 16 others by abstract, the preliminary sample comprised 132 publications for integral reading. Papers were then grouped into:

- Application: 74 articles indeed applied archetypes in their methods, but without expressing methodology guidance regarding archetype definition or classification; and
- Methodological detailing: 58 articles declared – to various detailing degrees – the methodological approach that led to the archetype classification used. From those papers, 21 articles explained in more detail the methods for defining variables and the grouping techniques for creating archetypes.

This 21-paper sample data extraction table (Table 1) summarizes answers to the five specific research questions posed to this SLR:

- How to define the ideal set of input variables to be used in the grouping algorithms?
- Which variables standardization methods are the most appropriate?
- Which grouping techniques should be employed and how to define the building sample?
- How to define the optimal number of clusters?
- How to evaluate the clustering quality?

*3.2. Methodological synthesis*

*3.2.1. Archetype development stages*
Certain publications provide an overview of the entire archetype development process. Stages vary slightly across publications: **Classification,** Characterization, Calibration [1,2]; **Classification,** Parametrization, Modelling [3]; **Data collection, Segmentation,** Characterization, Quantification and Modelling results [4]; **Preparation, Classification, Normalisation, Clustering,** Assessment, Augmentation [5]. Nonetheless, the general development structure of archetypes is very similar and can be summarised in two macro steps: creation and detailing (Figure 2).

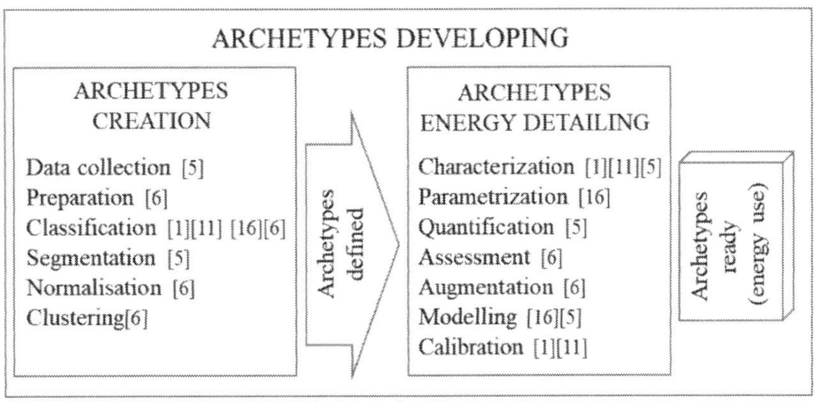


**Figure 2**. Archetype development macro steps

[1, 2, 6] presented robust and consistent studies on such archetypes development processes, but restricted to the 'Archetypes energy detailing' macro step, specifically to 'Characterization', in which archetypes have been already defined and relevant data for energy analyses are extensively specified. From those steps, the 'Archetypes creation' macro step is the most relevant for methodological transferring for LCA application, as it defines the parameters to be statistically worked on for sample partitioning. For estimating a building's potential impacts, its building materials (embodied impacts) and operational (water and energy) consumption are described in terms of mass and energy flows, and the corresponding parameters must be included in the archetypes defined. In particular, seven articles (see **Error! Reference source not found.**) stood out in terms of methodological detailing of archetype classification procedures.

IOP Conf. Series: Earth and Environmental Science **855** (2021) 012021 doi:10.1088/1755-1315/855/1/012021

***Table 1**. RSL data extraction table summarizing methodological aspects found in the sampled publications. (*A high-resolution version is available at https://drive.google.com/file/d/1jsJd4WA2yy6_PQkp61WBHDL1yIrT4GF_/view?usp=sharing)

| [21] | [20] | [8] | [19] | [6] | [16] | [11] | [10] | [18] | [15] | [14] | [3] | [2] | [13] | [1] | [12] | [9] | [5] | [4] | [7] | [17] | | Ref. |
|---|---|---|---|---|---|---|---|---|---|---|---|---|---|---|---|---|---|---|---|---|---|---|
| | | | | | | | |x| | | | | | | | | | | |x| | SRL |
|x|x|x|x| | |x|x| | | | | |x|x|x| |x|x|x| | Existing database | Data sources |
| | | | | | | | | |x| | | |x| | | | | | | | Specific created database | |
| | | | | | | | | | |x|x|x| |x|x| | | |x| | Geographic Information System (GIS) | |
| | | | |x| |x| |x| | | | | |x| | |x| | | Statistic data | |
| | | | | | | | | | | | |x| | |x| | |x| | | Building standards | |
| | | | | | | | | | | | |x|x| | |x| | | | | Field survey | |
| | | | | | | | | | | | |x| | | |x| | |x| | Image analysis | |
| | | | | | | | | | | | | | |x|x|x|x|x| | | Data filtering / processing | Normalization technique |
| | | | | | | | | | | | | | |x|x| | | | | | Automatizado /Inteligência Artificial | |
| | | | | | | | | |x| | | |x| | |x| |x|x| | Manual | |
| | | | | | | | | | | | | | | | | |x| |x| | rescaling / scaling | |
| | | | | | | | | | | | | | | | | | | |x| | standardising | |
| | | | | | | | | | | | | | | | | |x| | | | Centring | |
| | | | | | | | | | | | | | | | |x|x| | | | z-score | |
| | | | | | | | | | | | | | | | | |x| | | | PCA | |
| | | | | | | | | | | | | | | | | |x| | | | Box-Cox | |
| | | | | | | | | | | | | | | |x|x|x|x|x| | K-means | Clustering technique |
| | | | | | | | | | | | | | | | |x|x| | | | Partitioning around medoids | |
| | | | | | | | | | | | | | | |x| |x| |x| | Hierarchical clustering | |
| | | | | | | |x| | | | | | | | | | | | | | Cartesian product | |
| | | | | | |x| | | | | | | | | | | | | | | interpolating /extrapolated | |
| | | | | | |x| | | | | | | | | | | | | | | distribuição / função exponencial | |
| | | | | |x| | | |x|x|x| | | | |x| |x| | | centróide (virtual) | |
| |x|x| | | | | | |x| | | | | | |x| | |x| | amostra / medïodo (real) | |
| | | | |x| | | | | | | |x| |x| | | | | | | Bayesiana (para caracterização) | |
| | | | | | | | | | | | | | | |x|x| | | | | Model-based grouping - NbClust | Cluster ideal number |
| | | | | | | | | | | | | | | | |x| | | | | R Language | |
| | | | | | | | | | | | | | | | | |x| | | | Silhouette index | Cluster quality - Validation indices |
| | | | | | | | | | | | | | | |x| |x| | | | Dunn index | |
| | | | | | | | | | | | | | | |x| | | | | | Indice SD | |
| | | | | | | | | | | | | | | | | |x| | | | Calinski-Harabasz index | |
| | | | | | | | | | | | | | | | | |x| | | | Wb index | |
| | | | | | | | | | | | | | | | | |x| | | | Entropy index | |
| | | | | | | | | | | | | | | | | |x| | | | Dunn index variation: Dunn2 | |
| | | | | | | | | | | | | | | | | |x| | | | Goodman and Kruskal's coefficient | |
| | | | | | | | | | | | | | | | | |x| | | | G3 index | |
| | | | | | | | | | | | | | | | | |x| | | | Ratio of the number of elements | |
| |x|x|x| |x| | | |x| |x| | | | | | |x|x| | Construction typology | Cluster parameters |
|x|x|x|x| |x|x|x| |x|x|x| | |x|x| |x| | | | Construction typology | |
|x| | | | | | | | | | | |x| |x|x| | | |x| | Building footprint | |
| | | | | | |x|x| | | | | | |x| | | |x|x| | Floor area | |
| | | | | | | | | | | | | | | |x| | | | | | Total area to net area ratio | |
|x| | |x| | | |x| | |x| |x|x| |x| | |x| | | Construction year | |
| | | | | |x|x| | |x| |x| | | |x| |x| |x| | Construction year | |
| | | | | | | | | | | | | | |x| | | | | | | Year of the last renovation of the building | |
| | | |x| | |x| | |x| | |x| |x|x| | | |x| | Number of floors | |
| | | | | | |x| | | | | | | |x| | | | | | | Number of dwellings (buildings) / rooms | |
| | | | | | | |x| | | | | | | | | | | |x| | Heated floor area | |
| | | | | | | | | | | | | | | | | | | |x| | Heated volume | |
| | | | | | | | | | | | | | | | |x| | |x| | Compactness ratio | |
| | | | | | | | | | | | | | | | | | | |x| | Floor loss area (= heat loss area of the | |
| | | | | | | | | | | | | | | |x| | | |x| | Window area | |
| | | | | | | | | | | | |x| | |x| | | |x| | Window-to-wall ratio (WWR) | |
|x| | | | | | | | | | | |x| |x|x|x| | | | | Building height | |
| | | |x| | | |x| | | |x| |x| |x| |x| | | | End use / Function / Size class | |
| | | | | | | | | | | | | | | | |x| | | | | Aspect ratio | |
| | | | | | | | | | | | | | | |x| | | | | | Performance evaluation indexes | |
| | | | | | | | | | | |x| | | | | | | | | | Roof type (sloping and flat roofs) | |
| | | | | | | | | | | | | | |x|x| | | | | | Ground elevation | |
| | | | | | | | | | | | | | |x| | | | | | | Internal and external conditions | |
| | | | | | | | | | | | | | |x| | | | | | | Facade materials | |
| | | |x| | | |x| | | | | | |x| | | | | | | Fuels used | |
| |x|x|x| | |x|x| | | | | | |x| | | | | | | Heating system | |
| | | | | | | |x| | | | |x| | | | | | | | | Measured energy demand | |
| | | |x| |x| | | |x| | | | | | | | | | | | Climate zone | |
| | | | | | | |x| | | | | | | | | | | | | | Envelope shape | |
| |x|x| | | | | | | | | | | | | | | | | | | DHW cylinder insulation thickness | |
| |x|x| | | | | | | | | | | | |x| | | | | | U-values | |
| |x|x| | | | | | | | | | | |x| | | | |x| | multiple linear regression (MLR) | Selecting parameters Technique |
| | | | | | | | | | | | | | | | | | | |x| | Akaike Information Criterion (AIC) | Ideal num. parameters Technique |

**Table 2.** Outstanding articles regarding detailing of archetype classification methodological aspects

| Authors | Year | Approach |
|---|---|---|
| De Jaeger et al. [7] | 2020 | Data sources: Existing database, Geographic Information System (GIS), Image analysis<br>Normalization technique: rescaling, standardizing,<br>Clustering technique: K-means, Hierarchical clustering<br>Technique for selecting parameters: multiple linear regression (MLR)<br>Technique for selecting the ideal number of parameters: Akaike Information Criterion (AIC)<br>Support tools: CityGML (LOD 2),<br>Historical aerial photographs, Google Streetview |
| Tardioli et al. [5] | 2018 | Data sources: Existing database<br>Data filtering / processing: removal of missing data<br>Normalization technique: rescaling, centring, z-score, PCA, Box-Cox<br>Clustering technique: K-means, Partitioning around medoids, Hierarchical clustering,<br>Validation cluster indices: Silhouette index, Dunn index, Calinski-Harabasz index, Wb index, Entropy index, Dunn index variation: Dunn2, Goodman and Kruskal's coefficient, G3 index, Ratio of the number of elements<br>Support tools: PostgreSQL, R Language, Q-GIS. |
| Famuyibo, et al. [8] | 2012 | Data sources: Existing database<br>Technique for selecting parameters: multiple linear regression (MLR), frequency histograms |
| Ali et al. [4] | 2019 | Data sources: Existing database, Statistic data, Building standards<br>Data filtering / processing: complete missing data with average data<br>Clustering technique: K-means<br>Support tools: Dwelling Energy Assessment Procedure (DEAP), EnergyPlus, Geographic Information System (GIS) |
| Li et al. [9] | 2018 | Data sources: Field survey, Image analysis<br>Data filtering / processing: correlation analysis between variables<br>Normalization technique: Z-score<br>Clustering technique: K-means, Partitioning around medoids<br>Support tools: MATLAB/ Corner Shadow Length Ratio (CSLR), R Language, nbClust,<br>Urban Modeling Interface (UMI) |
| Pasichnyi et al. [10] | 2019 | Data sources: Existing database, Statistic data<br>Clustering technique: Cartesian product<br>Support tools: R Language / IDE RStudio |
| Nägeli et al. [11] | 2018 | Data sources: Existing database<br>Clustering technique: interpolating /extrapolated<br>Support tools: Python |

*3.2.2. Data source*

The selected publications used different sources to create an initial information base about the sample to be studied, which varied according to the availability and objective of each study. Existing databases [1,4,5,7,10,11,12,13], Geographic Information System (GIS) [1,2,7,12,14], statistical data [4,10,12,15,16], construction standards [2,4,12], field surveys [2,9,13], image analysis [2,7,9], and databases created specifically for the study [13,15] were used. In most cases, multiple sources were used for completeness sake.

*3.2.3. Set of input variables*

Definition of grouping parameters has not reached consensus among the studies presented. Use typology (residential, commercial and mixed use), Construction typology (twin, semi-grown, independent / isolated, continuous), Year/ period of construction and Number of floors are more commonly used in the classification phase (Figure 3), although each study uses certain data sets. The choice criteria is not always explicit and may vary from limitations on data acquisition to the result of in-depth statistical work. In studies which applied statistical methods, Multiple Linear Regression - MLR [1,7,8] was highlighted for defining the most significant variables to be used for classifying archetypes, and promoting a good trade-off between precision, ease of results interpretation and computational power [1]. MLR depends on prior identification of the optimal number of variables, which can be identified from the Akaike Information Criterion (AIC) metric, which estimates the information loss caused by a given model relatively to the others [7].

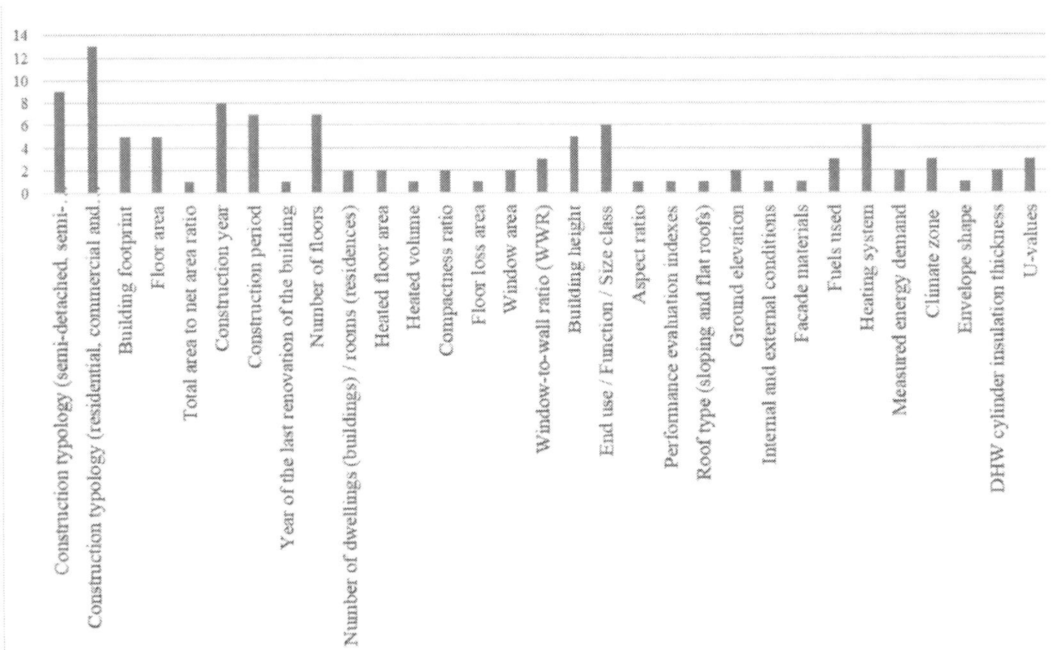


**Figure 3.** Most common variables used for creating construction archetypes.

*3.2.4. Data treatment: screening and normalization*

Studies point to the need for preparing data for statistical processing. Firstly, the raw data collected should be filtered, which corresponds to exclude the parameter [5]or to filling in the missing data with an average data [4]. Then, 'data normalisation techniques' are applied to standardise and make the data compatible with statistical treatment, as values for different construction characteristics display may differ by several orders of magnitude [5]. The data normalisation techniques addressed in the sampled studies are Reescaling or Scaling [7,5], Standardising [7], Centring [5], Z-score [5,9], Principal Component Analysis (PCA) and Box-

Cox [7]. Only one study compared the techniques and indicated Reescaling as the most suitable for residential buildings, PCA for commercial buildings and Z-score for mixed use buildings [7].

*3.2.5. Clustering techniques*

Clustering is a data mining approach that provides information on how elements should be grouped based on similarities and representative elements. The sampled studies used K-means clustering [7,4,5,9,12] K-medoids [5,9], Hierarchical clustering [4,7,5,12], partitioning around medoids (PAM)[5], Cartesian product [10], interpolation/extrapolation and distribution / exponential function [11]. Certain studies, predominantly on residential buildings, considered K-means as the best approach to archetype development [4,5,12], despite serious limitations found for discrepant input data values and clusters with different sizes and densities [4]. Conversely, others studies indicate that the K-medoids technique is more precise than K-means and that the K-medoids clustering process is more direct and convenient, as the cluster centres are real buildings rather than 'average' synthetic buildings, as when using K-means [9]. Hierarchical clustering is also praised, however, it is less used compared to K-means because it requires more computational resources [7]. For determining the best number of clusters in the data set, studies suggest the use of 'model-based clustering' supported by software such as the R Package "NbClust" [12][9]. Clustering quality metrics like Silhouette [5], Dunn index [5][12] and SD index [12] were applied in the RSL studies.

## 4. Conclusions

The SLR showed that using constructive archetypes for energy assessments at urban scale is a promising strategy, yet it still lacks methodological consensus. Data availability seems to be the major limitation for archetype creation in most studies, which rarely present in-depth information on their development. The few investigations which presented consistent methodological procedures in fact focus on detailing the initial classification process.

Simplifying the built stock may reduce model accuracy and introduce uncertainties regarding its representativeness. Guidance on improved modelling is a critical methodological need, which varies with the complexity and scope of the study at hand. Definition of grouping parameters depends on each research aims. For energy demand analysis of buildings, criteria like Use typology, Construction typology and Year/ period of construction seem to provide a good representation [19]. But, for archetypes to be consistently used to also support life cycle impacts estimation, a broadened parameters selection is needed (Figure 4).

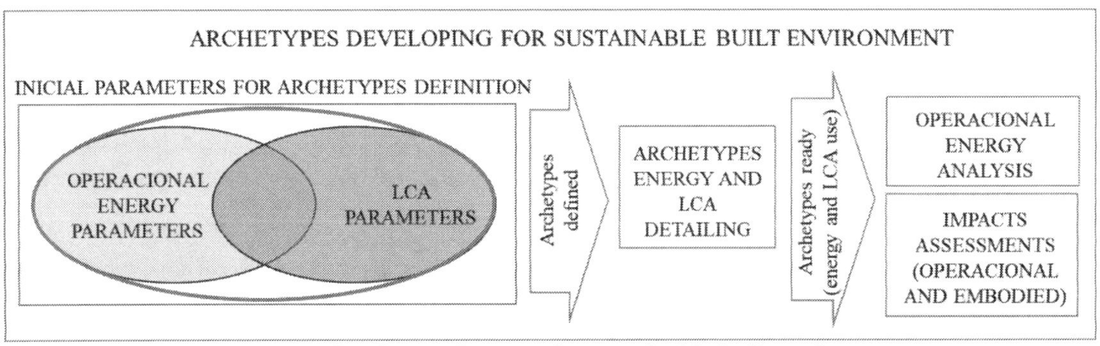


**Figure 4. Macro process to develop archetypes for sustainable built environment, merge energy analysis and impacts assessments**

Archetypes that include LCA parameters can support circularity challenges diagnosis, mapping and predictions, development of strategies to close material and energy loops, and monitoring their implementation within urban built environments. Yet, transposing the archetype strategy from energy assessment to LCA at urban scale faces practical limitations, such as:

- Although several databases support operational energy studies, the same does not apply to LCA. Supporting systems like GIS and statistical data rarely add consistent bill of construction materials. Thus, additional resources are required or LCA scope remains restricted;
- Urban building energy models typically do not model infrastructure, which is relevant for neighbourhood LCA; and
- Statistical tools, such as MLR, support selecting parameters for archetype definition, but data entry quality is key for robust variable selection, as little relevant data can be considered merely for being available, while high-relevance but unavailable information is left out. Thus, data quality must prevail over the adequacy of the tools and techniques, and extra efforts must be carried out for completeness sake.

Thus, developing representative archetypes to enable robust environmental assessments should consider building characteristics relevant for operational and embodied impact assessments, including building data that interferes with energy demand, consistent bill of construction materials and their quantitative. Optimal data collection and computational processing must be prioritized within financial, time, technological and resource reasonability.

## 5. References

[1] Sokol J, Cerezo Davila C, Reinhart CF. Energy Build. 2017;134:11–24.
[2] Cerezo C, Sokol J, AlKhaled S, Reinhart C, Al-Mumin A, Hajiah A. Energy Build. 2017;154:321–34.
[3] Monteiro CS, Pina A, Cerezo C, Reinhart C, Ferrão P. Energy Procedia. 2017. p. 817–25.
[4] Ali U, Shamsi MH, Hoare C, Mangina E, O'Donnell J. Energy Build. 2019;202.
[5] Tardioli G, Kerrigan R, Oates M, O'Donnell J, Finn DP. Build Environ. 2018;140:90–106.
[6] Kristensen MH, Hedegaard RE, Petersen S. Energy Build. 2018;175:219–34.
[7] De Jaeger I, Reynders G, Callebaut C, Saelens D. Energy Build. 2020;208.
[8] Famuyibo AA, Duffy A, Strachan P. Energy Build. 2012;50:150–7.
[9] Li X, Yao R, Liu M, Costanzo V, Yu W, Wang W, et al. Energy Build. 2018;169:417–29.
[10] Pasichnyi O, Wallin J, Kordas O. Energy. 2019;181:360–77.
[11] Nägeli C, Camarasa C, Jakob M, Catenazzi G, Ostermeyer Y. Energy Build. 2018;173:443–60.
[12] Ghiassi N, Mahdavi A. Energy Build. 2017;144:372–86.
[13] Hawkins D, Mumovic D. Energy Build. 2017;147:113–22.
[14] Cerezo Davila C, Reinhart CF, Bemis JL. Energy. 2016;117:237–50.
[15] Ballarini I, Corgnati SP, Corrado V. Energy Policy. 2014;68:273–84.
[16] Lavagna M, Baldassarri C, Campioli A, Giorgi S, Dalla Valle A, Castellani V, et al. Build Environ. 2018;145:260–75.
[17] Gassar AAA, Cha SH. Energy Build. 2020;224.
[18] Swan LG, Ugursal VI. Renew Sustain Energy Rev. 2009;13:1819–35.
[19] Mata É, Sasic Kalagasidis A, Johnsson F. Build Environ. 2014;81:270–82.
[20] Famuyibo AA, Duffy A, Strachan P. Build Environ. 2013;70:90–101.
[21] Stephan A, Athanassiadis A. Build Environ. 2017;114:187–202.
[22] Kitchenham B 2004 *Procedures for Performing Systematic Reviews* 2004 (Keele: Keele Univ.)
[23] Loga T, Stein B, Diefenbach N. Energy Build. 2016; 132: 4-12.

## Acknowledgments

The authors thank the National Council for Scientific and Technological Development - CNPq for the generous funding [GMC, PhD process # 140412/2018-2; and VGS, productivity grant #306048/2018-3].

IOP Conf. Series: Earth and Environmental Science **855** (2021) 012022 doi:10.1088/1755-1315/855/1/012022

# Bijlmermeer and Scampia: The potential of suburbs as centralities in sustainable and circular cities

Chiara Mazzarella P[1], Hilde Remoy[2]
P: Presenting author
1 Department of Architecture, University of Naples Federico II, 80134 Naples, Italy;
2 TU Delft / Faculty of Architecture and the Built Environment / Department of Management in the Built Environment / Julianalaan 134 – 2628 BL Delft

**Abstract.** In the sustainable paradigm of the circular economy, waste is a resource to be valued. Similarly, in the circular city, fringe areas and urban wastescapes are places full of potential. This study consider two modern suburbs in Amsterdam and Naples. The two cases aims to compare the characteristics, potential and problems of the two districts and the ongoing actions for sustainable or circular urban regeneration. The objective is to identify the potential of these two European suburbs in the wider process of development of fair and circular cities. Therefore, this paper aims to answer to the following research question: Based on the two case studies Bijlmermeer and Scampia, what is the potential for circular and socially sustainable urban transformation of the 1960s suburbs? Analysing the main changes that have taken place over the last ten years in these complex areas, the main drivers are identified with the historical value of modern architecture landmarks, the opportunities offered by big green public spaces in dense urbanized areas, the presence of active place-based communities and their position in nodal points of the metropolitan areas.

## 1. Introduction

Circular Economy (CE) is a recent and still contested concept (1). Many CE definitions have been developed over last years, and although there are a few explicit linkages with sustainability (2), it seems to be the most effective paradigm for sustainability (3). Many studies have highlighted existing links between CE and Sustainable Development Goals (SDGs) (4). The idea of circularity is mostly about closing supply chains, through the implementation of strategies for reduction, reducing, recovering and recycling, as much as possible, waste materials by optimizing resource management in metabolic systems. This is consistent with the SDG 12, that concerns precisely "responsible consumption and production".

In few years CE has gained momentum and its own conceptual autonomy, expanding the research field from industrial ecology to urban studies. Making cities sustainable, that correspond to the SDG 11, is explored in the idea of the circular city. A clear definition of what is a Circular City (CC) still under development (5), but it seems to mean more than a city where CE is implemented (6).

CC researches produced strategies for looping actions (7), indicators and monitoring (8,9), and adaptive reuse strategies of culture heritage buildings (10).

Several European cities are pursuing circular economy policies for the built environment, energy transition, sustainable mobility, industrial production, and local communities (11).

Because CE monitoring is mainly connected to urban metabolism study, urban studies included waste lands to be claimed and reused in the idea of *circularity*.

The research project REPAiR, *Resource Management in Peri-Urban Areas: Going beyond Urban Metabolism*, carried out a multidisciplinary research, aimed at developing tools and strategies to implement CE solutions in peri-urban areas (12). Amsterdam and Naples have been the pilot cases of the research, where urban metabolism and *wastescapes* have been co-explored through peri-urban living labs (13) with local stakeholders, in incremental processes of co-creation of innovative strategies (14). The Geodesign Decision Support Environments (15), based on the geodesign methodology (16) has been tested using a new open-source tool. Wastescapes have been analyzed in the REPAiR project as "patches of landscape related to waste-cycles both by functional relations and because they are 'wasted-lands': anomalous areas inconsistent with the peri-urban metabolism that become neglected spaces." (12), and seen as strategic places to recover to activate circular strategies (17). From literature on drosscape (18) and operational infrastructures of waste (19) wastescapes were investigated in the research project as spatial areas subject to risk and various abandonment phenomena.

In this study, we consider a specific category of urban wastescapes, seen not only as spatial categories of degraded and abandoned places, but also as those places where a negative perception has led to a rejection and, therefore, to a condition of exclusion, segregation and, finally, rejection and discarding of a part of the city.

An urban wastescape is therefore a part of a dysfunctional urban ecosystem, i.e. a place where the built environment, society and socio-economic dynamics are determined synergistically.

Emblematic of exclusion and rejection, two 1960s suburbs, famous for their iconic and deeply problematic architecture, are studied here as urban wastescapes.

Considering the regenerative and circular potential of wastescapes, this study attempts to identify the characteristic traits and the role that such neighbourhoods can play in the context of the SDGs (20) and urban circularity goals.

Section 2 presents the methodology of the study. Section 3 compares the two case studies in their respective urban contexts, identifying some salient features of recent urban regeneration processes. Finally, the discussions and conclusions in Section 4 propose answers to the research questions based on the analysis carried out.

The causes of the failures of planners, designers and public policy have been widely addressed in the literature. In the present study, the potential already expressed and that which these neighbourhoods have to give rise to more sustainable and circular city models is highlighted.

## 2. Methodology

The goal of the study is to identify the potential of modern urban districts that suffered (or are still suffering) the condition of urban wastescape. The methodology for comparison between Bijlmermeer and Scampia is based on the status quo of the two cases (Figure 1), considering three different key agents as drivers of transformation of the areas:

1. Public policies (PP) currently governing the sites,
2. The Built Environment (BE), considering architecture and public space,
3. Place Communities (PC) of residents, NGOs and active stakeholders.

**Figure 1.** Methodological framework.

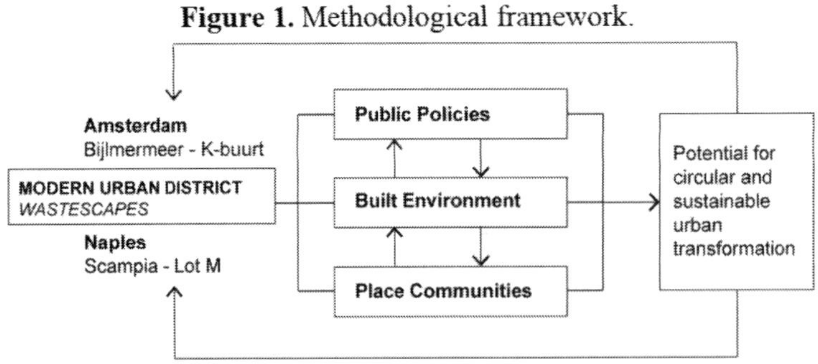

The common hallmarks of Public Policies, Built Environment and Place Communities are part of a unique system whose influences make them change each other in the interaction (21,22). The cause-effect of interaction of these agents defined in time the identity of the places.

## 3. Similar modern neighbourhoods in different European cities: the case of Amsterdam and Naples

The cities of Amsterdam and Napoli are both core of large metropolitan areas. The Amsterdam Metropolitan Area (AMA) is composed by a partnership of 32 municipalities, in the provinces of Noord-Holland and Flevoland. The Metropolitan Area of Naples (MAN) is one of the 14 Italian metropolitan cities instituted in 2001, and regulated by the Italian Law 56/2014 (23).

**Table 1.** The Metropolitan Areas of Amsterdam and Naples

| Indicator | Unit | AMA | MAN |
|---|---|---|---|
| Population | Inhabitants | 2,500,000 | 3,000,000 |
| Provinces | Name | North Holland and Flevoland | Province of Naples |
| Sub-zones | Number | 7 | 5 |
| Municipalities | Number | 32 | 92 |
| Area | km2 | 2,580 | 1,171 |
| Population | Million people | 2,5 | 3 |
| Density | Inh/km2 | 960 | 2,621.06 |

The two cities are both ports, daily crossed by large flows of people, goods, information and money. While Amsterdam is the capital and the main commercial and financial center of the Netherlands, Amsterdam urban population is growing, and the city is dealing with that (24). At the same time, Amsterdam is working to become a sustainable city the goal is to achieve a be fully circular by 2050. Construction, biomass and food, and consumer are the key issues of innovative circular actions to lead this transition (25). The recent *Doughnut economy model* (26) expands the seven circularity objectives of the DISRUPT framework (27) by introducing the social inclusion dimension at each stage. Also, the new Doughnut economy model substitutes the idea of *growth* with the one of *thriving* (26), implying recognition of the city's unexpressed (or wasted) potential.

Naples is one of the main southern Italian cities, but suffers from high unemployment and other socio-economic difficulties (28) like the most of the South of Italy. The First Strategic Plan for the 92 Municipalities of the MAN, *ImmagiNA,* (23) focuses on socio-economic development and environmental sustainability goals, looking at the 92 Municipalieties in a synergic approach, but without specific references to the transition to a circular economy model.

In these context, Bijlmermeer and Scampia are two suburbs of Amsterdam and Naples, famous wasted areas and no-go zones until a few years ago. The two neighbourhoods were designed almost at the same time, in the 1960s. Then, each one was realized and evolved differently, but both have been places of troubled and difficult events.

Since the 1990s, Bijlmermeer changed a lot, and today it is much a safer area and, also place of many vibrant and creative sports of the South-East district. The district is developing and transforming, and is doing so by implementing sustainable and CE solutions. In the Doughnut City model, tools and methods are defined as the *Doughnut Deals* have been conceived in Bijlmer (26).

The renaissance of Scampia is debated by public administration and stakeholders, but the district changed a lot over last twenty years. An important urban regeneration of the Scampia district is underway thanks to the Restart Scampia project (29), which has been funded by the Government as part of the 2016 national suburban plan (20).

IOP Conf. Series: Earth and Environmental Science **855** (2021) 012022 doi:10.1088/1755-1315/855/1/012022

In both cases architecture was blamed of being the origin of most issues, and the most significant actions for improvement started from redevelopment.
The K-buurt in Bijlmer and the lot M in Scampia (Fig.2) are the areas of the neighbourhood where still some original modern buildings are, and where the most of the debate about regeneration is focused.

**Figure 2.** In red the study areas areas in Bijlmer-Oost (Amsterdam) and Scampia (Naples).

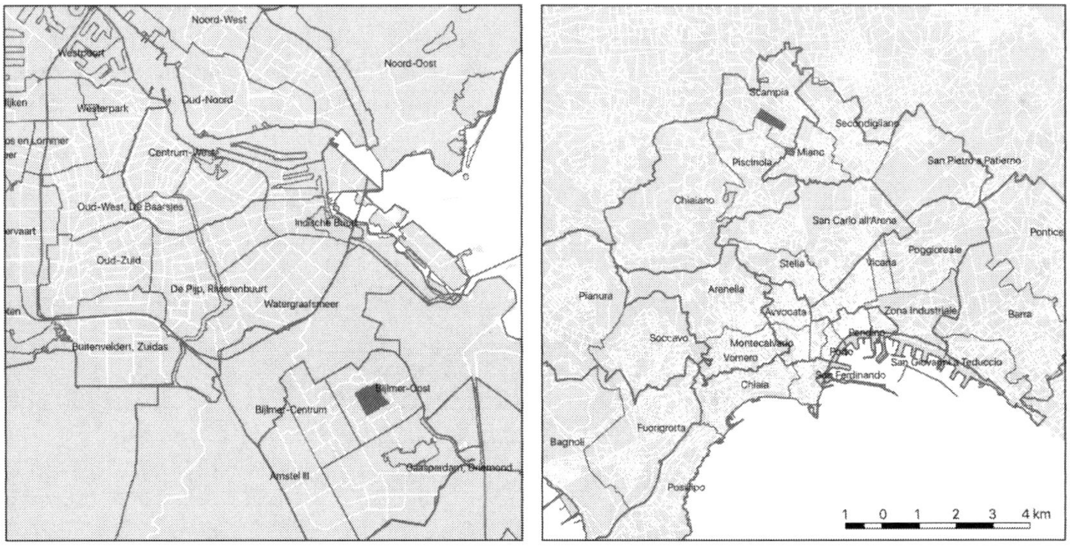


The comparison between these two similar even far areas allow at understanding the main hallmarks and enablers of the current regeneration processes in these two modern neighbourhoods, representatives of the 1960s suburbs.

## 4. Bijlmer and Scampia: Contested wastescapes and resources for the metropolitan areas

In the 1960s, many suburbs of Europe's major cities were built in response to population growth.
As many others in Europe and U.S., the districts of Bijlmermeer and Scampia were both born with similar assumptions: Megastructure design according the principles of the Modernist Utopia to cope with increasing housing demand. The modern architects dreamed of creating an egalitarian collective lifestyle for inhabitants, according CIAM's principles designing high-rise buildings, big public spaces with public green areas and collective facilities.
The result of modern planning and design systems of large housing estate (30), these parts of cities were not immediately absorbed by the existing fabric, and gave rise to phenomena of segregation and social exclusion, which led to scenarios of social and environmental degradation. Both districts had common problematic developments from the very beginning, albeit eventually they experienced different events.
Many incremental problems in time lead to their failure, i.e. the budget limitations that led to unfinish the facilities, the architecture typology, the isolation due to the lack of connections with the city centers, the cutting-edge aesthetic that did not like to people, and the concept of modern architecture based on utopic ideas of a daily life far from what people wanted (31).
Same planning background in different cities caused segregation and ghettos, where different socio-cultural phenomena led residents to similar conditions and problems that transformed the districts in no-go zones.
Renewal of Bijlmer high-rise (32) carried out over time have involved the demolition and reconstruction of some of the buildings, the construction of new public spaces and the completion of street furniture, a social renewal that has led to a reduction in unemployment rates and better control over crime, with the aim of facilitating the economic development of the area.

IOP Conf. Series: Earth and Environmental Science **855** (2021) 012022 doi:10.1088/1755-1315/855/1/012022

Bijlmermeer today is a lively, multicultural neighbourhood and part of the city of Amsterdam's extensive circular economy programme. Participants in the decades of renewal policies were the Municipality of Amsterdam, the district, the Dutch public housing authority, private operators and groups of inhabitants.
The Neapolitan district of Scampia is now in a deep process of transformation, but has not yet reached levels of economic well-being for its inhabitants such as to redeem it from the stigma it carries. After the first ten years of overcrowding and squatting, since the 1990s, Scampia has been a symbol of drugs, criminality and marginalization. In fact, Scampia was the biggest drug dealing place in Europe until about ten years ago, and today the problem is still not completely solved. In the mid-1990s, the buildings were declared unfit for human habitation and were therefore abandoned and not maintained by the public administration. The inhabitants organised themselves to provide the essential services for living, but with great inconvenience. A strong criminal control of the areas decreased along last years, but further and complex investigations would be required to clarify the current influence of this *agent* in the urban transformation dynamics. In the last ten years, the district has been experiencing a phase of regeneration through the action of many local NGOs and socially active groups in the area. The Municipality of Naples realized public works to improve the district, infrastructures, green care and a demolition and redevelopment plan.
Beyond the temporal asymmetries in regeneration policies, the main big difference in the two areas is in the approach of public administration and the residents to the conservation or demolition of the buildings. Urban regeneration processes affected both the physical environment and the socio-economic conditions of the areas, and involved public administrations residents and users. While public administration played a key role in the macro-scale planning and in new developments, residents willingness create conditions for the bottom-up community actions, that aimed at improving the daily life of inhabitants. People involvement and strong connection with their neighbourhood, determined also the fate of some spaces and modern buildings conservation or demolition.
In the perspective of circularity, planning policies, built environment and social disfunctions shaped waste areas and wastescapes, as unresolved territories procuring ecosystem disservices (33) and risks for environment, economy and society. Multiple dimensions of wasted areas have a regenerative potential (34).
Bijlmermeer and Scampia were born from a strong idea of city and residential architecture typology that immediately defined an image of a new urban landscape. These iconic landmarks have polarised the debate, and even if they did not work, their strong identity was blamed for producing all the social issues of the neighbourhoods, obviously caused by many different immaterial agents. Although a neighbourhood is defined by relationships between spaces and inhabitants, the synergy with the rest of the city and the socio-economic conditions of the employment opportunities offered to the inhabitants make a difference. In both Bijlmer and Scampia, areas where the original architecture has been completely modified have taken on a new urban identity (Fig.3).

**Figure 3.** On the left the perimeter of North and South Bijlmermuseum with the six original modern buildings. On the right the perimeter of lot L of the Vele di Scampia. In 2020 the Green Sail was demolished. (Source: Authors' elaboration on GoogleMaps top views)

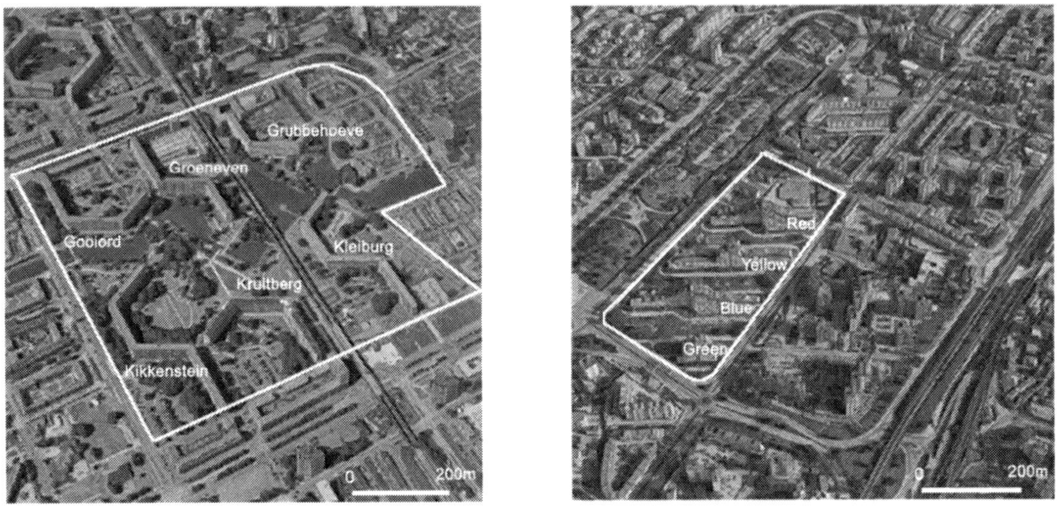


The K-buurt neighbourhood in Bijlmer and Lot M in Scampia have today the last significant bastions of the original conception and also of the problems that resulted. In these areas, regeneration processes are still more or less open. The value and potential of these architectures and places is being debated by researchers, writers, architects and residents. Today it seems evident that one of the strongest agent of change is the ability of public administration to interact with residents and place communities.

*4.1 K-buurt in Bijlmer: the Bijlmermuseum*

Massive investment in redevelopment in Bijlmer started after the disaster of 1992, when El Al cargo plane crashed into Kruitberg and Groeneveen buildings. In the point of the tragedy today the Growing Monument commemorate the 43 victims.

Of the 31 original honeycomb buildings, only seven remain today, including six in the area of the Bijlmermuseum in the K-buurt: Gooioord, Groeneveen, Grubbehoeve, Kikkenstein, Kriutberg, and Kleiburg (Fig. 3). Other elements of the original project by Siegfrid Nassuth the elevated road currently serving the metro line, some garages and the public park. Currently, there is no redevelopment or big transformation of space taking place in the area. The history of the K-buurt tells the story of the transition from a wastescape to a recognised cityscape and protected, the Bijlmermuseum since 2016. The last decade has marked several interesting moments in the history of the last original Bijlmermeer fragment. The Bijlmer-Spinoza Festival of the artist Thomas Hirshhorn took place in the middle of K-buurt in 2009, conceived for and with inhabitants in the public space (35) joining of a work of visual art with a series of activities like philosophy conferences, open reading zones and theatre and creative activities for children. The same year, the Grubbehoeve renovation of was awarded wuth South-East Architecture Prize. Also on 2009, Roshdale had decided to demolish Kleiburg, a vacant building whose renovation was estimated to be too expensive. But the last remaining original honeycomb flat building was defended by residents' groups, the Bijlmer pioneers. Rochdale then let the building for the symbolic price of €1 to the DeFlat Consortium, thus also saving the demolition costs. Using the Do It Youselft formula [36] Consortium DeFlat found the solution to restore the building at an affordable price and managed to sell all the flats. In 2017, NL Architects and XVW Architectuur's restoration project of Kleiburg won the Mies van der Rohe Award. It was the first time for a collective housing project. Revamping social housing also led to the prize in 2019 for the transformation of 530 dwellings is a renovation of three 1960s housing blocks in Bordeaux. The architects of the XVW and NLarchitects studios, through some minimal solutions to improve the functioning of the building and

the relationship with the public space on the ground floor, have revived the last modern building with an essential renovation operation. Thanks to Dutch legislation, however, it was not mandatory to meet many of the sustainability performance requirements that are imposed on new buildings today.
This made it possible to bear the expense of the renovation and save the building, but the environmental impact of the building would need to be investigated in order to assess its environmental sustainability over time. The new residents who own Kleiburg have formed neighbourhood ties and take care of an urban garden in the neighbouring public space. A chapel, the Kleiklooster, has also been established in the building, from which a brewery Brouwerij Kleiburg, with a community place and restaurant in Bullewijk, has also sprung up.
The multicultural Communities of Place are a strong component of the K-buurt identity. While Kleiburg was being completed, the cultural-historical significance of the area was recognised with the Bijlmermuseum, a protected cityscape where both public space, the trees, and the original buildings and remained infrastructure constitute an urban environment to which many are attached.
The Heart von de K-buurt group and Bijlmer pioneers opposed demolitions of the Kempering en Klusflat garages. As well as for the Bijlmermuseum *the value of the garage as a cultural heritage* was recognised by people. Originally, elevated roads such as Karspeldreef were connected directly to garages, which in turn led to buildings. Over time, however, the garages became unsafe places where people slept, gambled and engaged in illicit activities. Since the radical redevelopments of the 1990s, many streets have been redeveloped and many buildings and garages have been demolished.
Garages have been risky and characteristic places in Bijlmer.

**Figure 4.** The Bijlmermuseum today: On the right the public space and the original infrastructure, on the left part of Groeeneven building. (Photos: C. Mazzarella)

In June 2016, the municipality of Amsterdam had requested the redevelopment of the Kempering Garage. But last year the garage was demolished. To accompany the farewell of architecture, residents organised "The Last Nine Days of Garage Kempering" 600 candles in the garage every day between Ascension and Whitsuntide. Six hundred candles, one for each month that Kempering was allowed to stand in the Bijlmer.
Every building and modern architecture in the neighbourhood foundation is felt by the residents as a monument, so the garage also carries a history that the inhabitants would not want to erase. The International seminario *Negotiating Heritage in Emotion Networks: the Kempering Case* opened a

debate on the value of the garage as a symbol of collective memory. During the seminar, a large group of people visited the place, and Murat Isik[1] made a literary intervention.
Cultural heritage value as part of the *City of the Future*, as the Bijlmer was once called, where a leading role was claimed by high roads connected to the garages.
Today, the Bijlmermuseum is a liveable and safer area, (Fig.4) but despite major improvements in the area, public policies in the South-Eastern district, and in particular in K-buurt, still need policies that aim at improving youth development opportunities increasing security, making the living environment more sustainable, improving the position of vulnerable residents, reducing poverty, promoting participation and democratisation (36). This process is taking place on the basis of public recognition of the historical and cultural value of the area.

*4.2 The Lot M in Scampia: the Sails*
The modern iconic landmarks of the Scampia neigbourhood in Naples are the so called Sails (Vele) (Fig.5). These buildings have been dwelled in from the start of the 1980s and squatted while still unfinished after the 1980 earthquake, over-densified and lacking in services. These tent-like buildings, nicknamed Sails for their formal analogy, were built in an area north of Naples, that was originally agricultural. Still unfinished, by the early 1980s they had become a social condenser for displaced persons and people in difficulty, an isolated ghetto disconnected from the centre, where over time criminal activities and drug dealing found fertile ground in the absence of other job opportunities.
The Sails occupy only one of the twenty-one lots that make up Scampia, a record-breaking neighbourhood in Naples: the greenest and most feared in Naples.
The buildings designed by Franz di Salvo in the 1960s were not built as in the original project.
Neither people nor architecture fared well in housing units, such as Corviale in Rome (37), Forte Quezzi in Genoa, Unité d'Habitaciòn in Marseilles.
Although in an advanced state of degradation and a drug dealing place, the facades and interior walls of the Sails in Scampia are populated with street art, bearing messages, signs and drawings by foreign artists and inhabitants.
The Sails have often attracted those who do not live there. They have been observed, filmed and photographed by journalists and artists from all over Europe, as a recognisable dramatic icon of every conceivable problem in Italy's suburbs. In recent months they have returned to the centre of media attention.
Originally, there were seven buildings: four in the lot M and three in lot L. The four Sails built in the lot M are called: A, B, C, and D. Currently, they are also called by colours respectively: Green, Blue, Yellow and Red. Three Sails were built in the lot L: F, G and H (the E was never realized).
All the three Sails were demolished in lot L between 1999 and 2003, have been replaced by four-storey social housing, where hundreds of families have been moved. They found decent housing, but not excellent public space and the city has been architecturally impoverished. In this lot a new building circular plant has been realized to be a new University venue, designed by Gregotti, that should be opened in 2021.
The conditions around the whole neighbourhood, however, have changed profoundly in twenty years. Last architectural change is the demolition of the Green Sail began on February and was completed on July 2020. It has been celebrated by many and not appreciated by other residents of Scampia.
In the generic suburbs of northern Naples, the Sails have an architecturally strong image, they are a territorial landmark. Whether they are liked or not, they have become icons of the urban landscape and part of the local history and memories (38).
Having created their fame, today the last remnants of the stigma persist.
The social stigma has gradually overlapped with the formal one. *Comitato Vele*, the historic committee set up to demand the right to live in decent conditions, from 1989 struggle for better housing conditions

[1] He is the author of "Wees Onzichtbaar", the story of a Turkish boy in the Bijlmer that won the Libris award for literature in 2018.

for people of the Sails. Their main objective is the demolition of the buildings as a solution for obtain new houses to all the families who have been living in precarious conditions for decades.
The uninhabitable conditions have been blamed almost exclusively on architectural and structural configurations, the presence of toxic insulation materials, and general decay, but they cannot be examined without considering the social history of the place, the pressure due to very high density, the extreme lack of maintenance, the absence of services, and the strong criminal control.
The problem are not just the buildings. Without man, who uses them, things are dead bodies. In fact, men make and unmake places, judgements and prejudices.
In 2020, the first Dutch tourists guided by a *Local in Naples* tour came out to visit these buildings, going beyond the usual tourist circuits sold to visitors. Anyway the residents remain divided between those who want nothing more than demolition and those who feel an emotional bond with this urban landscape, due to collective history or personal memory. First group is organized with Comitato Vele, the most important association of inhabitants. Comitato Vele is collaborating with the Municipality of Naples in the Lot M redevelopment in the RESTART Scampia project. RESTART Scampia is implementing the urban regeneration strategy that aims to convert this area from an urban edge to a new centre of the metropolitan area. With respect to the Sails, the strategy includes the demolition of three of the four remaining Sails in Lot M and the construction of new houses. The Municipality's strategy is that only the Blue Sail will remain, as a symbol, memory and monument, following the renovation work, asbestos abatement and replacement of the famous concrete walkways with steel ones.
While the Sails are coming down and the built environment of the lot M is changing, what is happening in the human geography of the neighbourhood? A silent cultural renaissance is underway in Scampia. From the outside, it can be said that it is favoured by some major investments such as the new underground station, the Federico II Faculty of Medicine and "RESTART Scampia" project. Those who walk and live in the streets of the neighbourhood can find the vibration taking place in the activities of many NGOs, groups, activists and free citizens, who work for and with the community. Some socio-cultural reference points: *Comitato Vele* and *Cantiere 167*, the *Compare* Association that manages the *Mammut Territorial Centre*, *GRIDAS* - Gruppo Risveglio Dal Sonno (group wakening from sleep), *BandaBaleno - Murga Di Napoli*, *Dream Team - Donne in Rete*, *Centro Hurtado*, *CentroInsieme Onlus* with the project "Vela: Rendere Consapevoli" (Sail: making people aware) and L'albero delle storie (The tree of the stories) association, L'uomo e il legno (The man and the wood), I pollici verdi (The Green Thumbs), Chi rom e chi no, and many more.

**Figure 5.** The Sails today: On the left a view from North-East from the Ciro Esposito park (photo: S. Torrone), on the right the view on the Yellow Sail from the Red Sail (photo: C. Mazzarella).

Despite of many bottom-up initiatives and general improvements along time, what the inhabitants of Scampia and Naples lack most today is a labour. Even today, many young people who do not want to fall into the networks of the underworld or who seek independence from their families are forced to emigrate.

**5. Discussion and conclusions**

As defined by the donut economy model, EC and CC are not only about efficiency in the management of goods and materials, but also about welfare and social equity.

While the city of Amsterdam has set itself concrete objectives of thriving through CE models, the City of Naples does not yet have a plan aimed at implementing circular policies.

Starting from the concept of wastescape, developed in the h2020 REPAiR project, two urban wastescapes from Amsterdam and Naples have been considered. Bijlmermeer and Scampia have grown and deeply transformed over time; in this paper we have compared the two modern areas, that have suffered the condition of becoming wastescapes, in which there is still evidence of the original contested modern architecture. The so called *honeycomb* buildings of Bijlmer and the Sails of Scampia represent two cases in which modern architecture, born out of a utopia of egalitarianism and collective life, has come up against extraordinary needs and complex socio-cultural conditions. contributed to the failure of the districts. These neighbourhoods are recognised and characterised by the architectural landmarks of modern high-rise buildings, and have been stigmatised and excluded from the urban systems. The causes that led to the condition of abandonment and wastescape are found in the original rigid top-down planning approach, that did not consider the preferences and the needs of future inhabitants, the design choices that created some dysfunctional hallmarks of the modern built environment, and the unfinished condition of the areas, that made them the last choice and transformed the modernist utopia in a nightmare. Successive situations of socio-cultural exclusion due to the previous causes tuned these places in dangerous ghettos.

The policies implemented by the administrations of the two cities coped first with the physical evident problems of the built environment: Connections were set through infrastructure, demolition and reconstruction of many buildings, creation of public facilities points, and location of commercial areas. The socio-cultural actions implemented from the bottom up by NGOs and groups of citizens proved the importance of place-based communities in self-organization. Ties born of common daily difficulties created strong resilient communities, albeit with a different approach and link to modern architecture.

The main different elements in the two cases are the following. In Naples there has been a strong criminal control and the state of abandonment of the buildings was out of control since 1995. Since then, the ordinary maintenance has been self-managed by self-organised squatters and inhabitants. Most of resident are waiting for new social housing and for the demolition of the buildings, seen as icon of degradation and segregation mainly after the *Gomorra* effect.

On the one hand the historical area of Bijlmermeer has become a museum area. The cultural and historical value of the K-buurt with the Bijlmermuseum has been recognised by the Municipality in 2016. In this context, the case of Kleiburg shows also the value of a successful real estate operation of adaptive reuse and recognition of the cityscape from the conservation of modern buildings in the context.

While inhabitants of Naples asked for demolition of the buildings from the very beginning, the Bijlmer pioneer struggled for their conservation.

Currently the Sails are going through a phase of transformation, but the process seem to be slow and people still live in fragile conditions. Restart Scampia project and the strategic and planning actions put in place for the urban regeneration of the area are demolishing three of the four last buildings, and it is realizing new public and cultural functions for the social mix of the district.

In both cases, the regeneration processes were mainly oriented towards creating conditions of liveability, through architectural regeneration and the implementation of public services. The concept of sustainability has not been directly taken into account. So far, the idea of circular economy and circular city is present at a strategic level in the urban agenda of Amsterdam, but not in that of the Metropolitan City of Naples, nor specifically in the RESTART Scampia project. Some circular economy practices have been implemented by inhabitants through bottom-up urban regeneration projects, such as the urban garden of the inhabitants of Kleiburg or the recovery of Corto Maltese Park in Scampia by Pollici Verdi.

In a multidimensional and multi-scalar perspective, both neighbourhoods have great potential in the transition to the circular city:

1. the historical value of modern architecture, that is a landmarks, created identity, but it is a contested heritage due to the symbolic (positive and negative) value it assumed over time. It also need to be adapted to be more sustainable, flexible and liveable;
2. the opportunities offered by wide green public spaces in dense urban areas, provide ecosystem services and the space to implement nature-based solutions for circularity;
3. the presence of active place-based communities, committed in improving public and private space quality of their neighbourhoods, and interested in taking part in urban transformation decision-making processes. Co-governance of circularity at community scale is a big challenge of social circular economy;
4. their location in nodal points of metropolitan areas.

Thus, the relevance of urban regeneration in neighbourhoods like these -characterised by historical buildings of the modern movement- can be seen in the preservation of memory and the integration of new multiple place-based communities into neighbourhood decision-making processes to regenerate the built environment creating new labour. In this sense, actions to implement the CE and the CCs cannot be abstract and it is not possible to use the same strategies in different urban contexts. Each place must take into account local specificities.

These specificities are the identity of the areas and the meaning local communities gave them over the time.

If on a regional scale the CE has a great economic and environmental impact in urban metabolism of waste flows, on a local scale urban wastescapes regeneration can also assume a strong socio-cultural value.

**References**

1. Korhonen J, Nuur C, Feldmann A, Birkie SE. Circular economy as an essentially contested concept. J Clean Prod [Internet]. 2018;175:544–52. Available from: https://doi.org/10.1016/j.jclepro.2017.12.111
2. Kirchherr J, Reike D, Hekkert M. Conceptualizing the circular economy: An analysis of 114 definitions. Resour Conserv Recycl. 2017;127(September):221–32.
3. Geissdoerfer M, Savaget P, Bocken NMP, Hultink EJ. The Circular Economy – A new sustainability paradigm? J Clean Prod [Internet]. 2017 Feb;143:757–68. Available from: https://linkinghub.elsevier.com/retrieve/pii/S0959652616321023
4. Dantas TET, de-Souza ED, Destro IR, Hammes G, Rodriguez CMT, Soares SR. How the combination of Circular Economy and Industry 4.0 can contribute towards achieving the Sustainable Development Goals. Sustain Prod Consum. 2021;26:213–27.
5. Paiho S, Mäki E, Wessberg N, Paavola M, Tuominen P, Antikainen M, et al. Towards circular cities—Conceptualizing core aspects. Sustain Cities Soc. 2020;59(March).
6. Marin J, De Meulder B. Interpreting circularity. Circular city representations concealing transition drivers. Sustain. 2018;10(5).
7. Williams J. Circular cities: Challenges to implementing looping actions. Sustain. 2019;11(2):423.
8. Fusco Girard L, Nocca F. Moving towards the circular economy/city model: Which tools for operationalizing this model? Sustain. 2019;11(22):1–48.

9. Cerreta M, Muccio E, Poli G, Regalbuto S, Romano F. A Multidimensional Evaluation for Regenerative Strategies: Towards a Circular City-Port Model Implementation. In 2021. p. 1067–77. Available from: http://link.springer.com/10.1007/978-3-030-48279-4_100
10. Foster G. Circular economy strategies for adaptive reuse of cultural heritage buildings to reduce environmental impacts. Resour Conserv Recycl [Internet]. 2020;152(October 2019):104507. Available from: https://doi.org/10.1016/j.resconrec.2019.104507
11. Gravagnuolo A, Angrisano M, Girard LF. Circular economy strategies in eight historic port cities: Criteria and indicators towards a circular city assessment framework. Sustain. 2019;11(13):1–24.
12. Russo M, Amenta L, Attademo A, Cerreta M, Formato E, Remoy H, et al. D 5.1: PULLs Handbook [Internet]. 2017. Available from: http://h2020repair.eu/project-results/project-reports/
13. Amenta L, Attademo A, Remøy H, Berruti G, Cerreta M, Formato E, et al. Managing the transition towards circular metabolism: Living labs as a co-creation approach. Urban Plan. 2019;4(3):5–18.
14. Garzilli F, Mazzarella C, Vittiglio V. Integrated Approaches for Peri- Urban Wastescapes : Eco-Innovative Strategies of the REPAiR Project in the Naples Case Study. Int J Urban Plan Smart Cities. 2020;1(2):43–58.
15. Arciniegas G, Šileryté R, Dąbrowski M, Wandl A, Dukai B, Bohnet M, et al. A Geodesign Decision Support Environment for Integrating Management of Resource Flows in Spatial Planning. Urban Plan. 2019;4(3):32.
16. Steinitz C. A Framework for Geodesign: Changing Geography by Design. Vol. 7, Journal of Landscape Architecture. Redlands,California: esri press, redlands california; 2012.
17. Amenta Li, van Timmeren A. Beyond wastescapes: Towards circular landscapes. addressing the spatial dimension of circularity through the regeneration of wastescapes. Sustain. 2018;10(12).
18. Berger A. Drosscape, Wasting land in urban America. Princeton. New York, USA; 2006.
19. Brenner N. Implosions / Explosions Towards a Study of Planetary Urbanization " I'Ll Begin With the Following Hypothesis: Society Has Been Completely Urbanized. ". Berlin: Jovis; 2014.
20. Ministri P del C dei. Bando per la presentazione di progetti per la predisposizione del Programma straordinario di interventi di riqualificazione urbana e sicurezza delle periferie delle Città metropolitane e dei Comuni capoluogo. DPCM 25/5; 2016.
21. Hillier B. Space Is the Machine. A configurational theory of architecture. Space Synt. London; 2007. 1–109 p.
22. Heynen H. Space as Receptor, Instrument or Stage: Notes on the Interaction Between Spatial and Social Constellations. Int Plan Stud. 2013;18(3–4):342–57.
23. Del Prete G, Pascarella R, Pasquini S, Vanella V. ImmagiNA Napoli Metropoli 20/22. Napoli; 2020.
24. The Platform for Accelerating the Circular Economy (PACE). Circularity Gap Report [Internet]. 2019. Available from: https://www.circularity-gap.world/
25. Gemeente Amsterdam. Amsterdam Circular 2020-2025 Strategy. 2020;1–30. Available from: https://www.amsterdam.nl/bestuur-organisatie/volg-beleid/ambities/gezonde-duurzame-stad/amsterdam-circulair-2020-2025/
26. D.E.A.L., Biomimicry 3.8, Circle Economy, C40 Cities. The Amsterdam City Doughnut. A tool for transformative action. Amsterdam; 2020.
27. Circle Economy. DISRUPT framework [Internet]. Available from: https://www.circle-economy.com/resources/the-disrupt-framework
28. Punziano G (a cura di), Terracciano A, Lanzi M, Martone V, Gargiulo E, Sacco E, et al. Società, economia e spazio a Napoli. Esplorazioni e riflessioni (Society, Economy and Space in Naples. Explorations and Reflections). SSRN Electronic Journal. L'Aquila; 2016. Report

No.: 28.
29. De Magistris L, Piscopo M, Villani D, Borriello C, Pulli G, Santoro M, et al. RESTART SCAMPIA. Comune di Napoli. 2016.
30. Wassenberg F. Large social housing estates: From stigma to demolition? J Hous Built Environ. 2004;19(3):223–32.
31. Mentzel M. THE BIRTH OF BIJLMERMEER ( 1965 ): THE ORIGIN AND EXPLANATION OF HIGH-RISE DECISION MAKING. Netherlands J Hous Environ Res. 1990;5(4):359–75.
32. Wassenberg F. The integrated renewal of Amsterdam's Bijlmermeer high-rise. Informationen zur Raumentwicklung, H 3/4, 2006, pp 191-201. 2010;191–202.
33. Cerreta M, Mazzarella C, Spiezia M, Tramontano MR. Regenerativescapes: Incremental Evaluation for the Regeneration of Unresolved Territories in East Naples. Sustainability. 2020;12(17):6975.
34. Marini S. Returning to wasting away. 4th Int Conf Int Forum Urban [Internet]. 2009;249–58. Available from: http://newurbanquestion.ifou.org/proceedings/3 The Urbanized Society/full papers/B013_Marini_Sara_RETURNING TO WASTING AWAY.pdf
35. Hirschhorn T. An Electronic Conversation between Thomas Hirschhorn and Jacques Rancière: Presupposition of the Equality of Intelligences and Love of the Infinitude of Thought. Theory, Cult Soc. 2014;31(7–8):101–10.
36. Gemeente Amsterdam. Gebiedsagenda's Zuidoost 2019-2022. 2019.
37. Molinari L, Ingrosso C. The Corviale in Rome and the Vele in Naples: Can a 'monster' become an urban opportunity? In: Tostoes A, Ferreira Z, editors. Proceedings of the 14th International Docomomo Conference - Adaptive Reuse: The Modern Movement Towards the Future. Docomomo; 2016. p. 824–9.
38. Cerullo D. Diario di un buono a nulla. Scampia, dove la parola diventa riscatto. Società Editrice Fiorentina; 2016.

IOP Conf. Series: Earth and Environmental Science **855** (2021) 012023 doi:10.1088/1755-1315/855/1/012023

# Leadership of EU member States in building carbon footprint regulations and their role in promoting circular building design

**S Attia[1], M C Santos[2], M Al-Obaidy[1] and M Baskar[1,3]**
[1]Sustainable Building Design Lab, Dept. UEE, Faculty of Applied Science, University of Liege, 4000, Belgium
[2]Faculty of Architecture, University of Lisbon, Lisbon, 1000, Portugal
[3]EPF Graduate School of Engineering, Troyes, 10000, France

**Abstract**. European countries are working towards carbon neutrality of the building sector. Regulations and initiatives, including the European Green Deal, aim at promoting circular buildings and low carbon design. Therefore, this paper seeks to investigate the role of legislation in paving the way towards achieving the circularity of buildings design and construction. A systematic literature review is conducted to compare the current regulations in different EU member states that address carbon emissions and life cycle thinking to achieve circularity. The study aims to demonstrate how the low-carbon emissions regulations in leading countries can lead to making the construction sector's circularity. The research is focused on five leading EU member states in low carbon buildings, including Denmark, Finland, France, the Netherlands, and Sweden. The study compares the performance indicators, metrics, and target thresholds found in the five selected states' regulations and examines them across a circularity assessment framework developed earlier by the authors. This paper provides insights on low emission building regulations state-of-the-art. Moreover, it offers a better understanding of the relationship between low-carbon emissions regulations and building circularity. The article explains the role of the legislative landscape and its impact on circular building design practices. Key findings from the study will assist the European Commission to identify policy options to support the uptake of "Circular economy principles for buildings design" in European, national and local policies

## 1. Introduction

On Oct 7 2020, the European Parliament voted to update the EU's climate target for 2030 [1]. The vote is backing the decision of a 60% reduction in greenhouse gas emissions by 2030 compared to the emissions of 1990. Collectively, buildings in the EU are responsible for 36% of greenhouse gas emissions, mainly stemming from construction, operational emissions (heating, cooling, ventilation, and power), renovation, and demolition [2]. Therefore, the building sector has a vital role in responding to the climate emergency, and addressing upfront carbon is a critical and urgent focus [3]. Bringing embodied carbon upfront and the fast energy transition towards carbon neutrality is crucial. Simultaneously, the energy and carbon transition need to be following the principles of the circular economy. The European Commission is looking forward to promoting circular economy principles for buildings' design. However, coupling the carbon neutrality target with the building circularity target is challenging, and there are no established national and local policies within Europe.

Therefore, this study aims to assist the European Commission in identifying policies that support the uptake of "Circular Buildings "and "Carbon Neutral Buildings "to increase buildings' service life and ease the use of secondary materials improve resource efficiency throughout the building lifecycle. The main research questions are: To what extent are carbon neutrality and circularity concepts integrated into mandatory building regulations, and who are the EU leading countries in this regard?

The paper provides an overview of leading EU member states that integrated both concepts in their regulations based on a systematic literature review and expert interviews with relevant circularity and carbon neutrality experts. The research is focused on five leading EU member states in low carbon buildings, including Denmark, Finland, France, the Netherlands, and Sweden. Key findings from the study will be used as an input to define the policies to be implemented in other EU member states. The paper is the audience is mainly scientists and actors involved buildings value chain, namely building engineers, architects, contractors, and builders, including renovators, manufacturers of construction products, and government/regulators, including national, regional, and local municipal authorities.

## 2. Literature Review

In this section, we focus on two major concepts that are the focus of this study, namely, carbon neutrality and circularity. Both concepts are chosen because they are the most effective concepts that can reduce greenhouse gas emissions and other environmental impacts, such as end-of-life waste.

Several terms are used to express the carbon emissions associated with building construction that depend on the carbon neutrality period's emissions calculation method. From a life cycle thinking approach [4] and based on EN 15978 [5], there are two definitions associated with building carbon neutrality: namely the embodied carbon emissions stage (A1-5) and the carbon emissions in use stage (B1-7) (see Figure 1). Since the European Energy Performance of Buildings Directive (EPBD) introduction, building regulations requirements evolved until we reached nearly zero and net-zero energy building [6,7]. Several studies tried to associate energy neutrality targets with building materials emissions; however, very few countries addressed carbon emissions in the EPBD before the use stage [8]. The Paris agreement emphasized the need for deep decarbonization of the building stock beyond the energy neutrality concept. Carbon emissions from building materials have a significant climate impact. The Incorporated Carbon Review reports indicate that even by de-carbonizing the energy grid, buildings can continue to be a substantial generator of emissions in the long term due to the carbon incorporated in the materials used in emerging countries [9]. Research shows that energy efficiency has decreased fossil energy use and increased renewables; however, the embodied carbon increased [9,10]. The introduction of the Environmental Product Declaration in 2018 for building materials in most EU member states was the first step that was taken to quantify the carbon emissions of buildings [11]. However, the use of EPDs is voluntary in most countries without any emission thresholds. Figure 2 illustrates the European countries driving the demand for EPDs based on the Once Click Life Cycle Analysis (LCA) database.

| PRODUCT STAGE | | | CONSTRUCTION PROCESS STAGE | | USE STAGE | | | | | | | END OF LIFE STAGE | | | | BENEFITS AND LOADS BEYOND THE SYSTEM BOUNDARIES |
|---|---|---|---|---|---|---|---|---|---|---|---|---|---|---|---|---|
| Raw material supply | Transport | Manufacturing | Transport from the gate to the site | Assembly | Use | Maintenance | Repair | Replacement | Refurbishment | Operational energy use | Operational water use | De-construction demolition | Transport | Waste processing | Disposal | Reuse-Recovery-Recycling-potential |
| A1 | A2 | A3 | A4 | A5 | B1 | B2 | B3 | B4 | B5 | B6 | B7 | C1 | C2 | C3 | C4 | D |
| X | X | X | MND | MND | MND | MND | MNR | MNR | MNR | MND | MND | MND | MND | X | MND | X |

**Figure 1.** Description of the stages during the buildings' life, according to EN 15978.

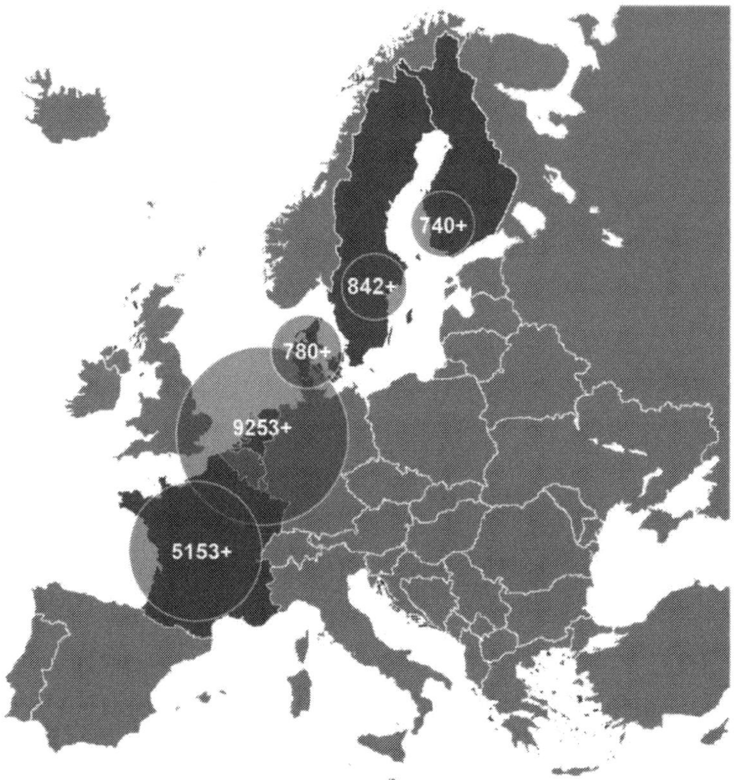


**Figure 2.** Numbers of European products EPDs available in One Click LCA's database.

The second concept investigated in this study is building circularity. A circular building seeks the highest efficiency in managing combined resources and the maximum generation of renewable resources. It seeks positive development to increase the carrying capacity to reverse ecological footprint [12]. Sustaining the material's value is the key to circular material use, and ways to harvest this value are at the center of the circular buildings. Few studies addressed the concept and definition of building circularity [13]. But several authors like Durmisevic and [14] Antonini et al. [15] tried to develop reliable indicators for reversibility and durability features of circular buildings. Cottafav et al. and Attia addressed the gap between embodied impacts and architectural and structural design aspects [12,16]. The difficulty of defining or building a circular building based on circularity indicators and technologies to enhance buildings' service life while closing material loops is still challenging [17].

Despite the evolution of the carbon neutrality and circular design of building, both concepts are not included in most European building standards and public procurements. The legislative requirements and compliance for circular construction remain excluded from most sustainable business models despite the climate urgency and resource scarcity. This literature review confirms the need to explore both concepts in national and local regulations to collect further insights on the leadership of EU member States in building carbon footprint regulations and their role in promoting circular building design.

## 3. Methodology

This study conducted a systematic literature review and expert interviews to build legislation concerning the two following concepts: circularity and carbon neutrality. Among the 27 EU member countries, a screening has been done to reduce the study countries to five: Denmark, Finland, France, the Netherlands, and Sweden. The inclusion criteria focused on countries that require a life cycle assessment and environmental product declarations (EPD) for building materials during building permit issuing.

The inclusion criteria included mandatory requirements or measures to increase the service life of buildings. The exclusion criteria included regulations that define circularity on an urban or territorial scale. The search focuses mainly on carbon neutrality and circularity on building scales. Carbon criteria and circularity were assessed based on the building design criteria for circular buildings developed by Attia et al. [12, 17, and 18]. The selected countries were the only countries fulfilling all criteria.

The study is mostly qualitative, based on the literature review of collected documents and articles and short interviews with experts representing the five countries. The principle of "triangulation" for findings was used. Three sources have been used, primary sources, secondary sources, and experts' knowledge to confirm results. Primary sources are official documents such as scientific articles and regulations; secondary sources are constituted by a literature review in the form of articles and magazines, eventually, and experts' short interviews. The texts concerning regulations have been carefully analyzed and summarized in tables and described in the results section. Thus, through this approach, the five countries' relevant regulations and publications are selectively reviewed.

## 4. Results

The results of the literature review and expert interviews are presented in the section. Firstly, an overall comparative analysis is presented. Secondly, a detailed analysis and facts check is provided individually for each country. Table 1 lists the most important publications that we selected during our review. Each publication is shortly described concerning building carbon footprint and circularity. Moreover, Table 2 compares the five investigated countries against a carbon neutrality and circularity checklist developed by Attia [17].

**Table 1.** Key publications in the five countries related to carbon neutrality and circularity criteria.

| Country | Reference | Description |
|---|---|---|
| Denmark | (19,20) | - Denmark outperforms EU28 on most selected resource and innovation metrics, such as share of renewable energy or Eco-innovation index. The improved utilization of assets and better use of waste or by-products as a resource has a potential value as a resource; however, recycled materials should be used in higher-value cycles, such as reuse or remanufacturing [19].<br>- Denmark has ambitious climate neutrality targets. Energy taxes stand in getting the electrification that can ensure a socio-economically beneficial integration of renewable energy in the total energy system. Even the reduced tax on electricity for space heating is almost twice as high as the energy tax on heating oil and natural gas. That inhibits the electrification of buildings heating and district heating, where heat pumps are both energy efficient and socio-economical efficient technology. [20] |
| Finland | (21,22) | - The goal is for Finland to build its competitiveness using sustainable material use. The demand for virgin materials will be minimized, the length of material and product life cycles maximized, and products designed so that they can be maintained and reused at the end of their first useful life [21].<br>- Finland developed a whole life carbon assessment of buildings method. The method is based on the European Commission's Level(s) method and European Standards [23]. A low-carbon building has a low carbon footprint and a giant carbon handprint. The method is intended |

| | | |
|---|---|---|
| | | to assess the carbon footprint and carbon handprint of new buildings and buildings undergoing extensive repairs [22]. |
| France | (24) | - The new French Building Regulation RE2020 introduces a new threshold of 4 $kgCO_2$ operational emissions equivalent/$m^2$ for new residential buildings and set at a level sufficiently ambitious to favor the low carbon-intensive energies [24]. This will eliminate the use of gas for heating. A conversion coefficient between primary energy and final energy of electricity of 2.3 will be used. It corresponds to the expected average value of this coefficient over the next 50 years, thus making it possible to consider the forecast evolution of the French electricity mix -over the life of new buildings.<br>-The new RE2020 is introducing a dynamic LCA calculation method that is ambitious, proposing a threshold for embodied carbon emission of 100 kg of CO2 per $m^2$, which favors biobased materials and timber [24]. |
| Netherlands | (25,26) | -In the Netherlands, the government and the business community work are currently using a method to measure the building's environmental performance via a life cycle analysis. LCA is used as a criterion for construction procurement. The Dutch recognize the limitation of their method regarding the treatment of demolished materials and their reuse [25,26]. |
| Sweden | (27,28) | -The Nordic construction section is witnessing acceleration towards circular construction. Experts suggest anchoring the requirements in the building regulation in Denmark, Finland, Norway, and Sweden in Environmental Product Declarations (EPD), Construction Products Regulation (CPR), Building Information Modelling System (BIM), and Material and building passports [27,28].<br>- Sweden has set the goals for achieving a climate-neutral value chain in the construction and civil engineering sector reaching by 2045: Net-zero emissions of greenhouse gases by 2040: 75% reduced greenhouse gas emissions by 2030: 50% reduction in greenhouse gas emissions [28]. |

### 4.1. Denmark

Denmark has adopted the German certification system, DGNB (*Deutsche Gesellschaft für Nachhaltiges Bauen*). A large part of the certification system refers to the Life Cycle Assessment (LCA). However, the use of LCA is voluntary. The promotion of sustainable solutions and green products is carried out through the Partnership for Public Green Procurement, a community of municipalities, regions, and public associations. The use of the Environmental Product Declaration (EPD) is voluntary in Denmark, except in Copenhagen, it is mandatory. However, there are strong incentives to use EPDs [19].

The Danes aimed that by 2030, 50% of all energy used in buildings should come from renewable sources, and it wants to reduce buildings total carbon emissions by 9.4 million tons by 2030 [20]. However, there is uncertainty to attain this target. In terms of energy consumption, the legislation provides that new buildings must be built as low-energy houses that consume up to 90% less energy. These requirements are implemented for the 2020 building regulations.

The current state of transition to a circular economy in the Danish construction sector is elaborated on to improve material recycling quality in the construction sector. Preferably, greater direct reuse and removal of hazardous substances in construction materials. The Danes developed their material passport tool, called "*Materialepas,*" to improve circular product policies [29]. Denmark is increasing existing building assets by encouraging multifunctional buildings such as schools and building reuse through building components' modular design. The construction industry had shown concerns in the disassembly process to enable the separations of materials after construction [28]. The main innovation area is industrial production processes to reduce waste during construction and renovation, including modular construction and 3D printing construction modules [19].

In terms of construction waste management strategies, the government plans to separate ten waste fractions: 1. natural stone, for example, granite and flint, 2. non-enameled tiles (bricks and tiles), 3. concrete, 4. mixtures of stone materials and unglazed tiles and concrete, 5. iron and metal, 6. plaster, 7. rock wool, 8. soil, 9. asphalt, 10. and mixtures of concrete and asphalt. Currently, 87% of building materials are recycled with low-quality applications, and less than 1% are made into components and materials for new construction [30]. However, large construction companies are going beyond their initiative's requirements - small builders can deliver different construction waste fractions to local recycling centers [31].

**Table 2.** Mapping of carbon neutrality and circularity criteria regulation landscapes in the five countries.

| | | Denmark | Finland | France | Netherlands | Sweden |
|---|---|---|---|---|---|---|
| Carbon Footprint Reduction Criteria | Carbon emission threshold | ✓ | ✓ | ✗ | ✗ | ✓ |
| | Green Public Procurement (GPP) | ✓ | ✓ | ✓ | ✓ | ✓ |
| | Embodied carbon in products | ✓ | ✓ | ✓ | ✓ | ✓ |
| | LCA | ✓ | ✓ | ✓ | ✓ | ✓ |
| | EPD & % | ✗ | ✓ | ✓ | ✓ | ✓ |
| | Energy efficiency improvements | ✓ | ✓ | ✓ | ✓ | ✓ |
| Circularity Criteria | Building Materials Passport | ✓ | ✗ | ✗ | ✓ | ✓ |
| | Dismantling Protocol | ✓ | ✗ | ✗ | ✗ | ✗ |
| | Carbon + water footprint | ✗ | ✗ | ✗ | ✗ | ✗ |
| | Competitiveness and innovation industrial | ✓ | ✓ | ✗ | ✗ | ✓ |
| | Waste management | ✓ | ✓ | ✓ | ✓ | ✓ |
| | Adapt Spaces | ✗ | ✗ | ✗ | ✗ | ✗ |
| | Reconfigure \ upgrade structure | ✓ | ✓ | ✗ | ✗ | ✗ |
| | Separate elements material | ✓ | ✗ | ✗ | ✗ | ✗ |

### 4.2. Finland

In Finland, the goal for a building's lifespan neighbors 80 years [7]. The plan will be to prolong this lifespan. This is where the idea of circularity comes into account. Construction parts must have net benefits and affect lowering carbon footprint. The first edition of the system for the lifetime carbon assessment of buildings was based on the Level(s) framework developed by the European Commission [23]. A project named "Circular Economy in the Built Environment" was launched by the Finnish Innovation Fund Sitra in early 2018 [21].

Material economics predicts that only cement, steel, aluminum, and plastic used for construction in Finland would result in emissions of 230 Mt $CO_2$ by 2050 if produced using today's manufacturing processes. The Ministry of the Environment released a plan for low-carbon construction in 2017: by the mid-2020s, a life-carbon assessment of buildings must be included in the building regulations. As Finland intends to achieve carbon neutrality by 2035, estimating buildings' carbon footprint and building-type specific emission limits are introduced in the building regulations of 2020. Furthermore, Finland also plans to reach an 80 percent reduction in greenhouse gas emissions by 2050 under the Climate Change Act, compared to the 1990 baseline figure. The Land Use and Construction Act, a recent reform, embraces low-carbon aspects and a life cycle thinking approach for a low-carbon construction sector [22,32].

### 4.3. France

The recently updated Energy-Climate Law promises to fight against climate change to reach a carbon-neutral France by 2050 [33]. The future French Environmental Legislation goals within this framework are to reduce the effect of new buildings on the environment by considering all of the building's pollution, right from construction, over its life cycle. A focus and special attention to bioclimatic design is (known as 'Bbio' indicator) is highlighted [34]. In the new building regulation law (RE 2020), new constructions carbon emissions are reduced substantially. Through the promotion of Energy Positive and Low Carbon Buildings, the regulation introduces a new threshold for CO2 emissions [24]. The CO2 emission factor is expressed in kg eq.CO2 /m$^2$ [24]. Unfortunately, according to the best available data, French regulations do not address circularity.

Moreover, according to the French Ministry of Energy Transition LCA methodology used to calculate the impact on global warming, buildings mandated by RE2020 will be based on a dynamic LCA methodology. The static LCA method proposed by the E + C- standard is going to be replaced. The calculation of global warming impacts via the dynamic LCA method will widen the gap already present between timber constructions and reinforced concrete construction. The new dynamic calculation will take into account benefits beyond the life cycle of building materials. It will be more adjusted to the reality of the materials integrated in projects, favoring bio-sourced construction solutions.

### 4.4. Netherlands

The Dutch Building Environmental Performance Declaration (DBED) is mandatory for every environmental permit [35]. The declaration indicates the environmental impact of the materials used in a building. This concerns new office buildings (larger than 100 m$^2$) and new-built homes. As of Jan 1, 2018, a maximum limit value of 1.0 applies to the Building Environmental Performance Declaration. The DBED is an essential measure of the sustainability of a building. The lower the declaration value the more sustainable the use of materials. The DBED is a tool that should be used during the design process, and it can be used in a Design Brief to record the result of a design process [26]. For example, recycled floor covering gives a sustainable value and is an essential means of communication. However, the DBED calculation shows that the floor slab's sustainability has a much more significant effect on the environmental impact than the floor finishing.

To determine the material's environmental impact, an LCA is performed. A qualified expert must perform the LCA. The LCA results are reported for 11 environmental indicators of a product. These 11 indicators are combined into one value: the environmental cost per unit of the product (kg, m3, m2, etc.). It is not necessary to perform an LCA of the same product/material over and over again. For the Netherlands, the characteristics of materials from the LCAs are collected in the National Environmental Database (NMD). The Building Quality Foundation manages this database (SBK). A producer or supplier must ensure that a product is included in the NMD.

Regarding circularity, no enforced building regulations exist. However, many guidelines and initiatives taken by municipalities and companies encourage green and circular building materials procurement. The Design for Modularity (DfM) or Design for Disassembly (DfD) concepts are implemented in several pilot projects. The renovation and adaptation of existing buildings, such as churches, and their transformation into dwellings, are examples of layouts circular design strategies in the construction, including adaptively, modularity, and design for reuse [25]. In 2019, 12.5 thousand homes were created by transforming existing buildings, such as offices, schools, and shops [36]. Netherlands is seeking to have a circular economy by 2050 and reduce the materials used by 50% [37].

### 4.5. Sweden

Under the Paris Agreement requirement, Sweden aims to eliminate its greenhouse gas emissions until 2045 [28]. The climate program legislation is made according to Ordinance (2015: 517) and is supported by local climate investors (Klimatklivsförordningen) [38]. Starting in 2022, all new buildings must reduce carbon emissions to meet a specific threshold. The approach is based on the LCA of buildings.

To ensure that the national energy target is 50% more efficient by 2030, Sweden has created several instruments that provide incentives to promote carbon efficiency and circularity. For waste management of the construction industry, Sweden aims to recycle 70% of materials [39]. The Swedish, Environmental Protection Agency requires recycling large amounts of construction residues guided through a manual to facilitate recycling in an environmentally safe and health-friendly manner. The Swedish Public Procurement Act states that contracting authorities should follow green procurement for public-funded projects [40].

## 5. Discussion

This research presented a literature review and expert interview results of five countries regarding building regulations associated with carbon neutrality and circularity. The five leading European countries are slowly increasing wood and biobased materials in construction to sequester carbon emissions. The use of EPDs and LCA with performance thresholds for carbon emissions is transforming the construction industry. However, the circularity concept is still not gaining momentum in regulations. As shown in Figure 3, our study indicates that Denmark is the most leading EU state to incorporate the circular economy's principles in the construction sector. The regulations landscape in Denmark is almost mature, with policies and tools that aim to waste reuse and materials separation. In second place comes Finland, with a multi-stakeholder approach to adopt "Circular Economy in the Built Environment" in the construction sector and achieve carbon neutrality by 2035. Sweden comes in third place by implementing a climate legislation that introduces and favors the execution of measures and supports the transition toward carbon neutrality in 2045, entering into effective January 2022. France is in fourth place in the qualification, leading to carbon footprint reduction measures and lagging regarding building circularity. The Netherlands comes forth with the mandatory LCA requirements for all new constructions.

IOP Conf. Series: Earth and Environmental Science **855** (2021) 012023 doi:10.1088/1755-1315/855/1/012023

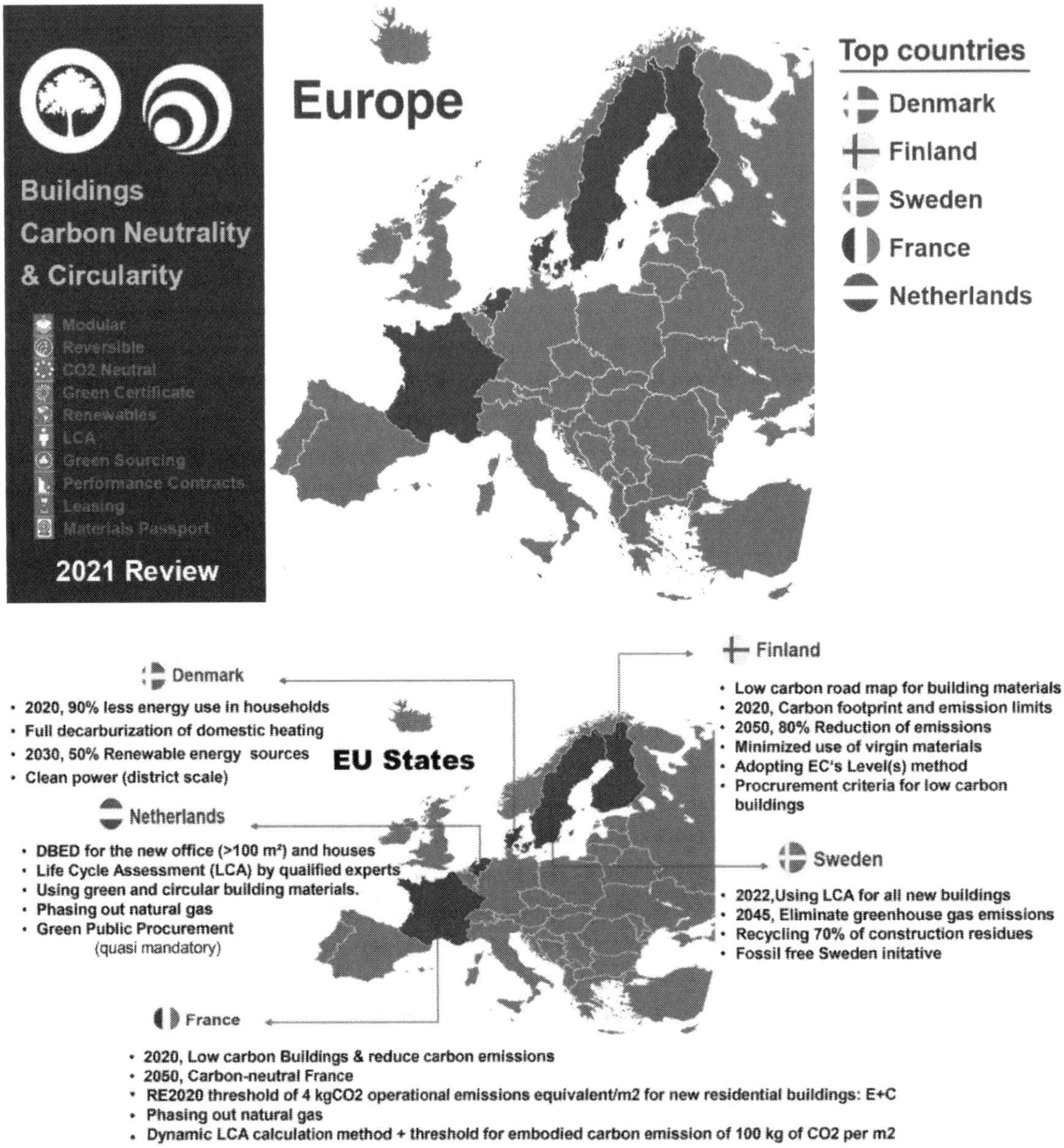


**Figure 3.** Ranking of EU leading member states that couple carbon neutrality and circularity requirements for new constructions.

Based on our review and analysis, we provide three main recommendations. Firstly, EU member states must adopt measures to limit carbon emissions and the carbon incorporated in products based on European standard EN 15978 and Level(s) framework [5, 25, 41]. The use of wooden buildings as carbon sinks or carbon sequestration is gaining momentum in the five investigated countries based on the evolving regulatory landscape [42]. Carbon emissions from heating are already very low in Sweden, Denmark, the Netherlands, and Finland due to district heating and heat pumps. France has recently adopted a new regulation that prohibits using gas as a source of heating in new households and is encouraging reversible heat pumps across the country.

Secondly, circularity criteria for buildings should not be treated in isolation from carbon neutrality. The circularity criteria must be coupled to carbon footprint reduction measures. As a consequence, timber construction materials and biosourced insulation materials will be in great demand. The current supply chain of timber is already under pressure, and the availability of wood sourced from sustainably managed forests will not be enough to meet the European demand. Moreover, the number of trained workers who can deliver timber construction projects is not enough. The training of professionals, including architects, masons, electricians, and plumbers, will require significant investments to prepare them to design and build timber constructions.

Thirdly, the EU member states must increase the share of recycled and reused construction waste and the possibility of separating the materials for future reuse. Timber and biosources materials will not cover the demand/supply during the coming 20 years. Also, none of the investigated countries addressed this matter thoroughly. Therefore, increasing the share of recycled and reused construction materials can help during the transition towards timber constructions. EPDs can facilitate the decision-making of architects and design teams while keeping the embodied carbon low.

Finally, this research's strengths are centered on reviewing carbon neutrality and the circularity of buildings together. The exhaustive analysis of the state of the art, regulation, and legislation resulted in selecting and ranking five leading member states incorporating the circular economy in buildings. At the same time, the research provides a snapshot and status quo analysis of a very dynamic field. The European housing market is under pressure to produce more and more housing units. There is already a housing supply gap in Europe. Most leading EU states are stopping the urbanization of land and setting limits for urban sprawl. The circularity and carbon neutrality paradigms will intensify this challenge of building sustainably.

We imagine it will be outdated very soon once this paper is published. Besides, the analysis is mainly quantitative and is not based on case studies or quantitative research approaches. However, this article aims to help researchers, the EU commission, and member states take the lead in legislation and regulation to embed circularity and carbon neutrality in buildings. The paper brings perspectives regarding the regulatory landscape of the construction sector in Europe. Future research should update our review and focus on case studies and a detailed dynamic LCA approach to assess carbon neutrality and circularity.

## 6. Conclusions

The three Nordic countries (Denmark, Finland, and Sweden) are leading the way in implementing policies and regulations that reduce the carbon footprint and incorporate the circular economy principles in the building construction sector. The five investigated countries implementing ambitious measures to reduce the environmental impact of buildings operation and materials use. However, our study found a separation between carbon neutrality targets and circularity targets. Integrating both targets in future building regulations is fundamental to reducing buildings' overall carbon and environmental footprint. The quantification of carbon neutrality and circularity targets through measurable indicators and critical performance thresholds is the key to accelerate the market uptake and inform policies and decision making.

## 7. Acknowledgments

The authors would like to express their thanks to Greg Vendena, Architect, and Danish Collaborative at the International Living Future Institute, for sharing his insights on Denmark's policies and best practices. Also, we acknowledge the Sustainable Building Design (SBD) Laboratory at the University of Liege for the operationalization of study variables and systematic reviews. The lab provided valuable support during the data collection and data analysis.

## 8. References

[1] eceee. EU Parliament votes in favour of cutting emissions 60% by 2030 [Internet]. European

Council for an Energy Efficient Economy. 2020 [cited 2020 Oct 23]. Available from: https://www.eceee.org/all-news/news/news-2020/eu-parliament-votes-in-favour-of-cutting-emissions-60-by-2030/

[2] EC. In focus: Energy efficiency in buildings [Internet]. European Commission; 2020 [cited 2020 Oct 23]. Available from: https://ec.europa.eu/info/news/focus-energy-efficiency-buildings-2020-feb-17_en#:~:text=Collectively%2C%20buildings%20in%20the%20EU,%2C%20usage%2C%20renovation%20and%20demolition.

[3] WGBC WGB. Bringing embodied carbon upfront: Coordinated action for the building and construction sector to tackle embodied carbon [Internet]. London: World Green Building Council; 2019 [cited 2020 Aug 25] p. 1–35. Available from: https://www.worldgbc.org/sites/default/files/WorldGBC_Bringing_Embodied_Carbon_Upfront.pdf

[4] Pombo O, Rivela B, Neila J. Life cycle thinking toward sustainable development policy-making: The case of energy retrofits. J Clean Prod. 2019 Jan; 206:267–81.

[5] EN 15978. Sustainability of construction works–Assessment of environmental performance of buildings–Calculation method. Bruss Belg Eur Comm Stand. 2011;

[6] Attia S. Net zero energy buildings (nzeb). 1st edition. Cambridge, MA: Elsevier; 2018.

[7] Belussi L, Barozzi B, Bellazzi A, Danza L, Devitofrancesco A, Fanciulli C, et al. A review of performance of zero energy buildings and energy efficiency solutions. J Build Eng. 2019 Sep;25:100772.

[8] Tallini A, Cedola L. A review of the properties of recycled and waste materials for energy refurbishment of existing buildings towards the requirements of NZEB. Energy Procedia. 2018 Aug;148:868–75.

[9] Pasanen P. The embodied carbon review Embodied carbon reduction in 100+ regulations & rating systems globally. 2018; Available from: https://www.oneclicklca.com/embodied-carbon-review/

[10] Tagliapietra S, Zachmann G, Edenhofer O, Glachant J-M, Linares P, Loeschel A. The European union energy transition: Key priorities for the next five years. Energy Policy. 2019 Sep;132:950–4.

[11] Passer A, Lasvaux S, Allacker K, De Lathauwer D, Spirinckx C, Wittstock B, et al. Environmental product declarations entering the building sector: critical reflections based on 5 to 10 years experience in different European countries. Int J Life Cycle Assess. 2015;20(9):1199–1212.

[12] Attia, S. (2018). *Regenerative and positive impact architecture: Learning from case studies*. Springer International Publishing, New York, USA. ISBN 978-2930909004, https://doi.org/10.1007/978-3-319-66718-8.

[13] Cambier C, Galle W, De Temmerman N. Research and Development Directions for Design Support Tools for Circular Building. Buildings. 2020;10(8):142.

[14] Durmisevic, E. Reversible Building Design: Reversible Building design guidelines [Internet]. University of Twente; 2018 [cited 2020 Oct 26]. Available from: https://www.bamb2020.eu/wp-content/uploads/2018/12/Reversible-Building-Design-guidelines-and-protocol.pdf

[15] Antonini E, Boeri A, Lauria M, Giglio F. Reversibility and durability as potential indicators for

Circular building Technologies. Sustainability. 2020;12(18):7659.
[16] Cottafava D, Ritzen M. Circularity indicator for residentials buildings: Addressing the gap between embodied impacts and design aspects. Resour Conserv Recycl. 2020;164:105120.
[17] Attia S, Al-Obaidy M. Design criteria for circular buildings. In: Crossing Boundaries Parkstad. Parkstadt, The Netherlands: Zuyd Uninversity; 2021.
[18] Attia S. Towards regenerative and positive impact architecture: A comparison of two net zero energy buildings. Sustain Cities Soc. 2016; 26:393–406.
[19] Sheppard R. Potential for Denmark as a Circular Economy: A Case Study From: Delivering the Circular Economy - a Toolkit for Policy Makers [Internet]. Ellen Macarthur Foundation; 2015 [cited 2020 Jan 11] p. 1–134. Available from: https://books.google.be/books?id=HgwBkAEACAAJ
[20] Blocks, B., & Society, L. (2017). Transition Towards 2030 NB ! Transition Towards 2030. June
[21] Karhu J, Linkola L. Circular Economy in the Built Environment in Finland - A case example of collaboration. IOP Conf Ser Earth Environ Sci. 2019 Sep 2;297:012–24.
[22] Kuittinen M. Method for the whole life carbon assessment of buildings. :60.
[23] Sánchez Cordero A, Gómez Melgar S, Andújar Márquez JM. Green building rating systems and the new framework level (s): A critical review of sustainability certification within Europe. Energies. 2020;13(1):66.
[24] RE2020 : Une nouvelle étape vers une future règlementation environnementale des bâtiments neufs plus ambitieuse contre le changement climatique. 2020.
[25] De Valk E, Quik JT. Eenduidig bepalen van circulariteit in de bouwsector. 2017 [cited 2020 Oct 21]; Available from: http://rivm.openrepository.com/rivm/handle/10029/620930
[26] Bepalingsmethode Milieuprestatie Gebouwen en GWW-werken. 2019; Available from: https://milieudatabase.nl/wp-content/uploads/2019/05/SBK-Bepalingsmethode-versie-3.0-1-januari-2019.pdf
[27] Høibye L, Sand H. Circular Economy in the Nordic Construction Sector [Internet]. Nordic Council of Ministers; 2018 [cited 2020 Oct 21]. (TemaNord). Available from: http://urn.kb.se/resolve?urn=urn:nbn:se:norden:org:diva-5166
[28] Wannerström A. Färdplan för fossilfri konkurrenskraft Bygg- och anläggningssektorn [Internet]. 2018. Available from: https://vpp.sbuf.se/Public/Documents/ProjectDocuments/80d42431-39fa-4c59-9e35 c6cbab45037b/FinalReport/SBUF%2013474%20Färdplan_Bygg_Anläggningssektorn.pdf
[29] Lewis M. Building Materials - Impacts and Transparency. 2020;31.
[30] Ejs L. Byg Cirkulært, Miniguide. 2019; Available from: https://www.circularitycity.dk/wp content/uploads/2019/11/Håndværker-Toolkit.pdf
[31] Klimarådet. Kendte veje og nye spor til 70 procents reduktion - retning og tiltag for de næste ti års klimaindsats i Danmark. Klimarådet;
[32] Bruce-Hyrkäs T, Pasanen P, Castro R. Overview of Whole Building Life-Cycle Assessment for Green Building Certification and Ecodesign through Industry Surveys and Interviews. Procedia CIRP. 2018;69:178–83.
[33] Gaëta R, Guldner L, Piton F, Priem L, Thiébaut A. Vers une réglementation environnementale pour les bâtiments neufs. Ann Mines - Responsab Environ. 2018;N° 90(2):55.
[34] Attia S, Eleftheriou P, Xeni F, Morlot R, Ménézo C, Kostopoulos V, et al. Overview and future challenges of nearly zero energy buildings (nZEB) design in Southern Europe. Energy Build.

2017;155:439–458.

[35] MilieuPrestatie Gebouwen - MPG [Internet]. 2020. Available from: https://www.rvo.nl/onderwerpen/duurzaam-ondernemen/gebouwen/wetten-en regels/nieuwbouw/milieuprestatie-gebouwen

[36] CBS. 12,5 duizend woningen door transformatie van gebouwen in 2019. 2020 Oct 29;1.

[37] van Oppen C, Bosch S, Circulair inkopen in 8 stappen, Copper8, Amsterdam, The Netherlands, 2020.

[38] Boverket. Miljöindikatorer. 2019;37.

[39] Riksdagsförvaltningen. "Förordning (2015:517) Om Stöd till Lokala Klimatinvesteringar Svensk Författningssamling 2015:2015:517 T.o.m. SFS 2019:526." Riksdagen [Internet]. 2014. Available from: https://www.riksdagen.se/sv/dokument-lagar/dokument/svensk forfattningssamling/forordning-2015517-om-stod-till-lokala_sfs-2015-517

[40] Høibye L, Sand H. Circular Economy in the Nordic Construction Sector [Internet]. Nordic Council of Ministers; 2018 [cited 2020 Oct 21]. (TemaNord). Available from: http://urn.kb.se/resolve?urn=urn:nbn:se:norden:org:diva-5166

[41] Kuittinen, M., & Häkkinen, T. (2020). Reduced carbon footprints of buildings: new Finnish standards and assessments. Buildings and Cities, 1(1).

[42] Amiri, A., Ottelin, J., Sorvari, J., & Junnila, S. (2020). Cities as carbon sinks—classification of wooden buildings. Environmental Research Letters, 15(9), 094076.

# Assessment of the circularity and carbon neutrality of an office building: The case of 't Centrum in Westerlo, Belgium

**M Al-Obaidy[1], M C Santos[2], M Baskar[1,3] and S Attia[1]**
[1]Sustainable Building Design Lab, Dept. UEE, Faculty of Applied Science, University of Liege, 4000, Belgium
[2]Faculty of Architecture, University of Lisbon, Lisbon, 1000, Portugal
[3]EPF Graduate School of Engineering, Troyes, 10000, France

**Abstract.** Circular building design is a concept that is gaining great interest from architects, construction professionals, and their clients but is still rarely adopted in practice. One of the earliest design decisions architects and developers should make to design a circular building is to determine the building's construction system. The choice of constructive and structural systems, such as columns, beams, and slabs, is crucial to upgrade the reuse cycles in the future. Flexible construction systems can make it easier to dismantle the structures and recover, upgrade, modify, or transform building materials. Therefore, this paper assesses the carbon emission impacts of two construction systems for an office building in Belgium using life cycle assessment (LCA) and circularity criteria. One-Click LCA software was used for the calculations. Parametric analysis took place for two construction systems scenarios involving a steel structure and a timber structure. Life Cycle Assessment and comparisons of the various construction systems are made based on ISO 14040, 14044, and CEN/TC 350 standards with a focus on carbon neutrality. The results show that using local plant-based materials such as wood can drastically reduce office buildings' carbon footprint. Based on the sensitivity analysis results, the overall global warming potential impact is mostly sensitive to the construction material's weight and reuse and dismantling ability. This paper provides a better understanding of building structural systems; to inform architects about the circularity potential of different construction systems.

## 1. Introduction

The circularity paradigm implicates a new era of building design that has been of great interest to architects, construction professionals, and their clients [1]. However, the paradigm is still not common, although it is present in practice and research [2,3]. Therefore, the paper aims to learn from a real case study of a circular office building under development. The case of *'t Centrum* in Westerlo, Belgium, is considered a pioneer project in circularity. The building is designed to last for 20 years and is designed, constructed, and operated within an experimental setting that allows the building's dismantling and reconstruction every five years. In this context, this study will answer three main questions:

- How can we evaluate the circularity of the building?
- What is the most sustainable structural system regarding carbon neutrality and circularity?
- To what extent is *'t Centrum* project circular and carbon neutral?

The study's scope is limited to the structural system and materials choice of *'t Centrum* building as a case study. Then, a parametric analysis will be performed for alternative construction materials (steel vs. timber). The research findings are used to inform architects about the circularity potential of different construction systems and their environmental impact.

## 2. Literature Review

The emergence of the building circularity concept is strongly related to environmental assessment methods applied in the construction industry. The proliferation of environmental product declarations (EPDs) and the introduction of the so-called "Material Passports" change the way designers materialize their buildings. The emergence of life cycle assessment tools and software allows for full building inventories and total building environmental evaluations. For instance, Heisel et al. [2] described the Madaster platform, storing the materials' details and assessing the building's circularity [4].

The climate emergency and energy transition goals are forcing the construction sector to embrace low-impact buildings. Several studies tried to associate energy neutrality targets with building materials emissions in embodied energy and embodied carbon. However, very few countries addressed embodied carbon emissions in their Energy Performance of Buildings Directive (EPBD) regulations [5]. The introduction of the EPDs in 2018 for building materials in most EU member states was one of the first large-scale steps to quantify buildings' carbon emissions [6]. But, the use of EPDs is still voluntary in most countries without any emission thresholds.

In parallel, the building circularity concept is gaining more and more attention, despite the few studies that address the concept and definition [7]. Several authors like Durmisevic and [8] Antonini et al. [9] tried to develop reliable indicators for circular buildings' reversibility and durability features. Cottafav et al. and Attia addressed the gap between embodied impacts and architectural and structural design aspects [1,3]. The difficulty of defining or building a circular building based on circularity indicators and technologies to enhance buildings' service life while closing material loops is still challenging [10].

The short review above confirms a need to investigate both concepts (carbon neutrality and circularity) and the need to couple them. Our case study approach aims to achieve that.

## 3. Methodology

The case study was chosen because it represents a circular office building that will be dismantled every five years over twenty years. The project's nature and brief make it one of the unique endeavors to learn about regenerative building's circularity. The case study's evaluation is focused mainly on the building structure during the construction, operation, and demolition phase. A parametric analysis took place for two different construction systems, namely steel and timber constructions. The analysis process was performed in two steps:

- First, perform a detailed LCA for the environmental impact of the two structural materials focusing on their carbon emissions.
- Second, perform parametric analysis for the project construction system and also for other different construction systems.

A site visit was conducted to learn about the project details from the design team and get acquainted with the project delivery process and design brief.

### *3.1. Life cycle standards and system boundary*

The building industry accounts for almost 40% of the total carbon emissions directly responsible for climate change and global warming [11]. This study included a life cycle assessment analysis for the materials and resources. A study pointed out that embodied energy can be up to 60% of the building life cycle [12]. A life cycle assessment to compare the environmental impact and $CO_2$ emissions of the structure has been done according to ISO 14040 and 14044 standards [13,14,15]. According to EN 15978 life cycle stages of the building divided into five stages: product stage (raw materials supply, transport, and manufacturing), construction stage (transport and construction installation on-site process), use stage (maintenance, repair, and replacement, refurbishment, operational energy use: heating cooling, ventilation, hot water and lighting and operational water use), end-life stage (deconstruction, transport, and disposal), and reuse stage (recovery and recycling potential) [16] (see Figure 1).

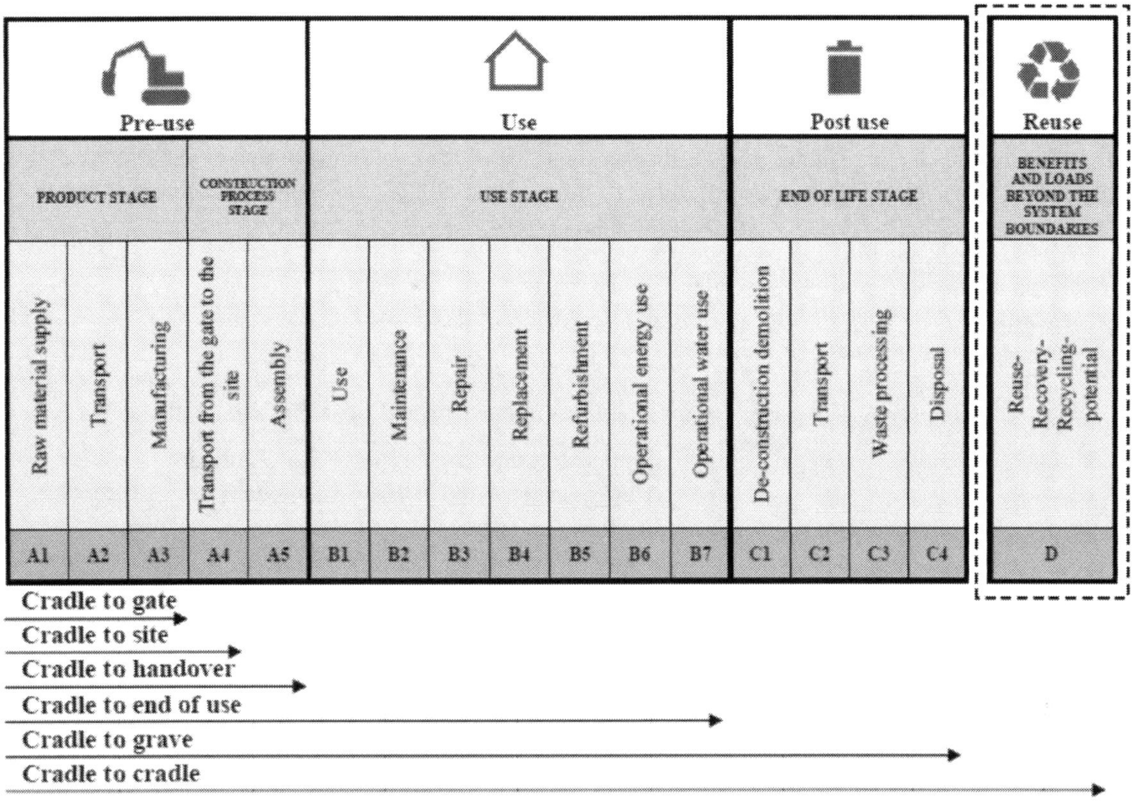


**Figure 1.** Description of the stages during the buildings' life, according to EN 15978.

### *3.2. One-Click LCA software*

One-Click LCA [17] is a life cycle assessment software that allows calculating life cycle assessment, life cycle costing, carbon footprint, and other environmental impacts. It is compliant with more than 40 Green Building certification schemes around the world. LCA considers several environmental impacts, including carbon footprint over the whole life cycle, which makes achieving real sustainability possible. One-Click LCA allows easy-to-use, entirely browser-based sustainability assessment for building projects and does not require LCA expertise from its users [18].

In terms of its built-in database, One Click LCA integrates data from nearly all of the available environmental product declaration EPD worldwide, including Belgium, of course. Customizations of the energy mix of the building are possible based on the project's geographical location.

### *3.3. Functional unit, year, tools, and indicators*

The functional unit to compare both buildings was 1kg/year. For the calculation model, the occupancy for 5 years to each phase of the project. The cradle to grave LCA was made based on collected data from the available information on the official websites of Kamp C and other consortium members of *'t Centrum* project. To calculate the environmental impact resulting from the biogenic $CO_2$ circulation, an approach of $CO_2$ storage in the buildings for five years and 20 years. According to Global Warming Potential (GWP), the LCA indicators were summarized as an indicator of greenhouse emissions, including biogenic carbon dioxide. Biogenic $CO_2$ is captured in biomass during a plant or tree's growth and, consequently, in a biologically-based product [1].

### *3.4. Life cycle inventory*

An inventory has been created, which referred to building materials of the five life cycle stages mentioned earlier, and according to EN 15978 [16]. The drawings were used to show the main difference between the structural system for the case study and the other structural system prepared, especially concerning materials, composition, and the various carbon emissions resulting from them.

By inspecting the project's structural system materials as shown in lists data concerning the weight of significant building materials (see Table 1 & Figure 2), we can see that timber is dominating the total building weight reaching almost 73% of the weight share. All the columns, beams, and the slabs of the ground, floors, and roof were made in timber. Cellulose insulation is the second most common material reaching almost 14%. It was used between the floor layers. In contrast, the concrete used in the building's foundation block reaching nearly 8% of the weight share.

We contacted some companies specialized in the manufacture of materials. We were provided with the weights of the different elements according to the design concept used in constructing the project and the steel structure proposal [19,20,21,22,23].

**Table 1.** Weight share of material groups in the analyzed two types of structure.

| Building Material Category | Timber structure | | Steel structure | |
|---|---|---|---|---|
| | Amount [kg] | Share [%] | Amount [kg] | Share [%] |
| **Pine timber (structure)** | 185005 | 73.5 | × | × |
| **Steel (structure / foundation bars)** | 1750 | 0.7 | 271576 | 27.5 |
| **Concrete (foundation / floors)** | 22000 | 8.74 | 682300 | 69 |
| **Cellulose / Rockwool (insulation)** | 36600 | 14.5 | 27450 | 2.77 |
| **Polyurethane & rubber (waterproof)** | 525 | 0.21 | 525 | 0.053 |
| **Recycled polyester(protective layer)** | 74 | 0.03 | 74 | 0.007 |
| **High-density polyethylene (drainage HDPE)** | 147 | 0.06 | 147 | 0.015 |
| **Polyester fibers (geotextile)** | 73 | 0.03 | 73 | 0.007 |
| **Polyethylene(filter layer)** | 25 | 0.01 | 25 | 0.002 |
| **Soil** | 5490 | 2.18 | 5490 | 0.56 |

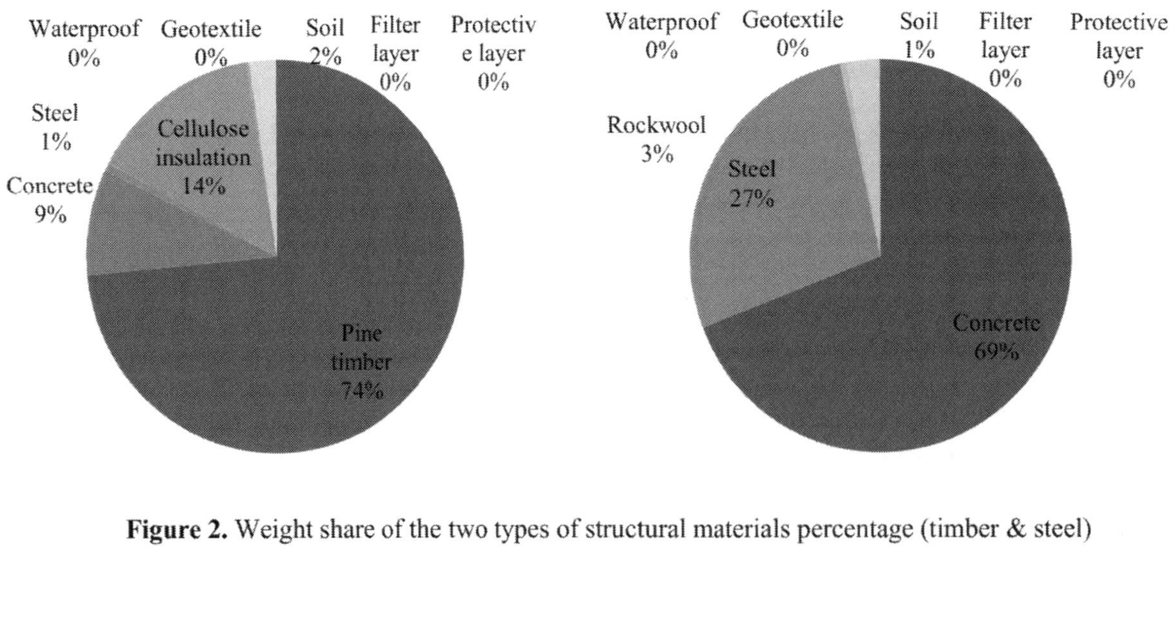


**Figure 2.** Weight share of the two types of structural materials percentage (timber & steel)

### *3.5. Operation carbon emissions*

According to Flanders requirements in 2020, the maximum energy-efficient (E-peil) for office buildings must be 55, and in 2021 the maximum E-peil will be 50 [24]. So this project should comply with these zero energy use requirements [25].

Use stage B1- B7: This stage should capture the carbon emissions associated with the activities and operation of the project from the end of construction stage to the end-of-life stage (see Figure 1) includes the following:

1. In-use emissions B1: This covers the carbon emissions from products and materials (e.g., refrigerants, paints, carpets) in the building's regular operation during the use stage.
2. Maintenance emissions B2: The carbon emissions of all maintenance activities, including cleaning and the carbon impacts from energy and water use associated to these activities.
3. Repair and replacement emissions B3–B4: This stage involves any emissions arising from the repair and replacement of relevant building components in line with realistic scenarios developed from the LCA. These should capture all emissions associated with the supply of new products.
4. Refurbishment emissions B5: The detailed LCA should incorporate any known refurbishment scenarios going forward in the building. This module would cover a planned future extension or change the spatial function of the building.
5. Operational energy use B6: All operational emissions from building-related systems should be included as assessed at the design stage. This should cover regulated energy consumption, including heating, cooling, ventilation, domestic hot water, lighting and auxiliary systems as projected over the project's life cycle, and all energy generating units such as solar thermal panels, etc. Data for this section is usually provided by the services consultants and should include an estimate of unregulated energy use.
6. Operational water use B7: All carbon emissions relating to operational water consumption, both supply and waste, throughout the building's life cycle should be included [26].

This study considers future uncertainty and realistic scenarios developed for the maintenance, repair, replacement, refurbishment, and building operation. This study based on the available information on the Kamp C website [27], and there is no technical information related to this stage.

### *3.6. Limitations*

This study excluded water installations and sewage installation, including roof gutter systems excluded from the study. Also, the damage categories were excluded - human health, ecosystem quality, climate change, resources, and impact categories(carcinogens, on-carcinogens, respiratory inorganics, ionizing radiation, ozone layer depletion, respiratory organics, aqua terrestrial ecotoxicity, terrestrial acidification/nitrification, land occupation, aquatic acidification, aquatic eutrophication, global warming, attic ecotoxicity.

## 4. Case study

The case study was chosen based on an extensive case study review conducted by Attia et al. (2021) [10]. Out of six case studies located in the Netherlands and Belgium, we identified the most promising building that embraces the circularity principles. The knowledge center of sustainable building Kamp C is currently erecting a fully circular building. The building ambitions aim to limit material impact, the reuse of building components, and the design's maximum adaptability to new and future needs. More importantly, the building will be dismantled and re-assembeled three times over 20 years. The project is designed and constructed by a consortium of Kamp C [27] (see Table 2). Construction works already started at the end of 2020 in Westerlo, Belgium (see Figure 3). The project is located at latitude N 51.13 and longitude E 4.86 and is 14 m above sea level.

IOP Conf. Series: Earth and Environmental Science **855** (2021) 012025 doi:10.1088/1755-1315/855/1/012025

**Table 2.** Kamp C consortium companies of *'t Centrum* project.

| Architect | Design & engineering | Structural engineering | Constructor | Constructing with green & natural elements | Geothermal energy | Research |
|---|---|---|---|---|---|---|
| West Architecture | TEN-agency | Streng-th | Beneens | Muurtuin | Tenerga | VITO |

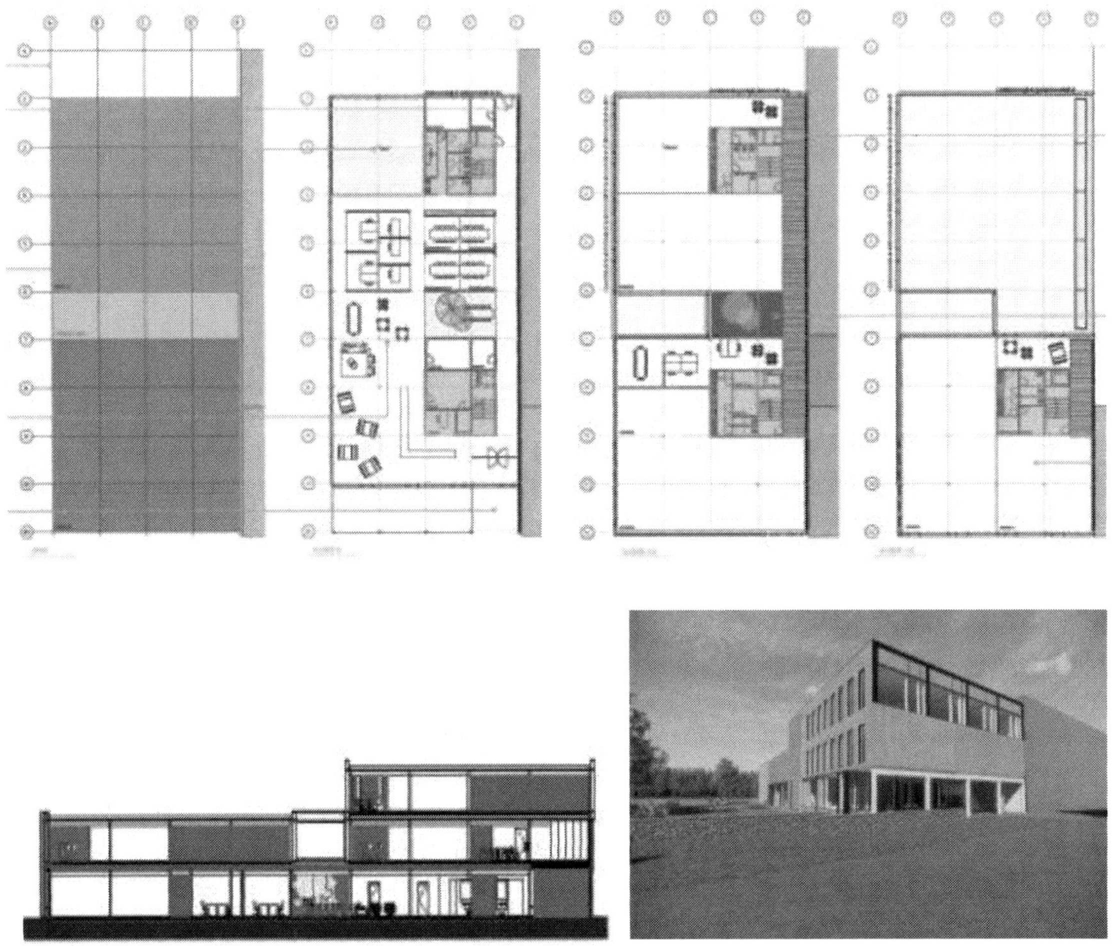

**Figure 3.** *'t Centrum* project drawings [28]

*'t Centrum* building has many advantages, not only for the environment but also for future users. The building will be sustainable and energy neutral, with lots of light and surrounded by greenery. It contains a modular office layout and transformable workplaces. Adjustments can be made quickly, without needing to add a lot of extra energy and new materials. The used materials are reusable, so when the building is outdated, it can be completely disassembled. The design is based on a fixed grid structure, in which two cores are used. The stairs, lifts, and sanitary blocks are located in these cores. Everything around the core is flexible and adaptable. What is an office one moment can become an open meeting space at another. The building can also be completely dismantled. Suppose modifications to the building are needed in the next 20 years. In that case, all elements can easily be disassembled and assembled in other places but in the same location every five years according to Kamp C scenario (see Figure 4).

Furthermore, sustainable and reusable materials have been used, and the building is a living lab for circular construction. A timber structure was chosen in *'t Centrum* building. This gave many more possibilities for dry fastening methods, associated dismount ability, and fire safety solutions.
The project will be built with a timber structure. Ceiling, floors, and interior partitions will be in timber as well.

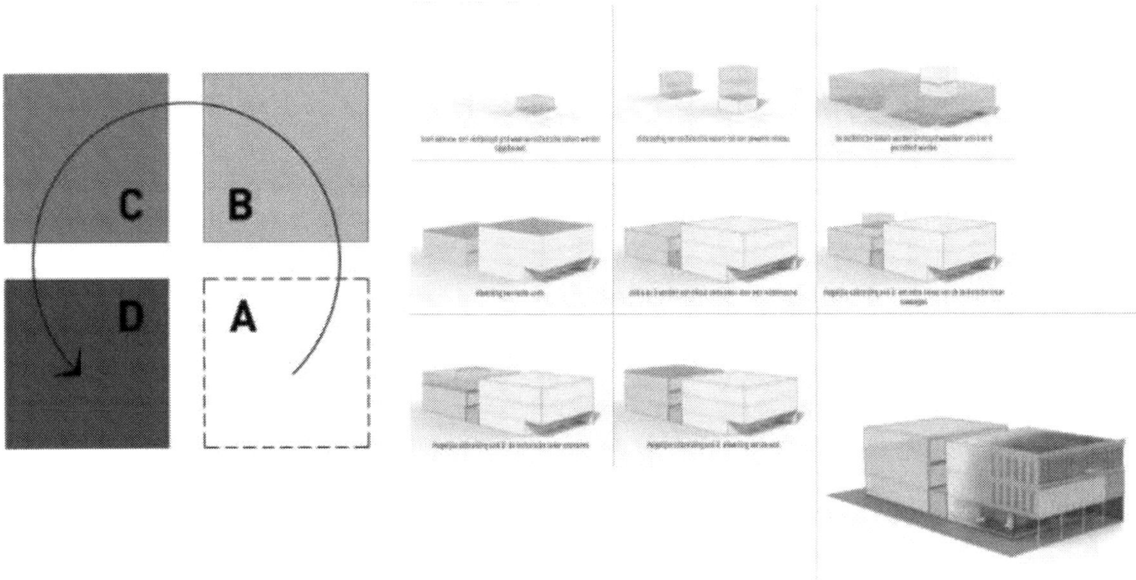


**Figure 4.** Circularity concept of *'t Centrum* project [28]

As a circular project, *'t Centrum* will be built in four buildings A, B, C, and D. They will be dismantled, rebuilt, and used in four phases, one by one:

**Phase 1**: Units A and B are being built. These units are connected using a green central zone and thus form one building. This building meets all aspects of circular construction. Materials are recruited as much as possible and recycled with as much value retention as possible.

**Phase 2**: Because the building is placed on a fixed surface, any extension is easy. For example, in a second phase, unit B can be bred with an extra taper. The same aspects of circular construction are used in building this level.

**Phase 3**: Based on a resource depot 2.0, materials technologies are continuously evaluated and, in this way, can also undergo improvements and transformations. Unit C can thus be built with the most sustainable, innovative raw materials, materials, and technology possible at that time. Heir, we speak of circular innovation. Unit C is connected to unit B with the same central zone.

**Phase 4**: Unit A is broken down so that the raw material depot 2.0 is physically replenished with reusable materials. Unit D can be built up in combination with the new, sustainable materials and techniques from the existing depot. Materials that cannot be used anymore in their current state are transformed into an application in which they can be used, recycled, or sold so that they can lead a new life elsewhere. In this way, we go one step further in circular innovation. Unit D is reconnected to unit C in the same way. The research has been based on the available information about the project on Kamp C website. The structure elements, dimensions, and details have been designed to start the simulation for timber and steel structural systems (see Figure 5 & Figure 6). Where the timber columns and beams were adopted dimensions 30cm, foundation precast bases with 70x70x20cm with a height of 50 cm, timber floors, Cellulose insulations, and roof garden has been used in this project as well. H beams have been used for the steel structural system with the same dimensions and precast foundation bases with

70x70x35cm but the same weight of the timber structure's foundation bases. Rockwool insulation has been used in the steel structure option.

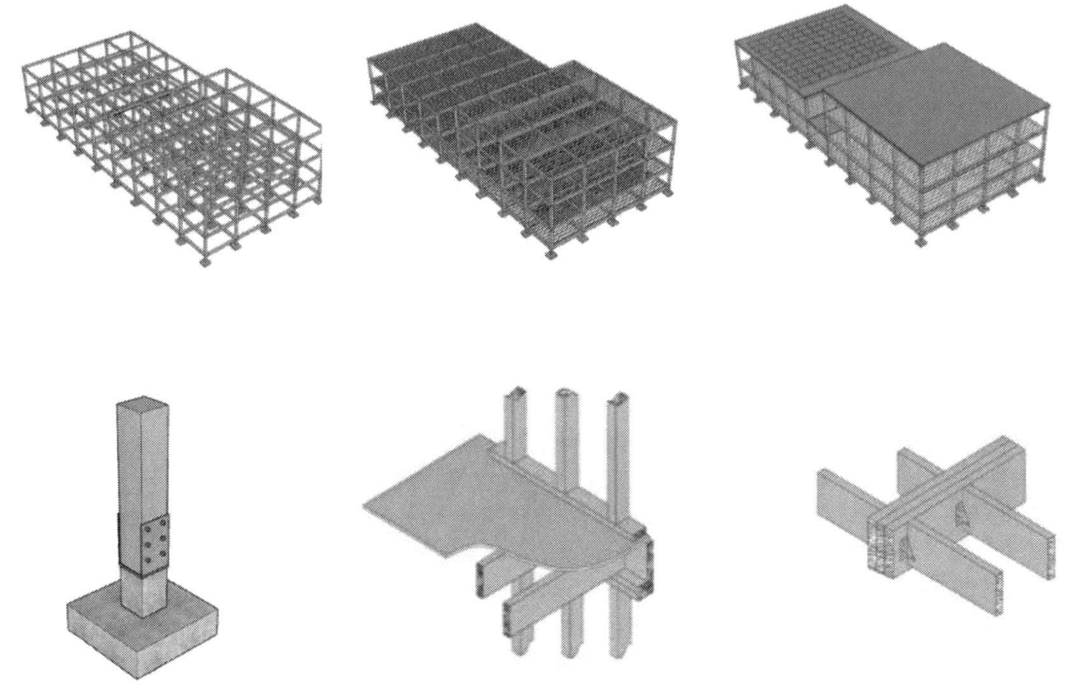

**Figure 5.** Visualization of timber structural system and timber details

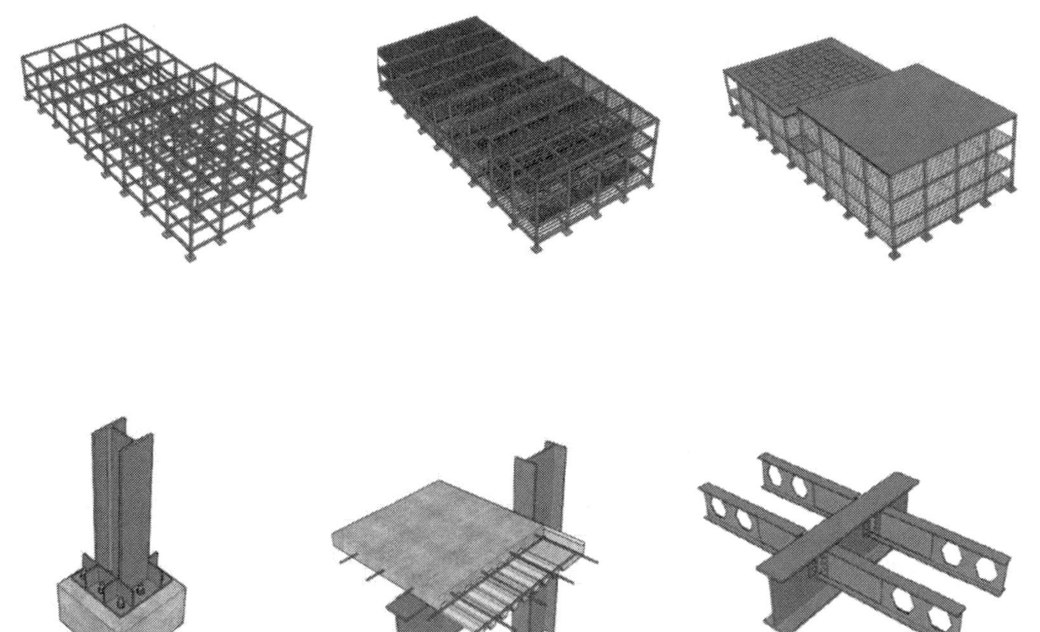

**Figure 6.** Visualization of proposed steel structural system and steel details

## 5. Results

### *5.1. Carbon neutrality*

From a life cycle thinking approach [29] and based on EN 15978 [16], there are two definitions associated with building carbon neutrality: the embodied carbon emissions stage (A1-A5) and the carbon emissions in the use stage (B1-B7) (see Figure 1). Table 6 shows the carbon emissions of building elements during the entire building life cycle and their reuse according to the proposed scenarios and focuses on the pre-use stage of structural system materials (see Figure 7 & Figure 8).

**Table 6.** kg $CO_2$ emissions during life cycle stages for different structure materials and other scenarios.

| Life cycle stages | Timber structure | | Steel structure | |
|---|---|---|---|---|
| EN 15978 | 5 years Kamp C scenario | 20 years Proposed scenario | 5 years Kamp C scenario | 20 years Proposed scenario |
| **Biogenic carbon storage** | -382101 | -382101 | -18 | -18 |
| **Product stage A1 - A3** | 211 030 | 211 030 | 1 210 963 | 1 210 963 |
| **Construction StageA 4 - A5** | 39211 | 39211 | 42589 | 42589 |
| **Use Stage B1 –B7** | 154817 | 125521 | 323845 | 125521 |
| **End-of-life Stage C1 –C4** | 11104 | 11104 | 12 180 | 12 180 |
| **Re-use D** | -134995 | -131303 | -488894 | -465515 |
| **Total life cycle** | -100933 | -126537 | 1100667 | 925722 |

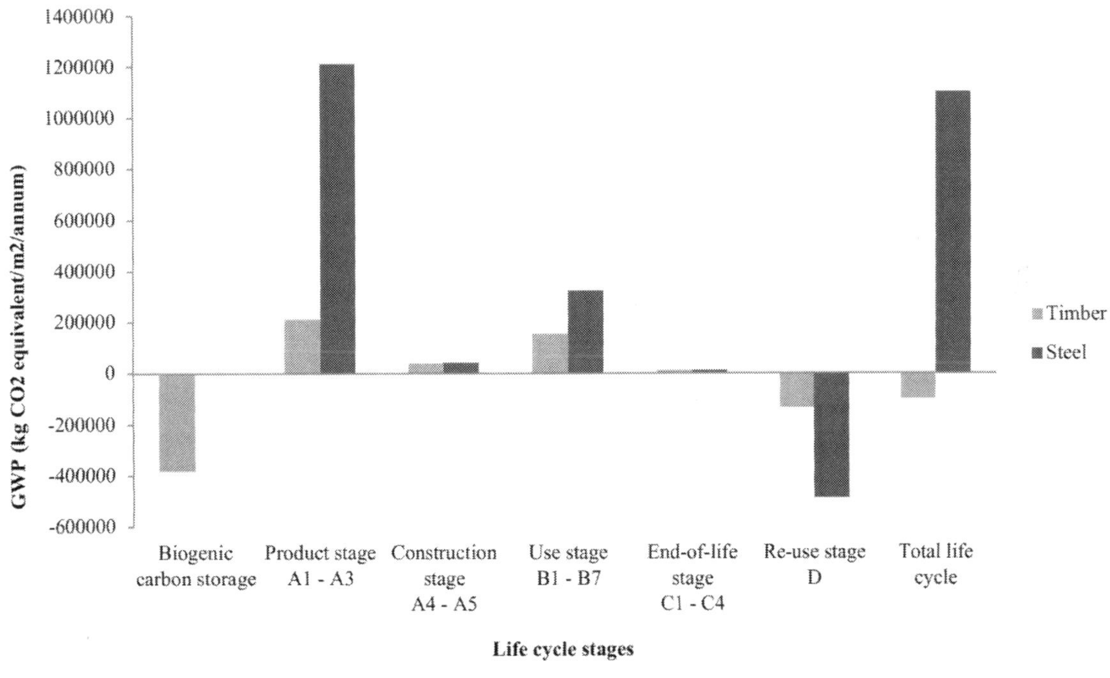


**Figure 7.** Comparison of the global warming potential for the different structure materials reuse every five years for 20 years.

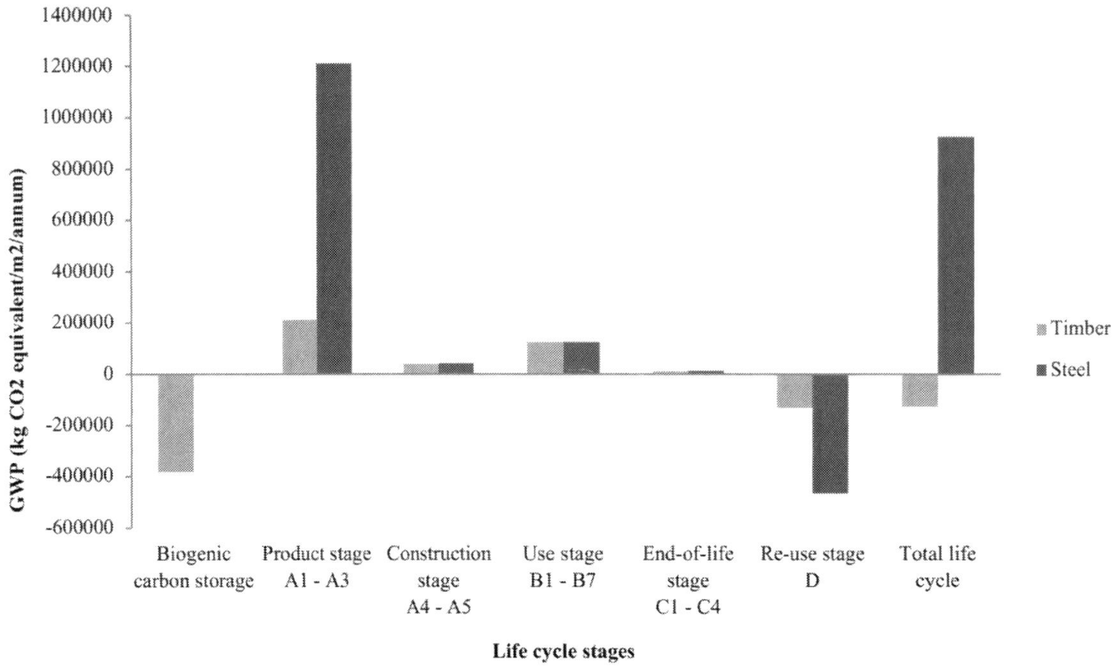


**Figure 8.** Comparison of the global warming potential for the different structure materials reuse every 20 years.

*5.2. Building materials inventory*

The results highlight the environmental impact of the building's structure during the building's life cycle and the reuse of materials several times according to the scenario targeted by Kamp C, and compare this to a hypothetical scenario for 20 years for timber and steel structure. Through an inventory of the main building materials for the structural system in the building (see Table 1 & Figure 2), we can see the following difference:

*5.2.1 Timber structural system*

Timber reaching almost 73% of the structural structure in the timber building. The study proved that the most carbon emissions during the life cycle of the building occur during the product stage (A1-A3) in general, as the study showed that carbon emissions in the timber structure are five times less of it in the steel structure during this stage (see Table 6).

*5.2.2 Steel structural system*

Steel reached almost 27% and concrete 69% of the steel building's structural system due to using concrete in the building floors, unlike the timber building. The study showed that carbon emissions in the product stage (A1-A3) are huge compared to the timber structure (see Table 6).

The carbon emissions were close during their calculation for the construction stage (A4-A5), use stage (B1-B7), and the end-of-life stage (C1-C4). The sum of carbon emissions during the life cycle of the whole building resulted in a remarkable difference between the use of timber structure and steel structure for 5 and 20 years scenario. On the other hand, the difference in carbon emissions is not significant between the 5 and 20 years scenarios for the same structural system material (see Figure 7 & Figure 8). Biogenic carbon storage had a substantial impact on carbon emissions. The difference between the timber structure and the steel structure leads to the conclusion that the building designed with a timber structure is considered.

### *5.3. Circular design criteria*

#### *5.3.1. Design criteria for circular building*

According to Attia et al. (2021), four primary critical design criteria should be used to assess the building's circularity, namely: (1) carbon footprint, (2) reused content, (3) disassembly potential, and (4) flexibility and adaptability design. They are the most effective design criteria that can help designers make strategic design decisions for a circular building design; construction system, spatial function, and building materials [10] (see Figure 9).

These design criteria are what Kamp C should adopt in designing and building *'t Centrum* project during the reuse phases. But in this study, we focused on the structural system, which will help develop future research addressing the other aspects of the circular building design.

Circularity criteria for buildings should not be treated in isolation from carbon neutrality. The circularity criteria must be coupled to carbon footprint reduction measures of the building materials [30].

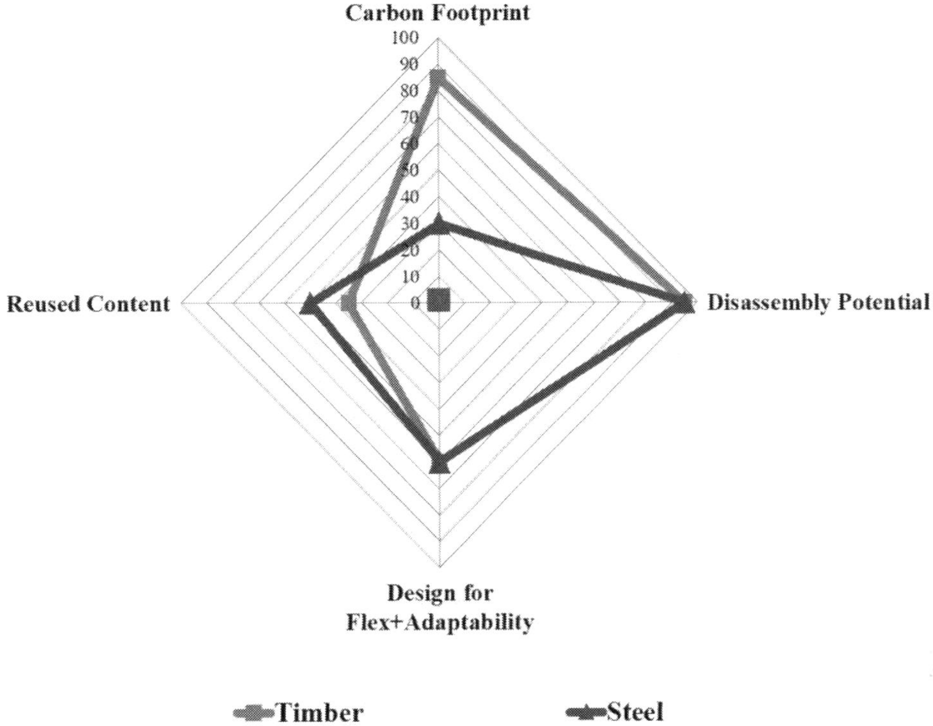


**Figure 9.** The circularity criteria of *'t Centrum* project for timber and steel structures

#### *5.3.2. Construction system*

The study's scope is limited to the construction system of *'t Centrum* Building as a case study. This paper will provide an assessment of the environmental impact of the timber and steel structural systems, and the extent to what it achieved circularity and carbon neutrality by focusing on the following:

1. Embodied carbon: By analyzing the project's timber structural system based on the available information, we find that the value of embodied carbon reaches -131860 kg CO2, making the project achieve the necessary carbon neutrality criterion.
2. Disassembly: The concept of this project depends on the disassembly of its components to be reused again.
3. Reuse cycle: After disassembling each time, the structural components will be reused about three times.

4. Recycled content: According to available information, this project's construction in the beginning will not be with recycled contents or maybe, but in a small proportion.
5. Usage adaptation: The building is an office building, and it will remain so throughout its use and reuse at every stage, and if a change in design occurs, it will not go outside the offices' framework.

Table 7 showed the durability of elements subject to reuse every five years for 20 years (Kamp C scenario). The table shows the building elements and how they are reused after the building was demolished and rebuilt. In contrast, some materials are not reused due to their damage after demolition. These materials that cannot be used anymore in their current state will be used, recycled, or sold so that they can lead a new life elsewhere. In this way, we go one step further in circular innovation, which will undoubtedly significantly impact evaluating these carbon emissions elements and achieving circularity. For the proposed scenario for 20 years, it was assumed that the building elements would not be replaced during the building's life.

**Table 7.** Durability of elements subject to reuse every five years for 20 years (Kamp C scenario).

| Building inventory elements | Timber structure | | Steel structure | |
|---|---|---|---|---|
| | Durability [years] | Number of reuse | Durability [years] | Number of reuse |
| **Structure elements** | 20 | 3 | 20 | 3 |
| **Foundation** | 20 | 3 | 20 | 3 |
| **Insulation** | 20 | 2 | 20 | 3 |
| **Waterproof** | 20 | 0 | 20 | 0 |
| **Protective layer** | 20 | 0 | 20 | 0 |
| **Drainage** | 20 | 0 | 20 | 0 |
| **Geotextile** | 20 | 0 | 20 | 0 |
| **Filter layer** | 20 | 0 | 20 | 0 |
| **Soil** | 20 | 0 | 20 | 0 |

## 6. Discussion

In this section, we discuss the answers to the three main research questions:

- How can we evaluate the circularity of the building?

This study adopted a multi-criteria evaluation approach developed by Attia et al. 2021 [10]. The evaluation approach combined four main criteria, namely: (1) carbon footprint, (2) reused content, (3) disassembly potential, and (4) flexibility and adaptability design. The four key indicators allowed to address circularity more consistently and involved spatial and qualitative evaluation methods. The strength of the suggested approach is mainly due to its comprehensiveness beyond the narrow scope of LCA. By applying the multi-criteria evaluation approach, the study informed the design team about *'t Centrum* buildings circularity. Despite the simplicity of the used radar graph, the design team of *'t Centrum* appreciated the evaluation results and changed the design to ultimaly increase the overall circularity score.

- What is the most sustainable structural system regarding carbon neutrality and circularity?

This article focused on the embodied carbon in structural steel and timber once its design has been optimized. Results indicate that timber structure is superior to the steel structure in achieving circularity and reducing carbon emissions. Carbon emissions in the timber structure are five times less than the steel structure during the product stage (A1-A3). There is a remarkable difference in carbon emissions between timber structure and steel structure during the whole building's life cycle (A1-D or Cradle to grave). The difference is up to ten times less in the timber structure in the five-year scenario and up to eight times less in the 20 years scenario. The biogenic carbon storage effect made the timber alternative negative in carbon emissions, unlike the steel structure.

- To what extent is *'t Centrum* project circular and carbon neutral?

As shown in Figure 9, the building scores high in carbon footprint and disassembly potential. The most influential factor in the environmental impact of the structural systems was the roof insulation. However, designers must increase reused content of materials as much as possible and prepare for occupancy alternative functional scenarios. As architects, we significantly influence the circularity and the amount of embodied carbon in buildings. Therefore our specifications must not be written in isolation of the circularity criteria applied in this study. Also, we advise the European Union legislators to combine the environmental impact assessment of materials with circularity indicators. LCA is a powerful method for building sustainability evaluations, but it is not enough to achieve long-lasting sustainability. Moreover, architects must engage with contractors, subcontractors, suppliers, and clients to make sure that, collectively, they realize the circularity compliance we all need to make. This may require us to change how we design and build but will ultimately leave us with circular and low-carbon buildings.

Finally, this study is initial and should be further developed. The boundary conditions should extend the study scope to include the building envelope and building services (HVAC systems). With other drivers such as durability and steel connections influencing timber members, we are often given far more strength in our timber structures than we assume in our designs. Could we engage with the contractor to agree on timber strengths aligned with the construction needs, leading to the dismantling of building construction elements?

## 7. Conclusions

The use of the building's steel structure is associated with high carbon emissions despite its circularity achievement. In contrast, the use of timber structural systems in the building is considered one of the best materials in the short and long term. It achieves a more significant benefit by combining circularity and achieving the desired goal of reducing carbon emissions. Timber can make the building have a negative environmental impact. The research recommends designers address the four circularity principles presented in this study to assess future buildings, namely: (1) carbon footprint, (2) reused content, (3) disassembly potential, and (4) flexibility and adaptability design. LCA's use in early design stages is not enough to inform the design decision-making to achieve circularity.

## 8. Acknowledgments

This publication is part of the research project 2020-2024 entitled: Design support for circular buildings: a guiding methodology for architects in early design. The authors would like to gratefully acknowledge the Kamp C for sharing information about the case study. We would also like to recognize the Sustainable Building Design (SBD) Laboratory at the University of Liege to use the circularity audits in this research and valuable support during the case study's observations and data analysis.

## 9. References

[1] Attia, S. (2018). Regenerative and positive impact architecture: Learning from case studies. Springer International Publishing, New York, USA. ISBN 978-2930909004, https://doi.org/10.1007/978-3-319-66718-8.

[2] Heisel, F., & Rau-Oberhuber, S. (2020). Calculation and evaluation of circularity indicators for the built environment using the case studies of UMAR and Madaster. Journal of Cleaner Production, 243, 118482.

[3] Cottafava, Dario & Ritzen, Michiel. (2021). Circularity indicator for residential buildings: Addressing the gap between embodied impacts and design aspects. Resources Conservation and Recycling. 164. 105120. 10.1016/j.resconrec.2020.105120.

[4] E. M. F. (EMF), Circularity indicators: An approach to measuring circularity, https://www.ellenmacarthurfoundation.org/assets/downloads/insight/Circularity-Indicators_Project-Overview_May2015.pdf, 2015. Online; accessed 19 March 2020.

[5] Tallini A, Cedola L. A review of the properties of recycled and waste materials for energy refurbishment of existing buildings towards the requirements of NZEB. Energy Procedia. 2018 Aug;148:868–75.
[6] Passer A, Lasvaux S, Allacker K, De Lathauwer D, Spirinckx C, Wittstock B, et al.Environmental product declarations entering the building sector: critical reflections based on 5 to 10 years experience in different European countries.Int J Life Cycle Assess. 2015;20(9):1199–1212.
[7] Cambier C, Galle W, De Temmerman N. Research and Development Directions for Design Support Tools for Circular Building. Buildings. 2020;10(8):142.
[8] Durmisevic, E. Reversible Building Design: Reversible Building design guidelines [Internet]. University of Twente; 2018 [cited 2020 Oct 26]. Available from: https://www.bamb2020.eu/wp-content/uploads/2018/12/Reversible-Building-Design-guidelines-and-protocol.pdf
[9] Antonini E, Boeri A, Lauria M, Giglio F. Reversibility and durability as potential indicators for Circular building Technologies. Sustainability. 2020;12(18):7659.
[10] Attia S, Al-Obaidy M. Design criteria for circular building. In: Crossing Boundaries Parkstad. Parkstadt, The Netherlands: Zuyd Uninversity; 2021.
[11] Grover, R. (2020). Towards Zero Carbon Buildings: Reducing the embodied carbon footprint of a construction (Master thesis); TU Delft. Available from: https://repository.tudelft.nl/islandora/object/uuid%3A98e2c46c-7321-4eab-9f88-c3d10912003a, Accessed: 28.02.2021.
[12] Huberman, N., & Pearlmutter, D. (2008). A life-cycle energy analysis of building materials in the Negev desert. Energy and Buildings, 40(5), 837–848.
[13] ISO I 14040 (2006): environmental management–life cycle assessment–principles and framework. London: British Standards Institution.
[14] ISO I 14044 (2006): environmental management—life cycle assessment—requirements and guidelines. International Organization for Standardization.
[15] Vogtländer, J. G. (2010). A practical guide to LCA for students, designers and business managers: cradle-to-grave and cradle-to-cradle. VSSD.
[16] EN 15978. Sustainability of construction works–Assessment of environmental performance of buildings–Calculation method. BrussBelgEurComm Stand. 2011.
[17] One Click LCA. Calculate your environmental impacts in minutes. Available from: https://www.oneclicklca.com/ , Accessed: 15.11.2020.
[18] One Click LCA. 90,000+ data points: up-to-date, verified and localised. Available from: https://www.oneclicklca.com/support/faq-and-guidance/documentation/database/ , Accessed: 15.02.2021.
[19] Engineeringclicks. Density of wood in kg/m3, g/cm3, lb/ft3 – the ultimate guide. Availablel from: https://www.engineeringclicks.com/density-of-wood/ , Accessed: 05.11.2020.
[20] Maiak. Calculator for Beams - HEB (IPB). Availablel from: https://www.maiak-m.bg/en/calculator/i15/Beams-HEB-(IPB).html , Accessed: 05.11.2020.
[21] Arcelor mittal. composite floor decking. Availablel from: https://constructalia.arcelormittal.com/files/AMC_floors_guide_EN_June2019_LR--b47b21285b39514432cf4da5d75b23f1.PDF , Accessed: 05.11.2020.
[22] Reinforcingsteelmesh. bs square mesh. Availablel from: http://www.reinforcingsteelmesh.com/products/bs-square-mesh.html , Accessed: 05.11.2020.
[23] Calculator.net. concrete calculator. 2020. Availablel from: https://www.calculator.net/concrete-calculator.html
[24] Vlaanderen is energie. E-peil: eisvoorniet-residentiëlegebouwen (2021). Available from: https://www.energiesparen.be/EPB-pedia/E-peil/eis-niet-residentieel-2021 , Accessed: 27.11.2020.
[25] Attia S. Net zero energy buildings (nzeb). 1st edition. Cambridge, MA: Elsevier; 2018.DOI: 10.1016/c2016-0-03166-2

IOP Conf. Series: Earth and Environmental Science **855** (2021) 012025 doi:10.1088/1755-1315/855/1/012025

[26] Sturgis, S. (2017). Embodied and whole life carbon assessment for architects. Riba Publishing. Available from: https://www.architecture.com/-/media/gathercontent/whole-life-carbon-assessment-for-architects/additional-documents/11241wholelifecarbonguidancev7pdf.pdf
[27] Kamp C. circulairbouwen; 't Centrum. Available from: https://www.kampc.be/tcentrum/circulair-bouwen-t-centrum , Accessed: 20.09.2020.
[28] Beneens. Wijbouwenkampcirculair. Availablefrom: https://www.beneens.be/circulair , Accessed: 20.09.2020.
[29] Pombo O, Rivela B, Neila J. Life cycle thinking toward sustainable development policy-making: The case of energy retrofits. J Clean Prod. 2019 Jan;206:267–81.
[30] Attia S, Santos M C, Al-Obaidy M, Baskar M. Leadership of EU member States in building carbon footprint regulations and their role in promoting circular building design. In: Crossing Boundaries Parkstad. Parkstadt, The Netherlands: Zuyd Uninversity; 2021.

# AUTHOR INDEX